FOSSILS, FINCHES AND FUEGIANS

By the same author

The *Beagle* Record:Selections from the Original Pictorial Records and Written Accounts of the Voyage of HMS *Beagle* (*editor*)

Nerve and Muscle (*author, with D.J. Aidley*)

Charles Darwin's *Beagle* Diary (*editor*)

Lydia and Maynard: Letters Between Lydia Lopokova and John Maynard Keynes (*editor, with Polly Hill*)

Charles Darwin's Zoology Notes and Specimen Lists from HMS *Beagle* (*editor*)

Fossils, Finches and Fuegians

Darwin's Adventures and Discoveries on the Beagle

Richard Keynes

UNIVERSITY PRESS

2003

OXFORD
UNIVERSITY PRESS

Oxford New York
Auckland Bangkok Buenos Aires
Cape Town Chennai Dar es Salaam Delhi Hong Kong Istanbul
Karachi Kolkata Kuala Lumpur Madrid Melbourne Mexico City Mumbai
Nairobi São Paulo Shanghai Singapore Taipei Tokyo Toronto

First published by HarperCollins Publishers, Great Britain, 2002

Published by Oxford University Press, Inc.
198 Madison Avenue, New York, New York 10016
www.oup.com

Library of Congress Cataloging-in-Publication Data
Keynes, R.D.
Fossils, finches, and Fuegians : Darwin's adventures and discoveries
on the Beagle / by Richard Keynes.
p. cm.
Originally published : London : HarperCollins, 2002.
Includes bibliographical references (p.).
ISBN 0-19-516649-3 (cloth : alk. paper)
1. Beagle Expedition (1831-1836) 2. Darwin, Charles,
1809-1882--Journeys. 3. Natural History. I. Title.
QH11.K49 2003
576.8'2'092--dc21
2002154176

1 3 5 7 9 10 8 6 4 2
Printed in the United States of America
on acid-free paper

Contents

Colour plates

Maps

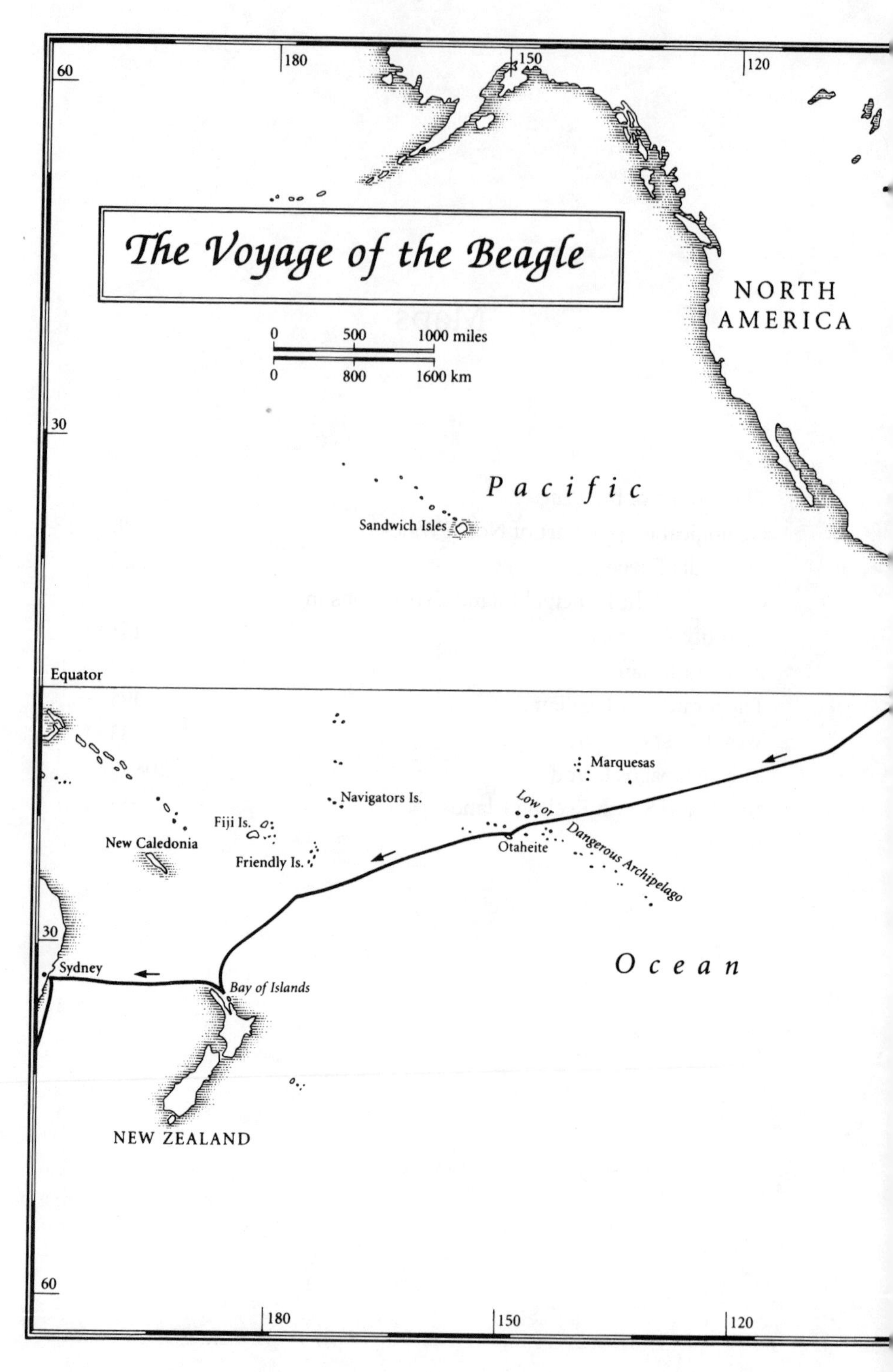

The Voyage of the Beagle
NORTH AMERICA
0
500
1000 miles
0
800
1600 km
Pacific
Ocean
Sandwich Isles
Equator
Marquesas
Navigators Is.
Low or Dangerous Archipelago
Fiji Is.
New Caledonia
Otaheite
Friendly Is.
Sydney
Bay of Islands
NEW ZEALAND
180
150
120
60
30
30
60
180
150
120

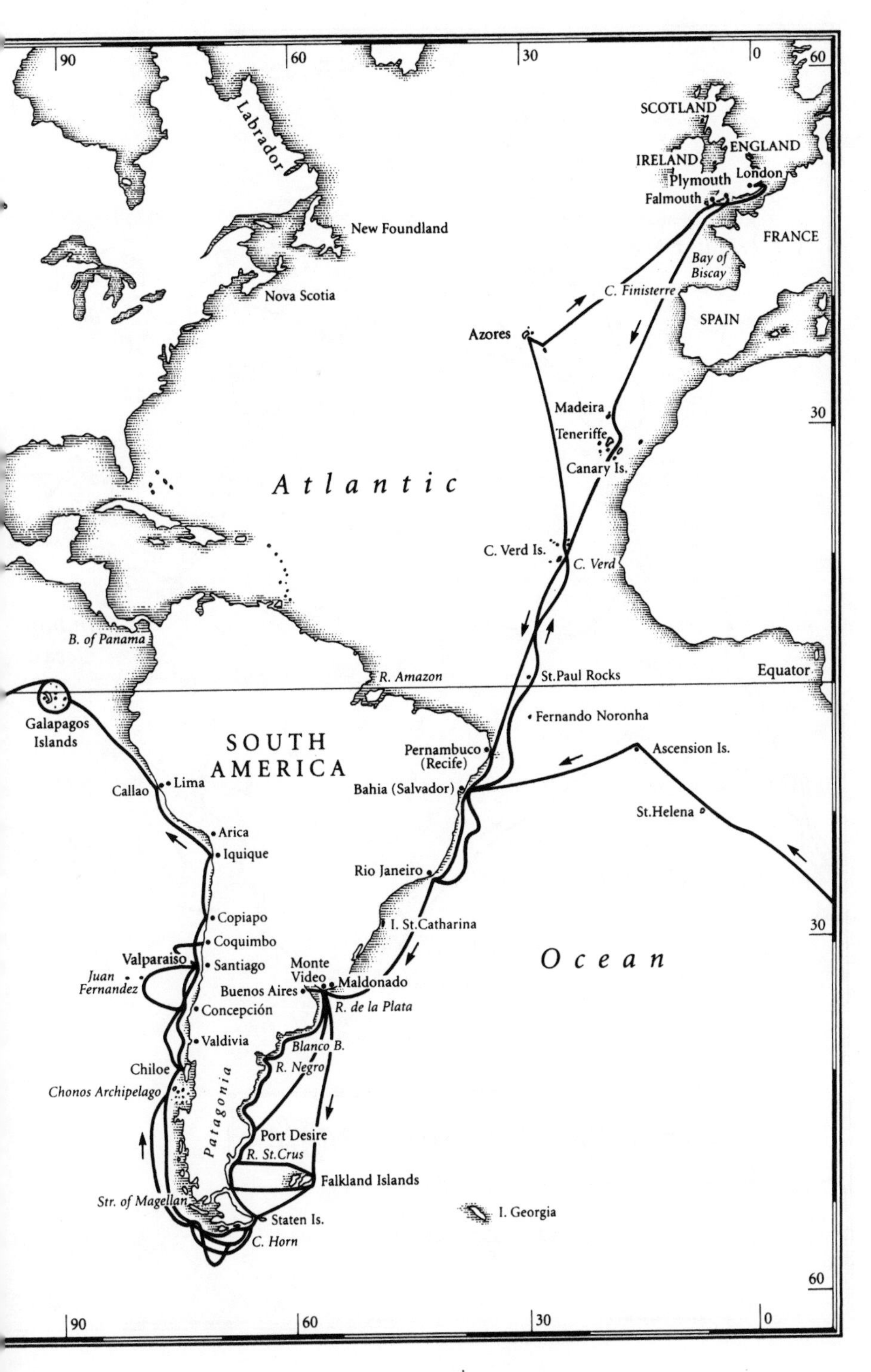

90
60
30
0
60
Labrador
SCOTLAND
IRELAND
ENGLAND
London
Plymouth
Falmouth
New Foundland
FRANCE
Bay of Biscay
Nova Scotia
C. Finisterre
SPAIN
Azores
Madeira
Teneriffe
Canary Is.
30
Atlantic
C. Verd Is.
C. Verd
B. of Panama
R. Amazon
St.Paul Rocks
Equator
Galapagos Islands
Fernando Noronha
SOUTH AMERICA
Pernambuco (Recife)
Ascension Is.
Callao
Lima
Bahia (Salvador)
St.Helena
Arica
Iquique
Rio Janeiro
Copiapo
I. St.Catharina
30
Coquimbo
Valparaiso
Santiago
Juan Fernandez
Monte Video
Maldonado
Ocean
Buenos Aires
R. de la Plata
Concepción
Valdivia
Blanco B.
R. Negro
Chiloe
Chonos Archipelago
Patagonia
Port Desire
R. St.Crus
Falkland Islands
Str. of Magellan
Staten Is.
I. Georgia
C. Horn
60
90
60
30
0

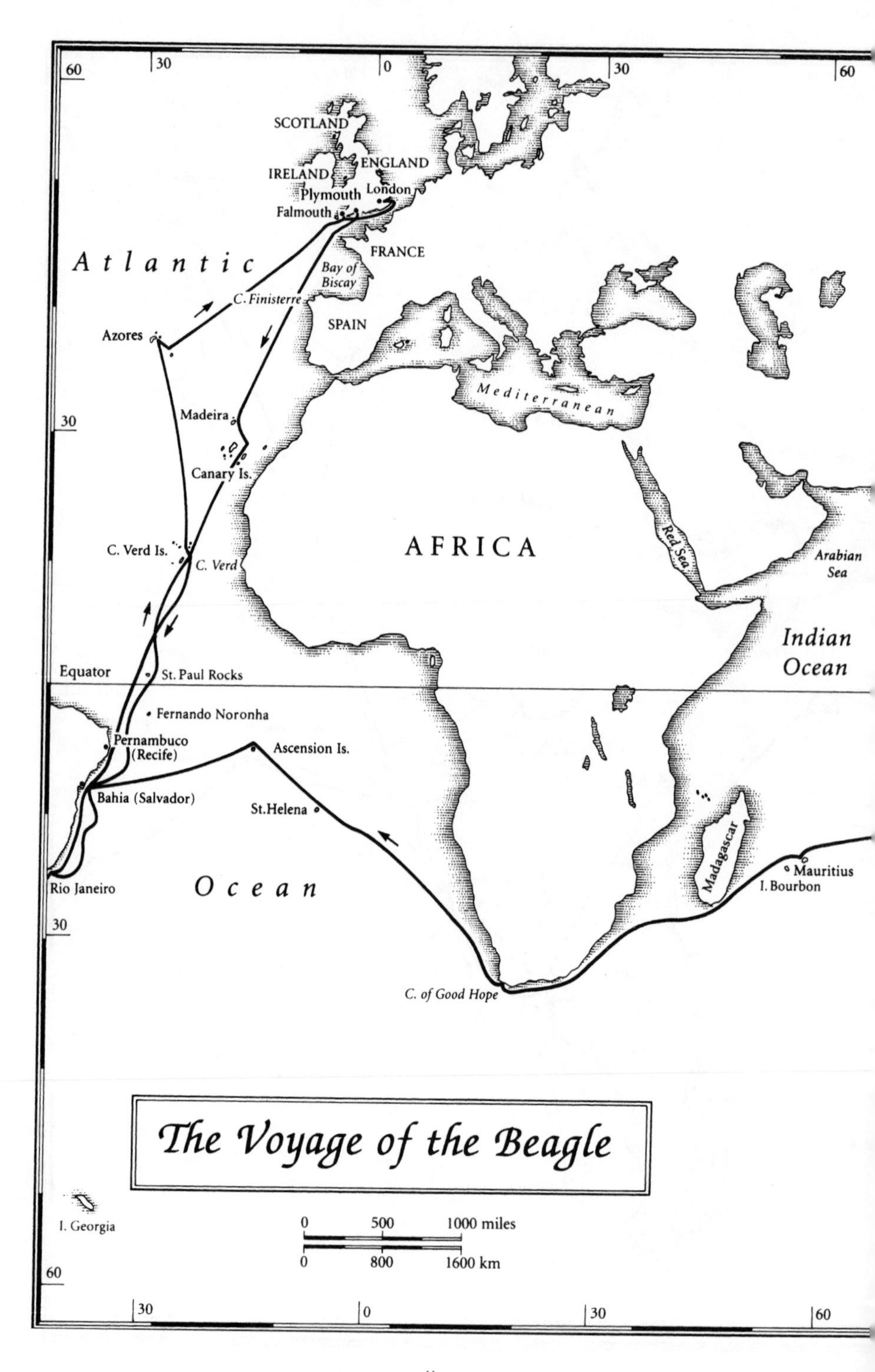

The Voyage of the Beagle
SCOTLAND
IRELAND
ENGLAND
London
Plymouth
Falmouth
FRANCE
Bay of Biscay
SPAIN
C. Finisterre
Atlantic
Ocean
Azores
Madeira
Canary Is.
C. Verd Is.
C. Verd
AFRICA
Mediterranean
Red Sea
Arabian Sea
Indian Ocean
Equator
St. Paul Rocks
Fernando Noronha
Pernambuco (Recife)
Bahia (Salvador)
Ascension Is.
St.Helena
Rio Janeiro
Madagascar
Mauritius
I. Bourbon
C. of Good Hope
I. Georgia
0 500 1000 miles
0 800 1600 km
60
30
0
30
60

90
120
150
60
ASIA
CHINA
30
INDIA
Pacific
Bay of Bengal
Philippines
Ceylon
Ocean
Equator
SUMATRA
Celebes
New Guinea
Java
Keeling Is.
New Caledonia
AUSTRALIA
30
Sydney
King George Sound
Bass Strait
Hobart
NEW ZEALAND
Kerguelen Land
60
90
120
150

Prologue

In the autobiography written by Charles Darwin near the end of his life for the benefit of his children and descendants,[1] he said:

> The voyage of the *Beagle* has been by far the most important event in my life and has determined my whole career; yet it depended on so small a circumstance as my uncle offering to drive me 30 miles to Shrewsbury, which few uncles would have done, and on such a trifle as the shape of my nose. I have always felt that I owe to the voyage the first real training or education of my mind. I was led to attend closely to several branches of natural history, and thus my powers of observation were improved, though they were already fairly developed.

There is no dispute with Darwin's own estimate of the importance of the voyage of the *Beagle* in the subsequent development of his scientific career. He himself provided a classical description of the voyage in his *Journal of Researches*,[2] but his biographers other than Janet Browne have not covered his scientific research on the *Beagle* in much detail, while Alan Moorehead's *Darwin and the Beagle*[3] was mainly concerned with the affair as the exciting adventure that it undoubtedly was, and gave a distinctly misleading picture of the relations between Darwin and FitzRoy. The purpose of this book is to retell the whole story, starting from the haphazard events in Tierra del Fuego that led Robert FitzRoy to take the *Beagle* there again, with a 'well-educated and scientific person' as companion. Then it is shown how FitzRoy's scientist had precisely the right talents to make highly effective use of the array of new scientific facts that were presented to him in South America and in the countries visited by the *Beagle* homeward bound round the far side of the world. And lastly it is explained how Charles Darwin's findings on the *Beagle* soon started him on the path that in due course led him to the discovery of the principle of Natural Selection and of the Origin of Species.

My interest in South America was first aroused in 1951, when I had the good fortune to be invited by Professor Carlos Chagas Filho to be Visiting Reader that summer at the Instituto de Biofisica in Rio de Janeiro. My ignorance about Brazil was profound, but I did know that Professor Chagas had a good supply of electric eels at his laboratory. Moreover, Alan Hodgkin, leader of our group working at the Physiological Laboratory in Cambridge on the mechanism of conduction of the nervous impulse, had just developed an important new technique for recording electrical activity in living cells that I could usefully apply to investigate the properties of what Charles Darwin had described as the 'wondrous' organs[4] in certain fishes that generated their powerful electrical discharges, though he had been puzzled about the manner of their evolution. So I spent two and a half happy months in Rio that summer, successfully unravelling the mystery of how the additive discharge of the electric organ was achieved.[5] The job complete, and having accumulated a pocketful of cruzeiros in payment for my efforts, I then took the recently established direct flight in a Braniff DC-6 from Rio to Lima over the Mato Grosso and the Andes, and made the first of many journeys to Peru, briefly calling on fellow physiologists in Chile and Argentina on the way home, and getting back to Cambridge for the Michaelmas term at the beginning of October.

In August 1968 I had been visiting a Chilean colleague at Viña del Mar, near Valparaiso, for discussions on a joint study of the biophysics and physiology of giant nerve fibres, and was flying home via Buenos Aires. Here the British Council representative, knowing me to be a member of the Darwin family, asked me whether I had seen the Darwin collection belonging to Dr Armando Braun Menendez. I had not previously taken any particular interest in the voyage of the *Beagle*, but fortunately the Argentinian professor who was showing me round knew Dr Braun Menendez, and took me to call on him. His impressive collection of papers and books was concerned with the exploration of the southern seas, and the Darwin item consisted of two little portfolios of pencil drawings and watercolours made on board the *Beagle* by Conrad Martens, of whom I knew vaguely as the second of the ship's official artists. I opened one of the portfolios to find the picture labelled by Martens as 'Slinging the monkey. Port Desire – Decr 25, 1833', which portrayed the ship's crew engaged on the Royal Navy's traditional celebration of Christmas Day, with the *Beagle* and *Adventure* anchored in the background. The picture bore the initials 'RF' at the top right-hand corner, though the *Beagle*'s Captain, Robert FitzRoy, evidently did not approve of every detail, because Martens had written below: 'Note. Mainmast of the Beagle a little farther aft – Miz. Mast to rake more'. This graphic document, and others in the portfolios, opened an exciting new window for me on to the voyage of the *Beagle*, and launched me on an entirely new and rewarding field of part-time study.

On my return home, I consulted Nora Barlow, my godmother and mentor, and like my mother a granddaughter of Charles Darwin. Through her pioneer editions of *Charles Darwin's Diary of the Voyage of H.M.S. "Beagle"* (1933), *Charles Darwin and the Voyage of the Beagle* (1945), Darwin's *Autobiography* (1958), his *Ornithological Notes* (1963), and *Darwin and Henslow: The Growth of an Idea. Letters 1831–1860* (1967), Nora Barlow was the true founder of what is often known nowadays as the Darwin Industry. Her wisdom and kindness had no bounds, and my debt to her is immense.

With Lady Barlow's encouragement, I set about assembling a catalogue of all the extant drawings and paintings made by Conrad Martens during his initial journey from England to Montevideo, where on 3 August 1833 he joined the *Beagle* as a replacement for the ship's first official artist, Augustus Earle, who had been taken ill. Then I listed the drawings and paintings that he made in Patagonia, Tierra del Fuego, the Falkland Islands, the Straits of Magellan, Chiloe and around Valparaiso, until in December 1834, there being no longer any space for him on the *Beagle*, he set sail for Australia. Here he established himself in Sydney as the leading artist of the period, and at the same time continued to paint developments of his *Beagle* drawings in watercolour, some of which he sold to Darwin, and others to Robert FitzRoy to be engraved as illustrations for the published accounts of the voyage. Selections from this material were reproduced in the first book that I edited, *The Beagle Record* (1979), to illustrate some of the places and people actually seen by Darwin and FitzRoy in South America. For the most immediate written records, I drew upon the vivid accounts in letters from Darwin to his sisters and to his mentor in Cambridge, Professor John Stevens Henslow, and also in what he himself called his 'commonplace journal', but which will be referred to here as the '*Beagle Diary*' to distinguish it from his subsequently published *Journal of Researches*. In addition I made use of FitzRoy's published account,[6] his few surviving diaries, and some of his letters.

Fifty years after its publication, Nora Barlow's edition of *The Beagle Diary* had long since gone out of print. Darwin's splendid description of his daily life on the *Beagle* and ashore, sent home at intervals to his family in Shrewsbury, is one of the major classics of scientific exploration. My next task was therefore to produce a new version revised according to modern standards of transcription, with Nora Barlow's very rare errors put right, and a new introduction and footnotes. This was published in 1988, and has recently been reprinted in paperback.[7]

Two substantial Darwin manuscripts still remained unpublished at the Cambridge University Library, namely the Zoology Notes[8] and the Geology Notes[9,10] made on the *Beagle*. Apart from a brief account of some observations made in Edinburgh in 1827,[11] these were his first scientific writings, containing a detailed record of all that he observed and collected during the four-and-three-quarter years of the voyage. Their importance is that here, and in his letters to Henslow,[12] it is possible to trace the first beginnings of Darwin's thinking about the evolution not only of the animal kingdom but also of the face of the earth.

After retiring in 1986 from administration and teaching in Cambridge, I had more time at my disposal, so in addition to visiting the marine laboratory at Roscoff in Brittany every autumn for continuation of some experimental work, I embarked on a transcription[13] of Darwin's Zoology Notes and Specimen Lists. They comprise some hundreds of quarto pages of notes, with descriptions of 1500 specimens preserved in Spirits of Wine, and some 3500 not in Spirits. In order to identify the insects and marine invertebrates collected by Darwin, I needed a great deal of help from experts on their taxonomy, but the vertebrates had mostly been covered in the five parts of *The Zoology of the Voyage of H.M.S. Beagle* published in 1839–1843.[14] Completion of the task has kept me busy for about ten years, but it has now at last been finished. The Geology Notes have not as yet been transcribed and published, occupying as they do around four times as many pages as the Zoology Notes. However, articles by Secord,[15] Herbert[16] and Rhodes[17] have discussed them helpfully, as has Janet Browne's account of the voyage.[18] Moreover, photocopies of the MSS are available at the Cambridge University Library for study by anyone well practised at reading Darwin's reasonably legible handwriting.

If any of my readers feel that they would like to become better acquainted with one of their forebears, I can recommend the course that I adopted thanks to that chance introduction in Buenos Aires in 1968. When you have transcribed several hundred thousand words of his writings, concerned with places a few of which you have seen for yourself not too greatly changed 160 years later, you may once in a while almost feel that you are talking to him. But it helps to be lucky enough to have a forebear who was as friendly to all men, and as constructively critical and honest about his own ideas, as Charles Darwin always was.

Richard Darwin Keynes
Cambridge, July 2001

CHAPTER 1

The Man who Walks with Henslow

Charles Robert Darwin was born at The Mount, Shrewsbury, on 12 February 1809, the second son and fifth child of Dr Robert Waring Darwin.

His genetic make-up in the male line was strongly scientific, although his father combined a position as one of the leading physicians in the Midlands not with science, but with acting widely as a private financial adviser to the gentry of the region. Charles would later have high praise for his father's powers of observation, and for his sympathy with his patients, but considered that his mind was not truly scientific. Robert's wealth was nevertheless as valuable an inheritance for his son as a gene for science, not only paying for Charles's board and lodging on the *Beagle* when he sailed as an unofficial scientist and companion to the Captain, but also enabling him to pursue a scientific career single-mindedly for the rest of his life, without ever having to earn his living.

However, his grandfathers, Dr Erasmus Darwin of Lichfield and Josiah Wedgwood I the potter, who built a model factory at Etruria near Stoke-on-Trent, were two of the leading scientists and technologists of their time. Both Fellows of the Royal Society, they were also founding members of the Lunar Society of Birmingham, which played the principal intellectual part in the establishment of the Industrial Revolution in England. Erasmus Darwin[19] was primarily a practising physician, but his prodigious energies overflowed in many other directions, as a poet, an inventor of mechanical devices of various kinds, a pioneer in meteorology and in the description of

A contemporary silhouette, c.1826, of Dr Robert Darwin

photosynthesis, and not least as author of an immense medical treatise entitled *Zoonomia*,* and of a fine poem, *The Temple of Nature*, in which as we shall see he made an important contribution to his grandson's Theory of Evolution. Josiah Wedgwood I made radical improvements in the handling of china clay, founded a famous pottery, and developed a canal system for the transport of his products. The chemists of Europe came to him for their glassware and retorts, and in due course his son Josiah Wedgwood II provided his nephews Erasmus and Charles with fireproof china dishes manufactured at

* *Zoonomia; or, the laws of organic life.* 2 vols. London, 1794–96.

the pottery, and an industrial thermometer for their private laboratory.

Charles recorded that at a very early age he had a passion for collecting 'all sorts of things, shells, seals, franks, coins, and minerals . . . which leads a man to be a systematic naturalist, a virtuoso or a miser'. Throughout his life he also exercised a scientifically useful taste for making careful lists, whether of his various collections, of the game that he had killed, of books that he had read or intended to read, of the pros and cons of marriage, or of his household accounts.

During his formal education at Shrewsbury School, where like his elder brother Erasmus he boarded for some years, he was taught mainly classics, ancient geography and history, and a little mathematics, but there was no place for science in the curriculum. However, their grandfathers' strong interest in chemistry managed to break through in both boys, first in Erasmus and then in Charles, and together they set up in an outhouse at home what they grandly called their 'Laboratory'. Here they could pursue a hobby fashionable at the time in the upper classes for investigating the composition of various domestic materials, sometimes after purification of their constituents by crystallisation, though they were seldom able to extract sufficient funds from their father to provide any really sophisticated chemical apparatus. For a while the application of elementary crystallography to his collection of rocks and stones was one of Charles's favourite occupations.

In 1822 Erasmus left school, and was sent to study at Cambridge, where he wrote a helpful series of letters to Charles with detailed instructions for further experiments. This encouraged Charles to examine the effect on different substances of heating them over an open flame, sometimes with a blow-pipe at the gaslight in his bedroom at school, earning him the nickname of 'Gas' and the strong disapproval of the headmaster. Over this period, Charles delighted in devising simple instruments for performance of his tests, and under the tuition of Erasmus served a useful initial apprenticeship in the art of scientific experimentation.

Robert Darwin now decided that Erasmus should proceed from Cambridge to Edinburgh University as he had done himself in order to take an M.B. degree, and that Charles should leave school at the age of sixteen and accompany his brother to Edinburgh in October 1825 with the same object. The plan did not quite work out, for although Erasmus did eventually pass the Cambridge M.B. exam in 1828, his poor health led to his retirement to London as a gentleman of leisure, and he never practised. Charles, on the other hand, having signed up for the traditional courses on anatomy, surgery, the practice of physic, and *materia medica*, the remedial substances used in medicine, which his father and grandfather had taken in their day, soon found that many of the lectures were now sadly out of date, and that conditions in the dissecting room and on two occasions in the operating room were so highly distasteful that he felt unable to continue on the course. It was not until the end of his second year that he was at last able to confess to his father his determination to abandon medicine as a career, but in the meantime Edinburgh provided other avenues to fill his time that assisted materially in his development as a scientist.

During his first year at Edinburgh, Charles took regular walks with Erasmus on the shores of the Firth of Forth, where he made his first acquaintance with some of the marine animals that later occupied him so intensively on the *Beagle*. At the same time he maintained his interest in ornithology, and arranged to have lessons on stuffing birds from a 'blackamoor' who had been taught taxidermy by the naturalist Charles Waterton. He and Erasmus also revived their knowledge of chemistry and related areas of geology by attending the stimulating lectures and demonstrations given by Thomas Charles Hope, Professor of Chemistry in the university from 1799 to 1843.

In 1826, Erasmus had remained at home, and Charles was left to fend for himself. He attended Robert Jameson's popular series of extracurricular lectures covering meteorology, hydrography, mineralogy, geology, botany and zoology, but said many years later that 'they were incredibly dull. The sole effect they produced on me was the determination never as long as I lived to read a book on Geology or in any way to study the science.' Although it was true that Jameson's style of lecturing did not inspire his audience, Charles's copy of Jameson's *Manual of Mineralogy* is heavily annotated, and provided him with a valuable source of practical information for his subsequent geological studies. He also benefited from exposure to the critical clash between Hope's Huttonian views and Jameson's preference for the Wernerian doctrine,* soon coming down firmly on Hope's side. In any case, any temporary prejudice that Charles may have had against geology did not last long, and in due course was banished by Professors Henslow and Sedgwick after his arrival at Cambridge.

In November, Charles became a member of the Plinian Society, named after Pliny the Elder, author of a famous account of the natural history of ancient Rome, at which a small group of undergraduates would meet informally for discussions of natural history or sometimes to go on collecting expeditions, but from participation in which the university professors were traditionally banned. He was also taken as a guest from time to time to the august Wernerian Society, whose membership was restricted to graduates, and whose proceedings were published in a series of learned memoirs. At a meeting of the Wernerian Society on 16 December 1826, he listened attentively to a paper on the buzzard in which the great American ornithologist and artist John James Audubon, who had recently arrived in Edinburgh to find an engraver for the first ten plates of his *Birds of America*, exploded the currently fashionable view of the extraordinary power of smelling possessed by vultures. When nine years later Charles was making observations in Chile on the behaviour of condors, he was happy to find himself in agreement with Audubon's conclusions.

The senior member of the Plinian Society was at that time Robert Grant, then aged thirty-three and a mere lecturer on invertebrate

* James Hutton (1726–97) was a Scottish natural philosopher and geologist who proposed that the earth's internal heat played an important part in the formation of strata of granite and basalt. Abraham Gottlob Werner (1749–1817) was a German mineralogist who held that the ordering of strata was determined by their precipitation in turn from a universal ocean.

animals at an extramural anatomy school, who had graduated as a doctor in 1814, travelled extensively on the Continent, and studied in Paris with the zoologist and anatomist Georges Cuvier (1769–1832). Soon Charles was taken by Grant to collect a variety of animals along the shores in the neighbourhood of Leith, and to go out with fishermen on the waters of the Firth of Forth. On these trips he was sometimes accompanied by another medical student, John Coldstream, who later advised him helpfully about fishing nets. Grant also taught Charles how to dissect specimens under sea water with the aid of a crude single-lens microscope, and gave him a valuable training in marine biology, with an emphasis on the importance of developmental studies on invertebrates, which was taken up with enthusiasm by a pupil who all too quickly outshone his master.

Among Grant's favourite subjects for research were the not very glamorous 'moss animals' of genus *Flustra* that encrusted the tidal rocks in bunches like a miniature seaweed, and which consisted of large numbers of microscopic polyps whose precise relationship with one another was unclear. There had long been controversy as to whether they should be classified as animals or plants, and the Swedish botanist and founder of the system of binomial nomenclature of species Carl Linnaeus (1707–78) had christened them Zoophyta, an intermediate form. By the beginning of the nineteenth century it had been widely but not yet universally accepted that these organisms were indeed sedentary aquatic animals, which formed colonies often containing millions of individual polyps or zooids with specialised functions. In 1830 the phylum* to which they belonged was termed Polyzoa by J. Vaughan Thompson, and nowadays the animals are classified as Bryozoa.

* Phylum is the rank immediately below kingdom in the hierarchy of classification of living organisms established by Linnaeus. The successively lower divisions are class, order, family, genus and species. The generic and species names of an animal or plant are customarily printed in italics. Thus, starting at the bottom, the European *Flustra foliacea* belongs to sub-order Anasca, in class Gymnolaemata, and in phylum Bryozoa.

1827

Charles set to work on the reproductive particles of *Flustra* and other marine animals, and to his great excitement confirmed Grant's observation that the eggs of *Flustra* were coated with fine cilia, hair-like vibrating organs whose coordinated movements endowed the ova with some degree of motility. He also noted that the 'sea pepper-corns' often found attached to old shells were not as previously assumed buttons of seaweed, but were the eggs of the marine leech *Pontobdella muricata*. This he duly reported in his first scientific paper, presented in a talk to the Plinian Society on 27 March 1827; but he had been seriously put out when three days earlier Grant read a long memoir to the Wernerian Society that included his pupil's findings without any proper acknowledgement of their source. What had happened was later described by Charles's daughter Henrietta:

> When he was at Edinburgh he found out that the spermatozoa [ova] of things that grow on seaweed move. He rushed instantly to Prof. Grant who was working on the same subject to tell him, thinking, he wd be delighted with so curious a fact. But was confounded on being told that it was very unfair of him to work at Prof. G's subject and in fact that he shd take it ill if my Father published it. This made a deep impression on my father and he has always expressed the strongest contempt for all such little feelings – unworthy of searchers after truth.[20]

At around the same time, as Charles recalled long afterwards, he had a significant conversation with Grant:

> He one day, when we were walking together, burst forth in high admiration of Lamarck and his views on evolution. I listened in silent astonishment, and as far as I can judge, without any effect on my mind. I had previously read the *Zoonomia* of my grandfather, in which similar views are maintained, but without producing any effect on me. Nevertheless it is probable that the hearing rather early in life such views maintained and praised may have favoured my upholding them under a different form in my *Origin of Species*. At this time I admired

> greatly the *Zoonomia*; but on reading it a second time after an interval of ten or fifteen years, I was much disappointed, the proportion of speculation being so large to the facts given.[21]

When in 1793 Jean Baptiste Pierre Antoine de Monet de Lamarck was appointed as Professor of the 'inferior animals' in Paris, he earned good marks for renaming them in a less uncomplimentary fashion as 'invertebrates', i.e. animals without backbones. He also came up with new and valid reasons for believing in the evolution of new species. But he then spoiled his case by endowing all animals with a special power to interact directly with their environment and acquire ever greater complexity or perfection, supposing for example that the length of a giraffe's neck was the result of the animal constantly reaching up for food, or that the length of an anteater's nose and loss of its teeth resulted from perpetual sniffing into anthills, and was inherited over many generations. In the absence of any good evidence for such an inheritance of acquired characteristics, the term 'Lamarckian' soon had pejorative connotations. The occasion to which Charles referred was possibly the first when Grant revealed his extreme views on transmutation in invertebrates, and metamorphoses in extinct fossils. At the end of 1827 Grant became the first Professor of Zoology and Comparative Anatomy at University College London,* and his strongly Lamarckian approach was more widely disseminated. He held this post until his retirement in 1874, and though he was reported by Charles's friend Frederick William Hope in 1834 to be 'working away at the Mollusca & Infusoria publishing at a great rate', in 1867 he was still teaching a defunct 1830s zoology in a frayed swallow-tail coat. Charles later noted that 'he did nothing more in science – a fact which has always been inexplicable to me'. Grant's excessively radical attitude, coupled with the disillusionment stemming from their falling out that March, may help to

* University College London was the 'godless institution' in Gower Street founded by nonconformists in 1827 to offset the religiously orthodox King's College in the Strand.

explain why their subsequent relations were never close, and there is no suggestion that Charles was ever subjected to the intimate approaches from Grant that may eventually have led to the nervous breakdown suffered by another of his students, John Coldstream.

Robert Darwin was far from pleased with Charles for giving up medicine, and told him angrily, and as Charles thought somewhat unjustly, 'You care for nothing but shooting, dogs and rat-catching and you will be a disgrace to yourself and all your family.'[22] After careful consideration, Robert decided that the only alternative for which there were several precedents in the Darwin and Wedgwood families would be for Charles to go up to Cambridge to take an ordinary Arts degree as the first step towards becoming an Anglican clergyman. In order to fulfil in due course the requirements for entry into the university, he had to brush up the Latin and Greek that he was supposed to have learnt at school, and a private tutor was therefore engaged for the last eight months of 1827. The period was not a very happy one for Charles, though he managed to escape to Uncle Josiah Wedgwood's house at Maer Hall in Staffordshire, seven miles from Stoke-on-Trent, for at least the start of the shooting season, and made his first and only visit to France to collect his youngest Wedgwood cousins from Paris. But he did not record whether he also fitted in a visit to Cuvier's famous Musée d'Histoire Naturelle.

Charles duly matriculated at Christ's College, Cambridge, in January 1828. Here he quickly fell in with a new circle of young men from his own background and sharing his own tastes, one of whom described him at the time as 'rather thick set in physical frame & of the most placid, unpretending & amiable nature'. Some years later, Charles advised his eldest son William at school:

> You will surely find that the greatest pleasure in life is in being beloved; & this depends almost more on pleasant manners, than on being kind with grave & gruff manners. You are almost always kind & only want the more easily acquired external appearance. Depend upon it, that the only way to acquire pleasant manners is to try to please *everybody* you come near, your school-fellows, servants &

everyone. Do, my own dear Boy, sometimes think over this, for you have plenty of sense & observation.[23]

Charles's own amiability and good relations with the rest of the world at every level were always among his most outstanding characteristics, and he had a true genius for friendship.

His new acquaintances included a number of schoolmates from Shrewsbury, and his cousin Hensleigh Wedgwood from Staffordshire, who had earlier seen a lot of Charles's brother Erasmus in Cambridge. By far his closest friend to begin with was a cousin from the other side of the family, William Darwin Fox, the only son of Robert Darwin's cousin Samuel Fox, who was then in his third year

Charles Darwin, drawn in 1840 by George Richmond

at Christ's and due to become a parson in Cheshire. William Darwin Fox's abiding passion, just as Charles's had been in his childhood, was for the collection of exotic natural history specimens that filled every cubic inch of his rooms. He took great pleasure in shooting and riding, and kept two dogs named Fan and Sappho at Christ's, about whose exploits, matching that of Charles's Mr Dash at Shrewsbury, they had a regular correspondence. Fox encouraged Charles to take an interest in both art and music, and together they visited print shops and the Fitzwilliam Museum, and attended concerts of choral works in college chapels. Charles later joined a musical set headed by another good friend, John Maurice Herbert.[24] They went regularly to King's College chapel to hear the anthem sung, and occasionally Charles hired the choristers to perform in his rooms. But as Herbert soon discovered, and as he himself freely admitted, Charles's pleasure in listening was not in fact accompanied by a good musical ear. The interests of the group were wide-ranging, and extended at one time to the formation of the 'Glutton' or 'Gourmet' Club at which they dined when not eating in hall, and consumed a range of animals that did not usually appear on the menu. The club was finally brought to an end by an attempt to eat an old brown owl whose flavour was considered by all to have been 'indescribable'. One wonders what the club would have made of some of the weird dishes later consumed in an experimental spirit by Charles on the *Beagle*.

The principal and most time-consuming occupation to which Charles was introduced by Fox was collecting and learning to identify beetles. Charles recalled later:

> No pursuit at Cambridge was followed with nearly so much eagerness or gave me so much pleasure as collecting beetles. It was the mere passion for collecting, for I did not dissect them and rarely compared their characters with published descriptions, but got them named anyhow. I will give a proof of my zeal: one day, on tearing off some old bark, I saw two rare beetles and seized one in each hand; then I saw a third and new kind, which I could not bear to lose, so that I popped the one which I held in my right hand into my mouth. Alas it ejected

> some intensely acrid fluid, which burnt my tongue so that I was forced to spit the beetle out, which was lost, as well as the third one. I was very successful in collecting and invented two new methods; I employed a labourer to scrape during the winter, moss off old trees and place [it] in a large bag, and likewise to collect the rubbish at the bottom of the barges in which reeds are brought from the fens, and thus I got some very rare species. No poet ever felt more delight at seeing his first poem published than I did at seeing in Stephen's *Illustrations of British Insects* the magic words, "captured by Charles Darwin, Esq."[25]

Public interest in natural history was at that time about to expand hugely, but few people pursued the new hobby with the passion, practical competence and competitiveness displayed by Fox and Charles. After breakfasting together daily, they scoured the fields and ditches closest to them at the 'backs' of the colleges, and the countryside further to the south of Cambridge, often accompanied by a bagman to carry their heavier equipment and their captures. When this man had learnt just what they were after, he would also collect for them when they were otherwise occupied. Returning to Charles's rooms they would go through his reference books, from Lamarck to Stephens, to identify any rarities they had secured, and pin them out on a board for all to admire. On one such occasion after Charles had, after a 'famous chace' in the Fens, caught an especially rare beetle, he was pleased when Leonard Jenyns, vicar of Swaffham Bulbeck, and one of his principal collecting rivals in Cambridge, quickly came round to inspect it. Some years later, it was Jenyns who identified the fishes brought back by Charles in the *Beagle*.

When in June 1828 Fox went down from Cambridge, Charles felt himself 'dying by inches, from not having any body to talk to about insects'. During the next three years, before the departure of the *Beagle*, he and Fox exchanged frequent letters, mainly concerned with entomology. They corresponded regularly during the voyage, and continued to keep closely in touch on family matters until Fox's death in 1880. On sending Fox a copy of *The Descent of Man* in 1870, in

John Stevens Henslow, Professor of Botany

which he had finally faced up to bringing Man into the picture, Charles added: 'It is very delightful to me to hear that you, my very old friend, like my other books.'

In the summer of 1828 Charles and a number of friends went to Barmouth on the coast of Wales as a reading party that was intended to brush up their mathematics, but in the event became more concerned with entomology. Even the unfortunate Herbert, who was severely lame thanks to a deformed foot, was dragged to the tops of the hills in search of beetles, though Charles made up for it by helping to carry him down. The enthusiasm and thoroughness with which Charles pursued his beetles had already become a legend among his contemporaries.

At the beginning of the Michaelmas term, Charles moved into the rooms at Christ's in which he lived for the next three years. A former occupant of the set had been the eighteenth-century theologian William Paley (1743–1805). Perhaps his influence still lingered there, for although Charles devoted very little of his time to theology, he said afterwards that he had greatly appreciated the clarity of Paley's language and the strength of his logic, and regarded his books on *Evidences of Christianity* and *Moral Philosophy* as the only part of the academic course which had helped to educate his mind. However, he now had plenty else to do, for he kept a horse in Cambridge for riding, and having persuaded his father and sisters to provide the funds for a powerful double-barrelled gun with percussion caps, could practise aiming it at a lighted candle in his rooms. His beetles were never neglected, and towards the end of his period in Cambridge he took up once more the study of the inner workings of living cells by microscopy that he had begun in Edinburgh with Robert Grant. This was made possible by a gift from a generous and initially anonymous donor, who later turned out to be Herbert, of the latest compound microscope designed by Henry Coddington, a mathematics tutor at Trinity.

Unquestionably the most important event in Charles's life while he was at Cambridge was his friendship with the Revd Professor John Stevens Henslow, of whom he had been told by his brother Erasmus, and to one of whose Friday evening soirées for scientifically inclined undergraduates and dons he was taken by Fox in 1828. Henslow had first been Professor of Mineralogy in Cambridge for five years, and then became Professor of Botany from 1827 to 1861. He and Adam Sedgwick, the Revd Professor of Geology and Senior Proctor in the University, had founded the Cambridge Philosophical Society in 1819. Together with William Whewell, polymath and later Master of Trinity, Charles Babbage, the designer of calculating engines, and George Peacock, mathematician and Professor of

Adam Sedgwick, Professor of Geology

Astronomy, Sedgwick and Henslow were the leading figures in the development of scientific research and teaching in the university during the first half of the nineteenth century. Henslow's wide-ranging lectures on botany, covering every aspect of the chemistry and biology of plants as well as the essential minimum of their taxonomic classification by Linnaeus, were attended annually by sixty or seventy undergraduates and several professors. The courses included field excursions, sometimes on foot or else in stagecoaches or on a barge drifting down the river to Ely, punctuated by talks on the variety of plants, insects, shells and fossils that had been collected. In the late spring there was always a trip to Gamlingay heath, twenty miles to the west of Cambridge, where rare plants and animals were to be found, and which ended with a convivial social gathering at a country inn. Charles signed up for these activities in 1829, 1830 and 1831. With at least some of Sedgwick's lectures that he also attended in 1831, they constituted the only formal instruction in science that he received at Cambridge.

Speaking of the last two terms at the university after he had passed his Bachelor of Arts examination in January 1831, tenth in the list of candidates who did not seek honours, Charles wrote of Henslow in his *Autobiography*:

> I took long walks with him on most days, so that I was called by some of the dons 'the man who walks with Henslow'; and in the evening I was very often asked to join his family dinner. His knowledge was great in botany, entomology, chemistry, mineralogy, and geology. His strongest taste was to draw conclusions from long-continued minute observations. His judgement was excellent, and his whole mind well-balanced; but I do not suppose that anyone would say that he possessed much original genius.

It is true that by modern standards Charles would not be regarded as having had an orthodox or adequate scientific training. But by the standards of 1831, and remembering the contacts with Grant and the lectures that he had attended in Edinburgh, he was by then as well educated in natural history as any student in the country. And although he had not passed any exams in the subject, he had greatly impressed some of the most eminent scientists in Cambridge with his practical ability as a collector, and with the high quality and purposefulness of his enquiring mind. 'What a fellow that Darwin is for asking questions,' said Henslow.

At around the same time Charles read two books that 'stirred up in me a burning zeal to add even the most humble contribution to the noble structure of Natural Science'. One was the classical account by the German naturalist, geophysicist, meteorologist and geographer Alexander von Humboldt (1769–1859) about his travels through the Brazilian rain forest to the Andes and beyond with the botanist Aimé Bonpland (1773–1858).[26] The second was the recent book by the astronomer and physicist John Herschel (1792–1871) on the study of natural science.[27] Charles insisted on inflicting long readings from Humboldt on his friends, and worked out plans for an expedition to the Canary Islands in July to inspect the volcanic cone of the Pico de Teide on Tenerife, whose summit had been closely inspected by Humboldt in 1799 on his way out to South America. Some of the requirements of his plan were tiresome to meet, such as taking 'intensely stupid' lessons in Spanish, though he was not to know how useful they would prove to have been when later on he was riding with gauchos across the pampas in Patagonia. A number of prospective participants were enlisted, but on enquiring about the sailing of passenger vessels to the Canaries, Charles found that his planning was already too late, for the boats were scheduled for departures only in June. The trip would therefore have to be postponed to 1832.

The combined compass and clinometer used by Charles on the *Beagle*

August 1831

It was pointed out by Henslow that such an enterprise would require a basic knowledge of geology. He therefore advised Charles on the purchase in London of the instrument for the measurement of the inclination of rock beds known as a clinometer, and showed him how to use it. Soon Charles could boast from Shrewsbury that 'I put all the tables in my bedroom at every conceivable angle & direction. I will venture to say I have measured them as accurately as any Geologist going could do.'[28]

Most significantly of all for Charles's training as a geologist, Henslow prevailed on Adam Sedgwick to take Charles with him for part of his usual field excursion during the summer vacation. Sedgwick was renowned as a field geologist, skilled at the recognition of regional patterns of strata from details that were strictly local, and in August he was planning to visit North Wales in continuation of a project to describe all the rocks in Great Britain below the Old Red Sandstone.* The first nights of the trip were spent by Sedgwick with the Darwins at Shrewsbury, where he made a great impression, especially on Charles's sister Susan, often teased by the accusation that 'anything in coat and trousers from eight years to eighty was fair game to Susan'. Charles had been practising his geology in the neighbourhood, and later related the story of the important scientific lesson that he learnt on that occasion:

> Whilst examining an old gravel-pit near Shrewsbury a labourer told me that he had found in it a large worn tropical Volute shell, such as may be seen on the chimney-pieces of cottages; and as he would not sell the shell I was convinced that he had really found it in the pit. I told Sedgwick of the fact, and he at once said (no doubt truly) that it must have been thrown away by someone into the pit; but then

* The Old Red Sandstone is a layer of sedimentary rock rich in iron laid down across England during the Devonian period 410–355 million years ago. It was succeeded by the rocks of the Carboniferous Limestone laid down during the next sixty-five million years, with above them the Coal swamps rich in carbon which existed during the Permian period 290–250 million years ago.

> added, if really embedded there it would be the greatest misfortune to geology, as it would overthrow all that we know about the superficial deposits of the midland counties. These gravel-beds belonged in fact to the glacial period, and in after years I found in them broken arctic shells. But I was then utterly astonished at Sedgwick not being delighted at so wonderful a fact as a tropical shell being found near the surface in the middle of England. Nothing before had ever made me thoroughly realise, though I had read various scientific books, that science consists in grouping facts so that general laws or conclusions may be drawn from them.[29]

Sedgwick had of course appreciated that the shell could not possibly be a genuine find in such a place. His scepticism taught Charles a valuable lesson, and brought home to him the importance of assembling plenty of mutually compatible observations to support any new scientific theory. Thereafter he would keep his mouth tightly shut until sufficient evidence had been accumulated.

Sedgwick's aim was to follow the line of contact along the Vale of Clwyd between the Carboniferous Limestone cliffs and the Old Red Sandstone, shown in the geological map with ribbons crossing the Vale at several points, starting at Llangollen and finishing at Great Ormes Head on the coast.[30] At a quarry near Ruthin they found a possible outcrop of the Old Red, and north of Henllan there was red sand and earth, but Sedgwick was not sure that this established with certainty the nature of the underlying strata. Charles was therefore dispatched on a traverse of his own from St Asaph to Abergele via Betwys-yn-Rhos, crossing a substantial band of Old Red shown on the map. Finding in some places a few loose stones and some reddish soil, he noted: 'It was in such points as these where the strata have been much disturbed, that I observed the greatest number of bits of Sandstone, but in no place could I find it in situ.'[31] Near Abergele the soil was indeed 'very red', but this he attributed 'entirely to the *very* ferruginous [rich in iron] seams in the rock itself, & not to the supposed sandstone beneath it'. That evening he told Sedgwick that there was no true Old Red to be seen, and to the end of his life could

August 1831

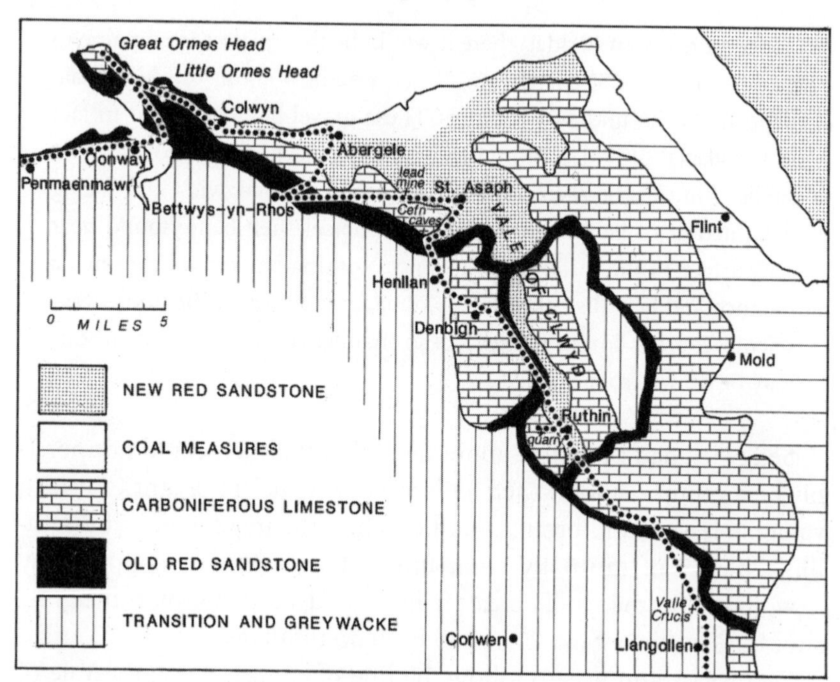

Geological map of part of North Wales, redrawn by Secord[15] after Greenough,[30] with Charles's route from Llangollen to Penmaenmawr as a dotted line. In the second edition of Greenough's map, published in 1839, the Old Red Sandstone had disappeared.

remember how pleased his teacher had been with this new evidence that the Vale of Clwyd did not have a complex structure as had been supposed, but was a simple trough-like syncline* resulting from a stretching of the strata. Although his experience of geology in the field was thus limited to just one week, Charles had sat at the feet of a master, and had solved his first problem with conspicuous success.

Among other sites of geological interest in the Vale were the famous caves in the limestone cliffs at Plas-yn-Cefn, above the River Elwy. Here the owner had excavated vertebrate fossils in the largest cave that included the tooth of a rhinoceros, and there were other bones in the mud. Charles's imagination was fired by the prospect of

* A large downward folding of the geological strata to form a trough.

making similar discoveries from past worlds in his projected trip to the Canaries, though his hopes were not in fact realised until his arrival in 1832 at the cliffs of Punta Alta in Patagonia. After a week, Charles and Sedgwick separated at Capel Curig in the neighbourhood of Bethesda, and Charles strode on across the central mountains of Wales, steering by map and compass, to join some Cambridge friends at Barmouth.

Reaching home at Shrewsbury on 29 August 1831 after two weeks of shooting with his Uncle Jos at Maer, 'for at that time I should have thought myself mad to give up the first days of partridge shooting for geology or any other science', Charles found the fateful letter from Henslow proposing that he should sail on the *Beagle*. The clock must next be turned back to explain its origin.

CHAPTER 2

The Strange Consequences of Stealing a Whale-Boat

On 25 September 1513 the Spanish explorer Vasco Núñez de Balboa crossed the narrow isthmus joining the two halves of the American continent to discover on the far side the Mar del Sur, later named the Pacific Ocean. The town of Panama was built on the Pacific shore, and became the base for a rapid expansion by the Spaniards. While Hernán Cortés was conquering the Aztec empire in Mexico, Francisco Pizarro was overcoming the Incas in Peru. During the next three hundred years prosperous Spanish colonies were established in the western and southernmost parts of South America, while in the east the Portuguese took over a large area in Brazil.

The English were jealous of their success, but for a long time could only benefit from it by robbery, following the example of Sir Francis Drake when he returned from his circumnavigation in 1580 with a rich cargo of treasures stolen from the Spanish colonies at Queen Elizabeth's behest. In 1806 Buenos Aires was attacked by a British force, which was successfully repelled, giving the Argentinians the confidence to join the other Spanish colonies in breaking away from Spain. By 1820 they were all independent countries, though not always at peace with one another.

The Hydrographic Office of the Royal Navy, founded in 1795, was initially responsible for looking after the Admiralty's collection of navigational charts, and of the 'Remark Books' about foreign shores and harbours that all naval captains were required to keep. In 1817 the second Hydrographer, Captain Hurd, was empowered to recruit some surveyors of his own, and had soon built up a programme of a dozen Admiralty surveys in home waters and abroad. Trade was quickly building up with the new governments of South America, and there was a need both for a British naval presence in South American waters, and for accurate charts of the coastline to assist shipping. Hence it came about that:

> In 1825, the Lords Commissioners of the Admiralty directed two ships to be prepared for a Survey of the Southern Coasts of South America; and in May of the following year the ADVENTURE and the BEAGLE were lying in Plymouth Sound, ready to carry the orders of their Lordships into execution. These vessels were well provided with every necessary, and every comfort, which the liberality and kindness of the Admiralty, Navy Board, and officers of the Dock-yards, could cause to be furnished.[32]

HMS *Adventure* was a 'roomy' ship of 330 tons, without guns, under the command of Captain Phillip Parker King. HMS *Beagle* was a smaller vessel of 235 tons, rigged as a barque carrying six guns, and commanded by Captain Pringle Stokes. On 19 November 1826, *Adventure* and *Beagle* sailed south from Monte Video, and until April 1827 carried out surveys in the south of Patagonia and in Tierra del Fuego, around the Straits of Magellan. In June 1827 they arrived back at Rio de Janeiro. Six months later, now accompanied by a schooner named *Adelaide* to assist in the surveys – for whose purchase Captain King had prudently obtained Admiralty approval in advance, unlike Captain FitzRoy when in 1833 he bought the second and smaller *Adventure* – they sailed south again. In January 1828 the *Adventure* was anchored for the winter at Port Famine. Captain Stokes was ordered in the *Beagle* 'to proceed to survey the western coasts, between the Strait of Magalhaens and latitude 47° south, or as much of those dangerous and exposed shores as he could examine', and to return to Port Famine (Puerto Hambre) by the end of July. Captain King allotted himself a more comfortable task in the *Adelaide*, charting the southern parts of the Strait relatively close at hand, and collecting birds and plants.

When at the appointed time the *Beagle* returned to Port Famine with her difficult assignment conscientiously completed, Captain

August 1828

Stokes was found to be in a state of acute depression thanks to the extreme privations and hardships that he and his crew had suffered from very severe weather, both stormy and wet, when working in the Gulf of Peñas. On 1 August 1828 he tried unsuccessfully to shoot himself, and although the surgeons thought for a while that he might recover, he died in great pain on 12 August. He was interred at the *Adventure*'s burial ground, the so-called English Cemetery two miles from Port Famine. (The tablet erected to his memory has since been moved to the Museo Saleciano in the modern town of Punta Arenas, some forty miles away along the Straits of Magellan.)

The *Adventure* and the *Beagle*, temporarily commanded by her First Lieutenant, William Skyring, sailed back to Rio de Janeiro in October for repairs and replenishment of their stores. Here Admiral Otway, Commander-in-Chief of the South American Station, appointed his young Flag Lieutenant, Robert FitzRoy,* to take over command of the *Beagle* in succession to Captain Pringle Stokes. His choice was successful, and FitzRoy had soon overcome the handicap of restoring the morale of a demoralised ship's company well enough to continue the charting of one of the world's most inhospitable coasts.

During 1829 the *Adventure*, *Beagle* and *Adelaide* conducted independent surveys at various points between Tierra del Fuego and Chiloe, coming together at Valparaiso in November. On 19 November the *Beagle* departed to survey more of the southern coasts of Tierra del Fuego before rejoining the *Adventure* at Rio de Janeiro for the final return to England.

---

* Robert FitzRoy was born on 5 July 1805 at Ampton Hall in Suffolk, seat of the Graftons. His family had naval as well as aristocratic connections, the first Duke of Grafton, son of Charles II, having died young as an English Admiral commanding his own ship against the French off Cork in the year 1690. There were two more admirals in later generations, the Lords Augustus and William FitzRoy, but Robert's father, son of the third Duke of Grafton, became a general, landowner and Member of Parliament. His mother, Lady Frances Anne Stewart, was sister of Lord Castlereagh. In February 1818, Robert joined the Royal Naval College at Portsmouth, and after nine years at sea had become Admiral Otway's well trusted Flag Lieutenant, ripe for promotion. See H.E.L. Mellersh. *FitzRoy of the Beagle*. Rupert Hart-Davis, London, 1968.

January 1830

Working among the Camden and Stewart Islands to the south of the mouth of the Cockburn Channel, FitzRoy found tiresome anomalies in his compass bearings, and wrote in his journal for 24 January 1830:

> There may be metal in many of the Fuegian mountains, and I much regret that no person in the vessel was skilled in mineralogy, or at all acquainted with geology. It is a pity that so good an opportunity of ascertaining the nature of the rocks and earths of these regions should have been almost lost. I could not avoid often thinking of the talent and experience required for such scientific researches, of which we were wholly destitute; and inwardly resolving, that if ever I left England again on a similar expedition, I would endeavour to carry out a person qualified to examine the land; while the officers and myself would attend to hydrography.[33]

A week later it was reported to FitzRoy that the ship's five-oared whale-boat, manned by Mr Murray, the Master, and a small crew, had been stolen during the night by the Fuegians near Cape Desolation, 'now doubly deserving of its name'. The bad news was brought to the *Beagle* by two of the sailors, paddling a basket-like canoe that they had thrown together for the purpose, and whose curious structure was commemorated in the names given both to the small island on which Cape Desolation was located, and to the first of the Fuegians taken hostage by FitzRoy.*

* Until the ending of the last ice age about 14,500 years ago, the southern tip of South America was wholly covered by ice and uninhabitable. A period of some thousands of years followed during which the nomad hunter-gatherers, who after crossing the Bering Straits into Alaska had spread steadily down the western coast of the American continent, found it possible to proceed still further to the south to occupy the desolate shores facing the Antarctic Peninsula. There was then a rising of sea levels eight thousand years ago that flooded the Straits of Magellan to leave the Fuegians trapped in the Isla Grande de Tierra del Fuego, battling for survival in what must have been the most hostile environment faced anywhere on earth by primitive man. Nevertheless, living almost naked in their temporary wigwams, and feeding mainly on birds' eggs, fish, shells and berries, survive they did. At the beginning of the nineteenth century, the total population of aborigines in Tierra del Fuego was probably close to ten thousand.

Robert FitzRoy's drawings of his Fuegians

May 1830

On the *Beagle*'s map of the Strait of Magalhaens (*sic*),[34] Basket Isle was inserted near the western end of Tierra del Fuego, with Thieves Sound to the north, and Whale Boat Sound to the east. In a modern map the area lies on the coast of Tierra del Fuego due south of Punta Arenas. FitzRoy responded to the theft with a campaign to capture hostages for return of the whale-boat, but the move failed, largely because the Fuegians showed no interest in exchanging their booty for their comrades, who remained quite happily on the ship. So he was left with the young girl Fuegia Basket, 'as broad as she was high', and the men York Minster, taken in Christmas Sound near the cliff of that name, and Boat Memory captured later nearby. He soon began to appreciate the practical difficulties that would arise in returning them immediately to their own peoples, and to consider the possibility of taking them back to England for a period of education before they were repatriated.

While a replacement for the whale-boat was being built at Doris Cove, situated on an island beside Adventurer Passage, Mr Murray was dispatched in the ship's cutter to explore the waters to the north and east of Nassau Bay. Not far to the north, but a long way to the east, he sailed through a channel little more than a third of a mile wide which became known as the Murray Narrow, and which 'led him into a straight channel, averaging about two miles or more in width, and extending nearly east and west as far as the eye could reach'. He had discovered the Beagle Channel, whose precise orientation on the map would provide grounds for legal dispute long afterwards in arguments between Argentina and Chile over territorial rights in the Antarctic. The new country was thickly populated, and on 11 May 1830, when the *Beagle* herself was in the Murray Narrow, some canoes full of natives anxious for barter were encountered. FitzRoy wrote: 'I told one of the boys in a canoe to come into our boat, and gave the man who was with him a large shining mother-of-pearl button.'[35] Jemmy Button, as the boat's crew called him, quickly settled down in his new surroundings, and there were now four Fuegians in FitzRoy's little group.[36]

At the end of June the *Beagle* sailed back to the Rio Plata. While in

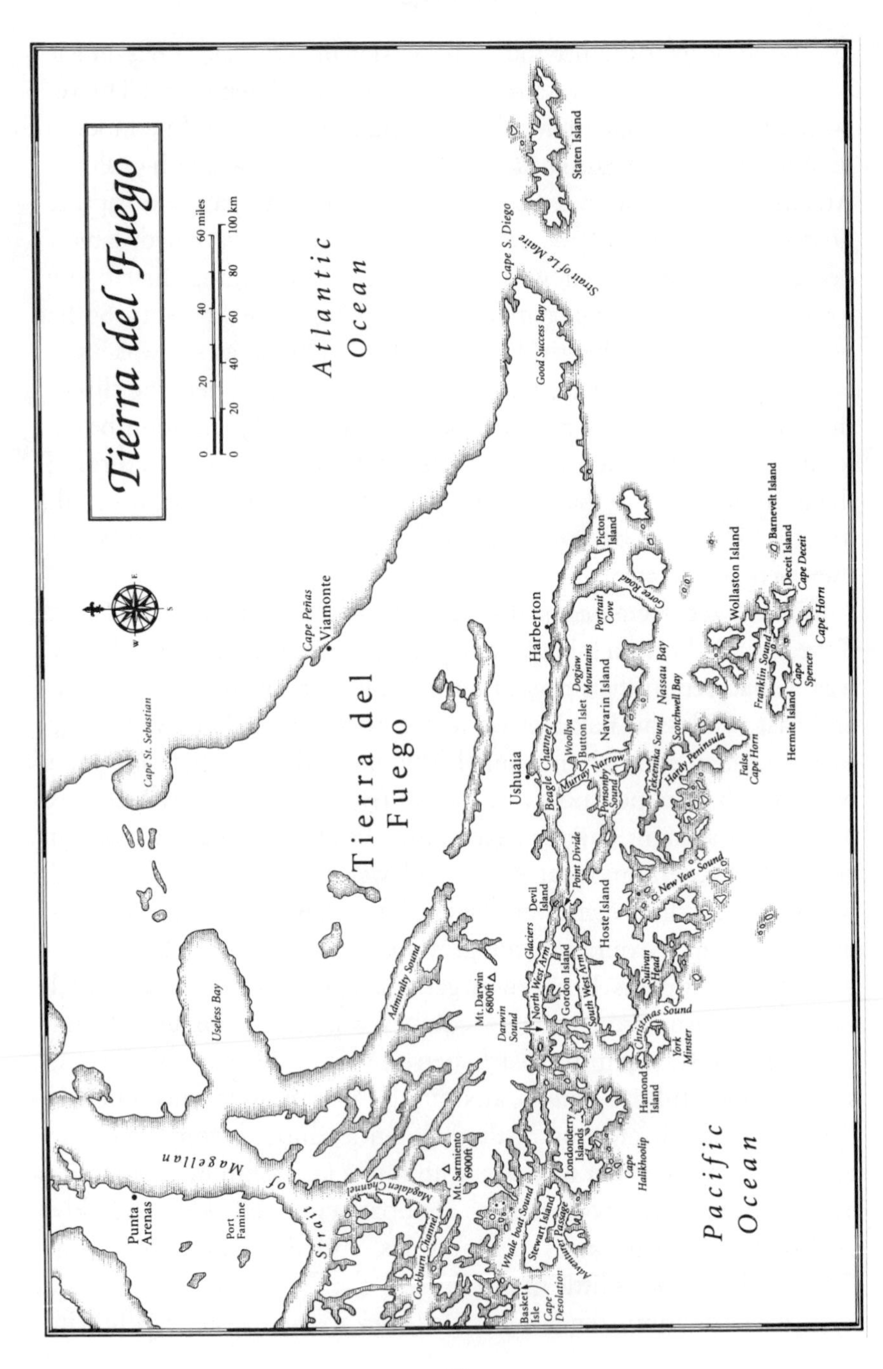
Tierra del Fuego
0 20 40 60 miles
0 20 40 60 80 100 km
Atlantic Ocean
Pacific Ocean
Tierra del Fuego
Cape St. Sebastian
Cape Peñas
Viamonte
Good Success Bay
Cape S. Diego
Strait of Le Maire
Staten Island
Harberton
Picton Island
Gorée Road
Portrait Cove
Ushuaia
Beagle Channel
Woollya
Button Islet
Dogjaw Mountains
Navarin Island
Murray Narrow
Ponsonby Sound
Tekeenika Sound
Nassau Bay
Scotchwell Bay
Hardy Peninsula
False Cape Horn
Wollaston Island
Barnevelt Island
Deceit Island
Cape Deceit
Cape Horn
Franklin Sound
Hermite Island
Cape Spencer
Point Divide
Devil Island
Glaciers
Hoste Island
New Year Sound
North West Arm
Gordon Island
South West Arm
Sullivan Head
Christmas Sound
York Minster
Mt. Darwin 6800ft
Darwin Sound
Admiralty Sound
Useless Bay
Hamond Island
Londonderry Islands
Cape Halikhoolip
Mt. Sarmiento 6900ft
Strait of Magellan
Magdalen Channel
Punta Arenas
Port Famine
Cockburn Channel
Whale boat Sound
Stewart Island
Adventure Passage
Basket Isle
Cape Desolation

Monte Video, FitzRoy tried to have the Fuegians vaccinated against smallpox, whose ravages were all too often fatal to unprotected natives, but the vaccination did not take. At the beginning of August the *Beagle* rejoined the *Adventure* in Rio de Janeiro, and together they made a 'most tedious' passage to Plymouth, where they anchored on 14 October.

FitzRoy's first thought was for the Fuegians. Landing after dark, they were taken to lodgings where next day they were vaccinated for the second time. With the *Beagle*'s coxswain James Bennett to look after them, they were then transferred to a farmhouse in the country near Plymstock, where they could enjoy the fresh air and hopefully avoid infection by other virus diseases, without attracting public attention. Meanwhile the *Beagle* was stripped and cleared out, and on 27 October her pendant was hauled down.

During the voyage home, FitzRoy had addressed through Captain King to John Barrow,[37] Second Secretary of the Admiralty, a long account of the manner in which he had taken the four Fuegians on board the *Beagle*,[38] and of his proposal to return them to their country after they had received some education. Mr Barrow's response, although it was negatively worded and predictably lacking in enthusiasm, said that their Lordships would not interfere with FitzRoy's benevolent intentions towards the Fuegians, would afford him facilities towards their maintenance and education, and would give them a passage home again. Their Lordships' promise was duly kept when early in November Boat Memory was taken ill with smallpox, and instructions were at once given for the Fuegians to be admitted to the Royal Naval Hospital at Plymouth for vaccination and treatment. Unhappily Boat Memory, who was FitzRoy's favourite among them, could not be saved, but the other three were successfully re-vaccinated. Fuegia Basket was in addition taken home by the doctor in charge of them in order to be exposed to measles with his own children. She duly had a favourable attack and quickly recovered with a strengthened immune system.

Through contacts with the Church Missionary Society, the Fuegians were next taken to Walthamstow just outside London for schooling in charge of the Revd William Wilson, and remained in his care until October 1831, still with James Bennett to keep an eye on them. Fuegia Basket and Jemmy Button were very receptive pupils, but the older man York Minster was not. He would reluctantly assist with practical activities like gardening, but firmly refused to learn to read. He also took what seemed to be an unhealthy interest in the ten-year-old Fuegia, following her everywhere, keeping her well away from other men, and treating her as if she was his personal possession. At this time there was no suggestion that anything sexual took place between them, though on board the *Beagle* later on she was deemed to be officially engaged to York in order to avoid embarrassment, and back in Tierra del Fuego she did become his wife. During that summer the Fuegians were taken to St James's Palace at King William IV's request, and Queen Adelaide honoured Fuegia Basket by placing one of her own bonnets on the girl's head and a ring on her finger, and gave her some money to buy clothes for returning home.

FitzRoy had been led by Captain King to suppose that the *Adventure* and *Beagle*'s surveys in South America would need to be continued by some other ship, giving him an opportunity to restore the Fuegians to their native land. But having in March 1831 completed his official obligations with respect to the *Beagle*'s 1826–1830 cruise, for which he was officially commended, FitzRoy discovered that the Admiralty's plans had for no stated reason been altered, and that their Lordships no longer intended to complete the survey. Feeling that he could not trust anyone but himself to return the Fuegians to the precise places from which they had been taken, he obtained twelve months' leave of absence from the Navy. In June he made at his own expense an agreement with the owner of a small merchant ship to take him with five companions, the Fuegians, and a number of goats to Tierra del Fuego, where he proposed to stock some of the islands with goats and deposit his protégée and protégés. This agreement did not, however, have to be put into effect, for FitzRoy happened one day to mention his problem to one of his aristocratic and politically influential uncles, the fourth Duke of Grafton, and the former Foreign Secretary Lord Castlereagh's half-brother Lord Londonderry. After some effective wire-pulling at the Admiralty, their Lordships were persuaded to appoint FitzRoy to command the *Beagle* once again for a second surveying cruise.

The greatest of hydrographers, Captain Francis Beaufort, who had taken charge of the Hydrographic Office in 1829, embraced with enthusiasm the opportunity of filling in some of the many blank spaces in the existing maps of the coast of Argentina and Tierra del Fuego, and extending the naval charts to cover not only Argentina and the Falkland Islands, but also more of the coasts of Chile and Peru as far north as Ecuador. FitzRoy would also be entrusted with the task of carrying a chain of meridian distances,

FIGURES

TO DENOTE THE FORCE OF THE WIND.

0	Calm.		
1	Light Air........	Or just sufficient to give steerage way.	
2	Light Breeze	Or that in which a man-of-war, with all sail set, and clean full, would go in smooth water from	1 to 2 knots.
3	Gentle Breeze ..		3 to 4 knots.
4	Moderate Breeze		5 to 6 knots.
5	Fresh Breeze	Or that to which a well-conditioned man-of-war could just carry in chase, full and by	Royals, &c.
6	Strong Breeze....		Single-reefed topsails and top-gall. sails.
7	Moderate Gale ..		Double-reefed top-sails, jib, &c.
8	Fresh Gale		Treble-reefed top-sails, &c.
9	Strong Gale......		Close-reefed topsails and courses.
10	Whole Gale	Or that with which she could scarcely bear close-reefed main-topsail and reefed fore-sail.	
11	Storm	Or that which would reduce her to storm stay-sails.	
12	Hurricane	Or that which no canvass could withstand.	

The Beaufort Scale

which measured the difference in longitude between an established location and a new one, all the way round the world by sailing back across the Pacific. The *Beagle* was therefore instructed to return via the Galapagos Islands, Tahiti, New Zealand, Australia – calling at Port Jackson observatory in Sydney, Hobart, and King George Sound – the Cocos Keeling Islands, Mauritius, the Cape of Good Hope, St Helena, Ascension, and so home. Beaufort's long Memorandum to FitzRoy,[39] carefully explaining this plan, included a note forbidding senior officers whom he might encounter to take from him any of his instruments or chronometers; instructions for sedulous observations of the eclipses of Jupiter's third and fourth satellites; and advice on the best way of handling natives. Lastly, the *Beagle* was the first ship in the Navy to be issued with Beaufort's list of the Figures, still in popular use today, to denote the force of the wind, based at the lower end on the speeds at which a man-of-war with all sails set would be driven, and at the upper end on what set of sails could just be carried safely at full chase. A second list of letters was drawn up to describe the state of the weather, but this has now fallen out of use.

While the *Beagle* was being extensively refitted at Devonport in preparation for her long voyage, FitzRoy, remembering his resolution to recruit a geologist should he pay another visit to Tierra del Fuego,[30] set about finding 'some well-educated and scientific person who would willingly share such accommodations as I had to offer, in order to profit by the opportunity of visiting distant countries yet little known'.[40] He began by consulting the most appropriate person at the Admiralty, Francis Beaufort, who being closely in touch with the scientific reformers at Cambridge and in the Royal Society was keen to modernise and bring more science into the Hydrographic Office, and was immediately sympathetic. Beaufort accordingly wrote to his mathematical friend George Peacock at Trinity College, Cambridge, telling him of the opening for 'a savant' on a surveying ship. Early in August, Peacock passed the news on to Henslow, although he had not perfectly interpreted the situation in speaking of a vacancy specifically for a naturalist, and in later correspondence placed greater emphasis on FitzRoy's need for a companionable and gentlemanly scientist:

> My dear Henslow
>
> Captain Fitz Roy is going out to survey the southern coast of Terra del Fuego, & afterwards to visit many of the South Sea Islands & to return by the Indian Archipelago: the vessel is fitted out expressly for scientific purposes, combined with the survey: it will furnish therefore a rare opportunity for a naturalist & it would be a great misfortune that it should be lost:
>
> An offer has been made to me to recommend a proper person to go out as a naturalist with this expedition; he will be treated with every consideration; the Captain is a young man of very pleasing manners (a nephew of the Duke of Grafton), of great zeal in his profession & who is very highly spoken of; if Leonard Jenyns could go, what treasures he might bring home with him, as the ship would be placed at his disposal, whenever his enquiries made it necessary or desirable; in the absence of so accomplished a naturalist, is there any person whom you could strongly recommend: he must be such a person as would do credit to our recommendation.
>
> Do think on this subject: it would be a serious loss to the cause of natural science, if this fine opportunity was lost.
>
> The ship sails about the end of Sept^{r}.
>
> Poor Ramsay!* what a loss to us all & particularly to you.
>
> Believe me / My dear Henslow / Most truly yours /
>
> George Peacock
>
> 7 Suffolk Street / Pall Mall East

* Marmaduke Ramsay, a Fellow and tutor at Jesus College, who had died.

August 1831

My dear Henslow

I wrote this letter on Saturday, but I was too late for the post. What a glorious opportunity this would be for forming collections for our museums: do write to me immediately & take care that the opportunity is not lost.

Believe me / My dear Henslow / Most truly yours /

Geo Peacock

7 Suffolk St. / Monday[41]

As has already been seen, Leonard Jenyns was another clerical naturalist, brother-in-law of Henslow and vicar of Swaffham Bulbeck near Cambridge. After a day's consideration of the offer, Jenyns decided regretfully that he could not leave his parish. Henslow therefore turned to Charles Darwin as the obvious alternative choice, and on 24 August wrote:

> My dear Darwin, Before I enter upon the immediate business of this letter, let us condole together upon the loss of our inestimable friend poor Ramsay of whose death you have undoubtedly heard long before this. I will not now dwell upon this painful subject as I shall hope to see you shortly fully expecting that you will eagerly catch at the offer which is likely to be made you of a trip to Terra del Fuego & home by the East Indies. I have been asked by Peacock who will read & forward this to you from London to recommend him a naturalist as companion to Capt Fitzroy employed by Government to survey the S. extremity of America. I have stated that I consider you to be the best qualified person I know of who is likely to undertake such a situation. I state this not on the supposition of y[r] being a *finished* Naturalist, but as amply qualified for collecting, observing, & noting anything worthy to be noted in Natural History. Peacock has the appointment at his disposal & if he can not find a man willing to take the office, the opportunity will probably be lost. Capt. F wants a man (I understand) more as a companion than a mere collector & would not take

* 'The bailiff that is close at the debtor's back, or that catches him in the rear.' – *O.E.D.*

> any one however good a Naturalist who was not recommended to him likewise as a *gentleman*. Particulars of salary &c I know nothing. The Voyage is to last 2 y^{rs} & if you take plenty of Books with you, any thing you please may be done – You will have ample opportunities at command – In short I suppose there never was a finer chance for a man of zeal & spirit. Capt. F is a young man. What I wish you to do is instantly to come to Town & consult with Peacock (at N° 7 Suffolk Street Pall Mall East or else at the University Club) & learn further particulars. Don't put on any modest doubts or fears about your disqualifications for I assure you I think you are the very man they are in search of – so conceive yourself to be tapped on the Shoulder by your Bum-Bailiff* & affect^{e} friend /J.S. Henslow[42]

This letter was reinforced in similar terms two days later by another from Peacock. Although both Peacock and Henslow said in their letters that FitzRoy was looking for a naturalist, it is evident that at some point in FitzRoy's original conversation with Beaufort a geologist had been mentioned, for Henslow's candidate for the post was described by FitzRoy himself as 'Mr. Charles Darwin, grandson of Dr. Darwin the poet, a young man of promising ability, extremely fond of geology, and indeed all branches of natural history.'[43] That FitzRoy thought he was primarily getting a geologist would be consistent with his gift to Charles on their departure from Plymouth of the first volume of Charles Lyell's *Principles of Geology*. Moreover in his first report to the Royal Geographical Society on the *Beagle*'s return to England in 1836 he said that 'Mr Charles Darwin will make known the results of his five years' voluntary seclusion and disinterested exertions in the cause of science. Geology has been his principal pursuit.'

CHAPTER 3

Preparations for the Voyage

Charles arrived back in Shrewsbury on Monday, 29 August from his trip in North Wales with Sedgwick, and was given Peacock's and Henslow's letters by his sisters. His immediate and joyful reaction was to accept, but finding next morning that his father was strongly opposed to the scheme, he wrote sorrowfully to Henslow:

> Mr Peacock's letter arrived on Saturday, & I received it late yesterday evening. As far as my own mind is concerned, I should think, *certainly* most gladly have accepted the opportunity, which you so kindly have offered me. But my Father, although he does not decidedly refuse me, gives such strong advice against going, that I should not be comfortable if I did not follow it. My Fathers objections are these: the unfitting me to settle down as a clergyman; my little habit of seafaring; the *shortness of the time* & the chance of my not suiting Captain Fitzroy. It is certainly a very serious objection, the very short time for all my preparations, as not only body but mind wants making up for such an undertaking. But if it had not been for my father, I would have taken all risks . . . Even if I was to go, my Father disliking would take away all energy, & I should want a good stock of that. Again I must thank you; it adds a little to the heavy, but pleasant load of gratitude which I owe to you.[44]

A letter in similar terms that he also wrote to Peacock has not survived.

All was not lost, however, for Robert had recognised the considerable compliment that had been paid to his son by the two eminent academics in Cambridge, and tempered his disapproval by telling Charles, 'If you can find any man of common sense who advises you to go, I will give my consent.' He well knew who that man might be, and wrote to Josiah Wedgwood II on 30 August: 'Charles will tell you of the offer he has had made to him of going for a voyage of discovery for 2 years. – I strongly object to it on various grounds, but I will not detail my reasons that he may have your unbiassed opinion on the subject, & if you feel differently from me I shall wish him to follow your advice.'[45]

Charles himself rode straight over to Maer, where he found his uncle and cousins full of enthusiasm for his embarking on the voyage, and by the evening all were urging him to reopen the case with his father. On 31 August, with Uncle Josiah at his elbow, he wrote an extremely apologetic note to his father, ending on a separate piece of paper with the list of objections to be answered:

> I am afraid I am going to make you yet again very uncomfortable. I think you will excuse me once again stating my opinions on the offer of the Voyage. My excuse and reason is the different way all the Wedgwoods view the subject from what you & my sisters do . . . But pray do not consider that I am so bent on going, that I would for one single moment hesitate if you thought that after a short period, you should continue uncomfortable.

(1) Disreputable to my character as a Clergyman hereafter
(2) A wild scheme
(3) That they must have offered to many others before me, the place of Naturalist
(4) And from its not being accepted there must be some serious objection to the vessel or expedition
(5) That I should never settle down to a steady life hereafter
(6) That my accomodations [*sic*] would be most uncomfortable
(7) That you should consider it as again changing my profession
(8) That it would be a useless undertaking[46]

August 1831

Enclosed with this letter was one from Josiah to Robert:

My dear Doctor I feel the responsibility of your application to me on the offer that has been made to Charles as being weighty, but as you have desired Charles to consult me I cannot refuse to give the result of such consideration as I have been able to give it. Charles has put down what he conceives to be your principal objections & I think the best course I can take will be to state what occurs to me upon each of them.

1— I should not think that it would be in any degree disreputable to his character as a clergyman. I should on the contrary think the offer honorable to him, and the pursuit of Natural History, though certainly not professional, is very suitable to a clergyman. 2— I hardly know how to meet this objection, but he would have definite objects upon which to apply himself, and might acquire and strengthen habits of application, and I should think would be as likely to do so in any way in which he is likely to pass the next two years at home. 3— The notion did not occur to me in reading the letters & on reading them again with that object in my mind I see no ground for it. 4— I cannot conceive that the Admiralty would send out a bad vessel on such a service. As to objections to the expedition, they will differ in each mans case & nothing would, I think, be inferred in Charles's case if it were known that others had objected. 5— You are a much better judge of Charles's character than I can be. If, on comparing this mode of spending the next two years, with the way in which he will probably spend them if he does not accept this offer, you think him more likely to be rendered unsteady & unable to settle, it is undoubtedly a weighty objection. Is it not the case that sailors are prone to settle in domestic and quiet habits. 6— I can form no opinion on this further than that, if appointed by the Admiralty, he will have a claim to be as well accommodated as the vessel will allow. 7— If I saw Charles now absorbed in professional studies I should probably think it would not be advisable to interrupt them, but this is not, and I think will not be, the case with him. His present pursuit of knowledge is in the same track as he would have to follow in the expedition. 8— The undertaking would be useless as regards his profession, but looking upon him as a man of enlarged curiousity, it affords him such an opportunity of seeing men and things as happens to few.

You will bear in mind that I have very little time for consideration & that you and Charles are the persons who must decide. I am / My dear Doctor / Affectionately yours / Josiah Wedgwood[47]

Uncle Jos had thus with calm sense disposed effectively of Robert's rather exaggerated qualms, and his description of Charles as 'a man of enlarged curiousity' was truly prophetic.

The two letters were dispatched to Shrewsbury early on 1 September, leaving Josiah and Charles to take out their guns for what would normally have been a specially enjoyable occasion for them, the opening day of the partridge season. They both had much on their minds, and Charles had only brought down a single bird when at ten o'clock Josiah did what 'few uncles would have done', bundled him into a carriage, and whisked him off to Shrewsbury to do battle in person with Robert. But there they found to their relief that Robert had already changed his mind, and was now ready to give 'all the assistance in my power'.

On this same day Beaufort had conveyed the news to Robert FitzRoy that 'I believe my friend Mr Peacock of Trin^{y} College Camb^{e} has succeeded in getting a "Savant" for you – A Mr Darwin grandson of the well known philosopher and poet – full of zeal and enterprize and having contemplated a voyage on his own account to S. America.' And that afternoon Charles himself sat down at Shrewsbury and wrote for the first time directly to Beaufort, explaining that the situation had changed since he had sent his refusal to Henslow and Peacock, and that if the appointment was still unfilled 'I shall be very happy to have the honor of accepting it'.

On Saturday, 2 September, Charles returned to Cambridge, and spent Sunday closeted with Henslow, 'thinking what is to be done'. Henslow explained how he himself had nearly accepted the appointment, but turned it down because of his wife's unhappiness at the

Robert FitzRoy sketched at the time of the voyage (P.G. King)

prospect, while Leonard Jenyns had done the same because he could not desert his parish at Swaffham Bulbeck. At Henslow's, Charles met Alexander Charles Wood, a cousin and good friend of FitzRoy, and currently an undergraduate at Trinity College, Cambridge, and pupil of Peacock. At Peacock's urging, Wood had written to recommend Charles to FitzRoy, which he had dutifully done, taking the precaution of warning his Tory cousin that Charles was politically a Whig and decidedly liberal in outlook. FitzRoy had just replied in a letter which Charles felt was '*most* straightforward & *gentlemanlike*, but so much against my going, that I immediately gave up the scheme'. However,

Henslow firmly urged him not to make up his mind until he had had serious consultations with Beaufort and FitzRoy, so on Monday morning he took the coach to London, and had his first encounter with his sometimes erratic 'beau ideal of a captain'.

The two young men, FitzRoy being then only twenty-six years old, and Charles four years younger, at once put themselves out to be agreeable. Charles wrote to his sister Susan, 'Cap. Fitzroy is in town & I have seen him; it is no use attempting to praise him as much as I feel inclined to do, for you would not believe me.' FitzRoy proposed that they would mess together, with no wine and the plainest dinners, fully sharing such limited working space as was available in so small a vessel. The trip would inevitably be stormy and uncomfortable, though if Charles found it too much for him he would always be at liberty to withdraw, and during the worst weather he might spend two months ashore in some healthy, safe and nice country. Such openness quickly restored all Charles's enthusiasm for the voyage, leaving only a slight doubt – soon laid to rest by Beaufort – as to whether the *Beagle* would indeed sail back across the Pacific and thus circumnavigate the world. Charles wrote later that:

> On becoming very intimate with FitzRoy I heard that I had run a very narrow risk of being rejected, on account of the shape of my nose! He was an ardent disciple of Lavater,* and was convinced that he could judge a man's character by the outline of his features; and he doubted whether anyone with my nose could possess sufficient energy and determination for the voyage. But I think he was afterwards well-satisfied that my nose had spoken falsely.

To his sister Charles concluded, 'There is indeed a tide in the affairs of men, & I have experienced it, & I had *entirely* given it up until 1 to day.'

There ensued some frenzied activity in London, while Charles followed up introductions given him by Henslow to experts for advice

* Johann Kaspar Lavater (1741–1801), Swiss theologian and mystic, chiefly known for his work on determining character from facial characteristics.

of all kinds, from naturalists knowledgeable on what to collect and how best to preserve the specimens, to suppliers of instruments, glassware, paper, books and guns. His sisters looked after his wardrobe, buying him strong new pairs of shoes and shirts marked with his name for the ship's laundry. His friend John Coldstream advised him to use an oyster-trawl *of the ordinary size* for collecting marine animals, to provide himself with a few lobster pots, and when at anchor to 'shoot' some deep-sea fishing lines baited with small pieces of worm-eaten wood and hooks. He met Captain P.P. King, FitzRoy's senior officer on the previous expedition, and, unasked, King said that FitzRoy's temper was perfect, and that he was sending his own son Philip Gidley King on the *Beagle* as a midshipman. Charles lashed out and bought an expensive portable dissecting microscope made by Bancks, of the type recommended to him by the eminent botanist and microscopist Robert Brown. With the help of the bookseller and ornithologist William Yarrell he ordered a rifle and a brace of pistols for £50, flattering himself with the thought that FitzRoy would have spent £400. They might be needed, he told a friend, 'for we shall have plenty of fighting those d— Cannibals'. On 8 September, when all the shops were closed, he found an excellent seat from which to watch the procession at the coronation of King William IV, who 'looked very well, & seemed popular: but there was very little enthusiasm, so little that I can hardly think there will be a coronation this time 50 years'. In other spare moments he worked at astronomy, 'as I suppose it would astound a sailor if one did not know how to find Lat & Long'.

On 11 September Charles embarked with FitzRoy on the three-day passage in a steam packet along the south coast to Plymouth, where his first view of the *Beagle* in the dockyard at Devonport was an unflattering one, 'without her masts or bulkheads, & looking more like a wreck than a vessel commissioned to go round the world'. The trip gave FitzRoy an initial chance to evaluate Charles's sea legs, and to indoctrinate him more fully in the Admiralty's plans for the *Beagle* and what had been achieved during her previous voyage. The ship was disconcertingly small, ninety feet long and twenty-four-and-a-half feet wide amidships, with a displacement of 235 tons. Inside the poop cabin, measuring ten feet by eleven, and filled mainly by a large chart table and three chairs, Charles would work with Midshipman Philip Gidley King and the Assistant Surveyor John Lort Stokes (no relative of the *Beagle*'s former Captain), and sleep in a hammock slung above the table. Stokes's bed was in a cubicle outside the door, and King's was on a lower deck. They would eat in the gunroom, but Charles would have the run of the Captain's cabin on the deck below, where they would take their meals together. Sixty years later, King drew from memory for an edition of Charles's *Journal of Researches* a picture of the internal layout of the ship.

Charles sped back to London by coach, 'wonderful quick travelling, 250 miles in 24 hours'. Here he had a useful talk to Beaufort, who told him that the normal naval practice was for any collections made by the ship's surgeon automatically to become the property of the government. But as Charles's appointment was not an official one, he would be best advised to retain for himself the disposal of his collection amongst the different bodies in London. Which of those bodies would be the most suitable ones was an issue that greatly concerned him in subsequent correspondence with Henslow. But the first step was to spend a couple of days in Cambridge, making detailed arrangements for Henslow and a brother of his in London to receive, and keep safely in store until the return of the *Beagle*, consignments of specimens that in due course would be shipped from South America. Henslow's parting gift to Charles was a copy of an English translation of the first two volumes of Humboldt's *Personal narrative of travels to the equinoctial regions of the new continent*, inscribed 'J.S. Henslow to his friend C. Darwin on his departure from England upon a voyage round the world. 21 Sept. 1831.'

Back at home in Shrewsbury for his last ten days, Charles packed up his clothes and books, and settled his complicated and sometimes overstrained financial affairs. He wrote to Henslow that his father was much more reconciled to the idea of the voyage now that he had become accustomed to it, and was in fact treating him with a generosity wholly belying the reputation he once had for being an

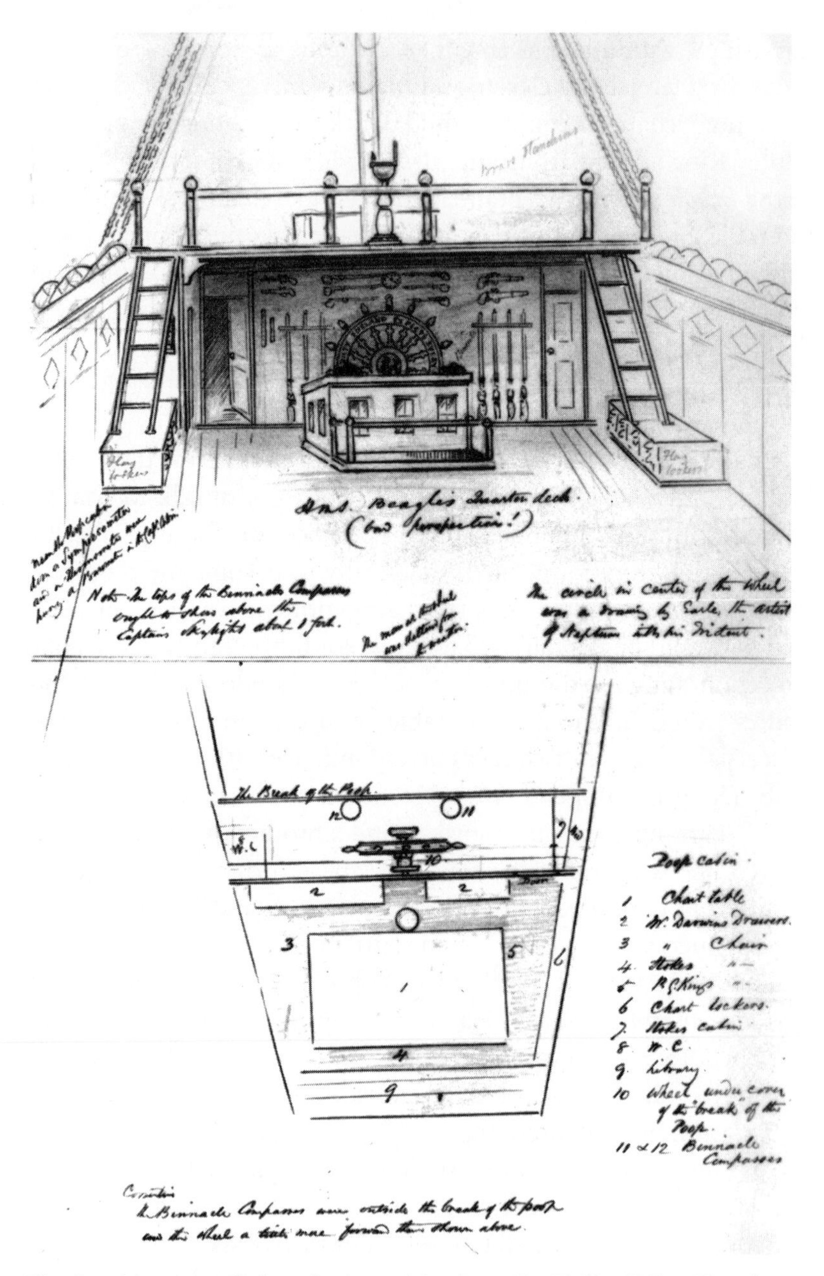

The *Beagle*'s quarterdeck and poop cabin, drawn by Philip Gidley King from memory in 1890 for Mr Hallam Murray

extremely severe parent. Earlier in the summer 'the Governor' had handed over a £200 note, no less, to meet Charles's debts at Cambridge. The cost of equipping him for the voyage amounted to some £600, not far from that of two years' support at Cambridge, while the Admiralty was exacting £50 per annum for his board and lodging. Charles tried to console his father by saying that he would have to be 'deuced clever to spend more than my allowance whilst on board the *Beagle*', to be answered with a smile, 'But they all tell me you are very clever.'

On 2 October, Charles said goodbye to his father and sisters. But his best-loved neighbour Fanny Owen, 'the prettiest, plumpest, Charming personage that Shropshire possesses', to whom in their correspondence he was 'Dr Postillion' and she was 'the Housemaid', was away from home in Exeter, and could only write two long letters expressing her grief that she would not see him for two years or more, 'when we *must* be grown *old* & steady'. Charles went to London, and thanks to further delays in the *Beagle*'s readiness to sail imposed by the dockyard, remained there for three weeks. During this period he wrote to FitzRoy to apologise for the bulkiness of the luggage that he had dispatched to Devonport, and was assured that it was acceptable. In a second letter accompanying yet another parcel containing some talc, which Charles may have required for taxidermy, he enquired whether FitzRoy had a good set of mountain barometers, for 'Several great guns in the Scientific World have told me some points in geology to ascertain, which entirely depend on their relative height.' FitzRoy's reply is not recorded, but Charles did obtain a set of aneroid barometers, of which he made extensive use during the voyage in his investigations of the rise and fall of the land on either side of the Andes. His copy of *Jones's companion to the mountain barometer & tables* survives among his papers in the Cambridge University Library.

After a pleasant drive from London, Charles arrived in Devonport on 24 October, and next day found the *Beagle* 'moored to the *Active* hulk & in a state of bustle & confusion'. The carpenters were busy fitting up the drawers in the poop cabin, and 'My own private corner

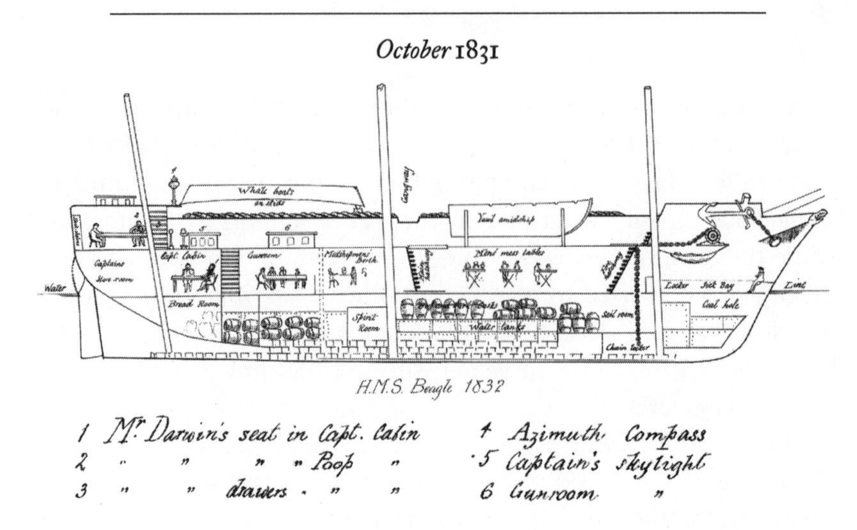

Press copy of Philip Gidley King's drawing of the *Beagle*'s layout

looks so small that I cannot help fearing that many of my things must be left behind.' Later on he was wont to refer a little unfairly to the next two months as spent at 'that horrid Plymouth'. From his letters to his sisters and the entries in the private journal that he kept from 24 October 1831 to 4 October 1836, it is evident that the principal defect of Plymouth in his eyes was its not infrequent storms and rain at the best of times, for 'It does not require a rain gauge to show how much more rain falls in the Western than in the Central & Eastern parts of England.' Even more aggravating had been the long series of south-westerly gales in December that forced the abandonment of several attempts to sail. Apart from this, however, Charles's chief complaint was having had more social engagements than he wished, although they included dinners with the Commander-in-Chief Admiral Sir Manley Dixon at which everyone except for him was a naval officer and of course the conversation was almost exclusively nautical. However, he confessed that this made the evening very pleasant to him.

He dined with the ship's nine gunroom officers for the first time on 29 October, and was regaled with horrific accounts of the manner in which he would be treated by Neptune on crossing the Equator. The following day he had lunch with the midshipmen, and after being taken for a sail to Millbrook, went for a long scrambling walk with Stokes and Charles Musters, a young Volunteer First Class hoping to become a midshipman. On other occasions he walked with various friends to Cawsand, Rame Head and Whitesand Bay, but his favourite walk, on which he took his brother Erasmus during a week's farewell visit at the beginning of December, was to the park at Mount Edgcumbe, with what he described as its 'birds eye view' of Devonport, Stonehouse and Plymouth.

On 31 October Charles went with Stokes to the Plymouth Athenaeum, where space had been reserved to set up an 'astronomical house' for the *Beagle* in which observations were to be carried out with a dipping needle for the determination of the angular depression of the earth's magnetic field. During the next weeks he had several sessions with FitzRoy at the needle, which he first described as a 'very long & delicate operation', though later he said less enthusiastically that he had been 'unpleasantly employed in finding out the inaccuracies of Gambey's new dipping needle'. At the Athenaeum one evening he attended a lecture by a Mr Harris on the virtues of the new system of lightning conductors with which the *Beagle* was fitted, and whose utility would be tested. He was taken out by FitzRoy to the breakwater protecting Plymouth Sound, where bearings were being taken to connect a particular stone, from which Captain King had based his longitudes for the previous voyage, with the quay at Clarence Baths where the true time was then taken. And he was assigned a regular task every morning of taking and comparing the differences in the *Beagle*'s barometers. His education in navigation and meteorology therefore proceeded apace.

Charles's relations with FitzRoy were mostly very cordial, though he later recalled an example of the storms that could suddenly arise, and as quickly be quelled:

> At Plymouth before we sailed, he was extremely angry with a dealer in crockery who refused to exchange some article purchased in his shop: the Captain asked the man the price of a very expensive set of china and said 'I should have purchased this if you had not been so disoblig-

> ing.' As I knew that the cabin was amply stocked with crockery, I doubted whether he had any such intention; and I must have shown my doubts in my face, for I said not a word. After leaving the shop he looked at me, saying You do not believe what I have said, and I was forced to own that it was so. He was silent for a few minutes and then said You are right and I acted wrongly in my anger at the blackguard.

On 14 November FitzRoy moved his twenty-two chronometers on to the ship, but the paint was still wet in the poop cabin, so his books could not yet be arranged. A week later, Charles carried all his books and instruments on board the *Beagle*, after which two hard days' work with Stokes reduced the poop cabin to 'very neat order'. The *Beagle* had now sailed from Devonport to her moorings at Barnett Pool under Mount Edgcumbe ready for her final departure. On 28 November FitzRoy gave a magnificent luncheon for about forty people as a ship's warming, with waltzing that continued until late in the evening.

On 4 December, in the first journal entry written on board ship, Charles wrote:

> I intend sleeping in my hammock. I did so last night & experienced a most ludicrous difficulty in getting into it; my great fault of jockeyship was in trying to put my legs in first. The hammock being suspended, I thus only succeeded in pushing it away without making any progress in inserting my own body. The correct method is to sit accurately in centre of bed, then give yourself a dexterous twist & your head & feet come into their respective places. After a little time I daresay I shall, like others, find it very comfortable.

The next morning was tolerably clear, and sights were obtained, so now the *Beagle* was ready for her long-delayed moment of starting. But at midday a heavy gale blew up from the south, making the ship move so much that Charles was nearly sick, and ruling out any escape from the harbour. 'In the evening dined with Erasmus,' he wrote. 'I shall not often have such quiet snug dinners.' However, gales from that unlucky point south-west recurred daily, and Charles had more last dinners and one more walk to Mount Edgcumbe with Erasmus. On 10 December the wind dropped more hopefully, and the *Beagle* sailed at ten o'clock with Erasmus on board, dropping him off after doubling the breakwater. But in the evening the barometer gave notice of yet another gale, and after a wild and very uncomfortable night it was determined to put back to Plymouth and there remain for a more fortunate wind. Charles reflected ruefully that although he had done right to accept the offer of the voyage, 'I think it is doubtful how far it will add to the happiness of one's life. If I keep my health & return, & then have strength of mind quietly to settle down in life, my present & future share of vexation & want of comfort will be amply repaid.'

Two days later, while the weather still showed little sign of improvement, Charles made some serious resolutions:

> An idle day; dined for the first time in Captain's cabin & felt quite at home. Of all the luxuries the Captain has given me, none will be so essential as that of having my meals with him. I am often afraid I shall be quite overwhelmed with the numbers of subjects which I ought to take into hand. It is difficult to mark out any plan & without method on ship-board I am sure little will be done. The principal objects are 1st, collecting observing & reading in all branches of Natural history that I possibly can manage. Observations in Meteorology. French & Spanish, Mathematics, & a little Classics, perhaps not more than Greek Testament on Sundays. I hope generally to have some one English book to hand for my amusement, exclusive of the above mentioned branches. If I have not energy enough to make myself steadily industrious during the voyage, how great & uncommon an opportunity of improving myself shall I throw away. May this never for one moment escape my mind, & then perhaps I may have the same opportunity of drilling my mind that I threw away whilst at Cambridge.

The wind remained obstinately in the wrong point, until on 21 December there was a light north-westerly which encouraged the

Beagle to try again to depart. After going aground off Drake's Island and taking several hours to get off again, the ship sailed out of the harbour, and once in the open sea Charles was soon overcome by sickness and retreated first to the Captain's cabin and then to his hammock. But during the night the wind strengthened and shifted back yet again to the south-west, and Charles awoke to find himself once more back in Plymouth Sound.

On Christmas Day, Charles went ashore to church, to find an old Cambridge friend preaching. He then lunched in the gunroom, where the dullness of the conversation made him 'properly grateful for my good luck in living with the Captain'. In the meantime the crew exercised their traditional custom of making themselves so drunk that Midshipman King was obliged to perform the duty of sentry. Boxing Day was greeted with an excellent wind for sailing, but the ship was still in a state approaching anarchy, with the worst offenders in heavy chains. At long last, on 27 December, the *Beagle* tacked with some difficulty out of the harbour, accompanied by the Commissioner Captain Ross in what Charles described as his 'Yatch'. After lunching with Captain Ross on mutton chops and champagne, Lieutenant Sulivan and Charles 'joined the *Beagle* about 2 o'clock outside the Breakwater, & immediately with every sail filled by a light breeze we scudded away at the rate of 7 or 8 knots an hour – I was not sick that evening but went to bed early'. The voyage of the *Beagle* had begun.

CHAPTER 4

From Plymouth to the Cape Verde Islands

During his first week at sea there was a heavy swell in the Bay of Biscay, and Charles suffered severely from sea-sickness, as he did throughout the voyage, noting gloomily that 'I often said before starting that I had no doubt I should frequently repent of the whole undertaking. Little did I think with what fervour I should do so. I can scarcely conceive any more miserable state, than when such dark & gloomy thoughts are haunting the mind as have today pursued me.' It did not help that his thoughts were also most unpleasantly occupied by his having had to witness the flogging of several members of the crew in punishment for their drunkenness on Christmas Day. Fifteen years later he wrote to FitzRoy:

> Farewell, dear Fitz-Roy, I often think of your many acts of kindness to me, and not seldomest on the time, no doubt quite forgotten by you, when, before making Madeira, you came and arranged my hammock with your own hands, and which, as I afterwards heard, brought tears into my father's eyes.[48]

On 31 December the weather was milder, and Charles's spirits were raised by his first sight of a shoal of porpoises dashing round the vessel, and a stormy petrel skimming over the waves. He spent the afternoon lying on the sofa in the Captain's cabin, and reading Humboldt's glowing accounts of tropical scenery, concluding that nothing could be better adapted for cheering the heart of a sea-sick man.

January 1832

On 6 January 1832 the *Beagle* arrived at the Canary Islands, and anchored off the port of Santa Cruz on Teneriffe. Although some geographers had adopted the Peak of Teneriffe as a zero point from which to reckon longitude, FitzRoy considered that such a procedure was unsatisfactory because of a lack of data on the precise position of starting points in the neighbourhood of Teneriffe with respect to the Peak itself. However, immediately the anchor had been lowered, a boat carrying the British Vice-Consul and some quarantine officers came alongside with the news that because of reports on the occurrence of cholera in England, nobody would be permitted to land without first undergoing a rigorous twelve days of quarantine. FitzRoy felt that his objective could not be achieved without making observations on shore, and that such a delay was not acceptable, so up came the anchor and the *Beagle* made sail for the Cape Verde Islands. Knowing of Charles's unfulfilled plans for an expedition to Teneriffe, FitzRoy wrote:

> This was a great disappointment to Mr Darwin, who had cherished a hope of visiting the Peak. To see it – to anchor and be on the point of landing, yet be obliged to turn away without the slightest prospect of beholding Teneriffe again – was indeed to him a real calamity.[49]

The first of Charles's zoology notes was made off Santa Cruz that day, when he wrote:

> The sea was luminous in specks & in the wake of the vessel of an uniform slight milky colour. – When the water was put into a bottle it gave out sparks for some few minutes after having been drawn up. – When examined both at night & next morning, it was found full of numerous small (but many bits visible to naked eye) irregular pieces of a (gelatinous?) matter. – The sea next morning was in the same place equally impure.[50]

Four days later the sea was again very luminous from the presence of myriads of tiny shrimps giving out a strong green light. Charles wrote in his journal:

January 1832

> I proved today the utility of a contrivance which will afford me many hours of amusement & work. It is a bag four feet deep made of bunting; & attached to a semicircular bow this by lines is kept upright, & dragged behind the vessel. This evening it brought up a mass of small animals, & tomorrow I look forward to a greater harvest.[51]

This was only the second recorded use, following that of the Irish zoologist J. Vaughan Thompson a few years earlier, of a net specifically designed for the capture of plankton, the name adopted sixty years later for the many kinds of small plants (phytoplankton) or animals (zooplankton) found floating or drifting at various depths in the ocean. Unlike John Coldstream's oyster trawl, whose lower bar was dragged along the bottom of the sea so that it gathered up the organisms that lived there, Charles's net was intended to collect from the surface of the water. His notes continue:

> 11th. I am quite tired having worked all day at the produce of my net. The number of animals that the net collects is very great & fully explains the manner so many animals of a large size live so far from land. Many of these creatures so low in the scale of nature are most exquisite in their forms & rich colours. It creates a feeling of wonder that so much beauty should be apparently created for such little purpose.

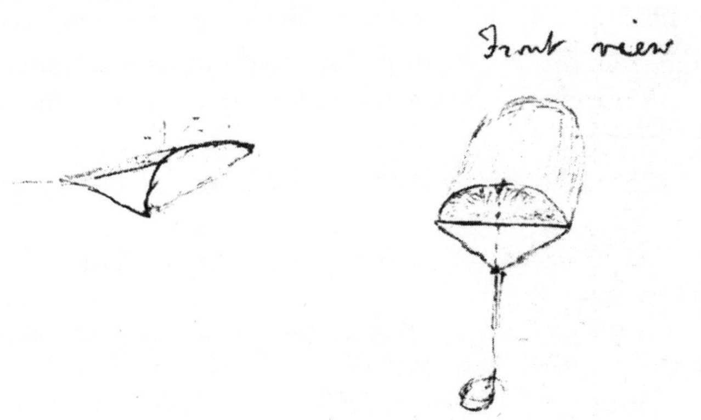

Charles's plankton net

January 1832

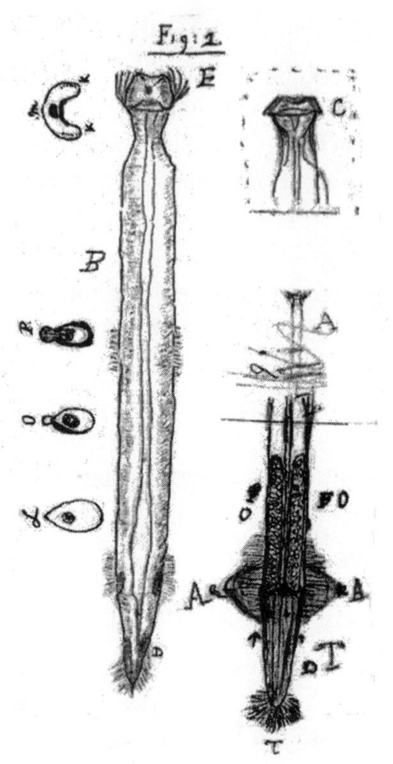

'A very simple animal' – Charles's drawing of an arrow-worm

This was indeed a most auspicious beginning, for Charles was far ahead of his time in his instant perception of the significance of plankton in what nowadays would be called the oceanic food web.

During the next few days a number of interesting animals were captured in the net. One of the first was a Portuguese man-of-war, *Physalia*, a colonial hydrozoan of the order known as siphonophores,*

* The first animals to possess a primitive nervous system, that evolved in the sea about 550 million years ago, belonged to the phylum known as Cnidaria, which contains many familiar creatures such as jellyfish, sea anemones, corals and sea fans, and less familiar ones such as siphonophores. Cnidarians take their name from their characteristic possession of stinging structures called cnidae or nematocysts. Many cnidarians are, like bryozoans (see pp.99–100), colonial animals consisting of large colonies of individual polyps or zooids with specialised functions.

The Bancks microscope used by Charles on the *Beagle*

which have a large horizontal float to hold them at the surface of the sea and long tentacles for capturing their prey. This particular species is well known to swimmers in the warm parts of the North Atlantic for its capability of inflicting a painful sting, as Charles quickly found when he got some of the slime on to his fingers, and on accidentally putting them into his mouth felt the disagreeable sensation, familiar to him, that biting the root of the arum lily produces. There were other hydrozoans such as sea butterflies like the By-the-Wind Sailor *Velella*, with a small sail on its upper surface, some salps growing in long chains, and 'a very simple animal' of which Charles produced an excellent picture on the first page of the twenty plates that he drew under his Bancks microscope to illustrate his Zoology Notes. He later found more of these creatures off the coasts of Brazil and Patagonia, where he described their anatomy in greater detail, but he had still not succeeded in classifying them when the *Beagle* returned to England in 1836. Today they are instantly recognisable as arrow-worms of the genus *Sagitta*, powerful carnivorous predators on other planktonic animals, which are seized by grasping spines located on either side of the head. They are plankton common in all tropical seas, but they had only been formally named in 1827, in a paper that was not in the *Beagle*'s library.

On the *Beagle*'s arrival at St Jago (São Tiago on a modern map) in the Cape Verde Islands on 16 January, Charles divided his time between geology in the mornings, collecting the animals that he found on the seashore in the middle of the day, and examining his specimens and writing his notes in the evening. With FitzRoy and the First Lieutenant John Wickham he visited St Jago's famous baobab tree of legendary age, whose height was measured with naval accuracy both by triangulation and by being climbed by the Captain and letting down a string from the top. The marine wildlife included a variety of sea slugs (*Doris*), sea hares (*Aplysia*), sea urchins, sea anemones, shells, turbellarian flatworms, and some corals, but the highlight was Charles's encounter with an octopus, of which he wrote:

> Found amongst the rocks West of Quail Island at low water an Octopus. When first discovered he was in a hole & it was difficult to perceive what it was. As soon as I drove him from his den he *shot with great rapidity* across the pool of water, leaving in his train a large quantity of the ink. Even then, when in shallow place it was difficult to catch him, for he twisted his body with great ease between the stones & by his suckers stuck very fast to them. When in the water the animal was of a brownish purple, but immediately when on the beach the colour changed to a yellowish green. When I had the animal in a basin of salt water on board this fact was explained by its having the Chamælion like power of changing the colour of its body. The general colour of animal was French grey with numerous spots of bright yellow. The former of these colours varied in intensity, the other entirely disappeared & then again returned. Over the whole body there were *continually* passing clouds, varying in colour from a "hyacinth red" to a "Chesnut brown". As seen under a lens these clouds

January 1832

5

5   1832   St Jago

Jan 28th 14 Octopus

Found amongst the rocks. West of Quail
Island at low water an Octopus. — When
first discovered. he was in a hole. & it was
difficult to perceive what it was. —
As soon as I drove him from his den
he shot with great rapidity across
the pool of water. — leaving in his train
a large quantity of the ink. — even then
it was difficult to catch him. for
he twisted his body with great ease
between the stones & by his suckers
stuck very fast to them. — When in
the water the animal was of a
brownish purple. but immediately when
on the beach the colour changed to
a yellowish green. — When I had the
animal in a basin of salt water. on
board — this fact was explained. by its
having the Chameleon like power of changing
the colour of its body. — The general
colour of animal was French grey. with numerous
spots of bright yellow. — the former of these
colours varied in intensity. — the other entirely
disappeared & then again returned. —
Over the whole body there were continually
passing clouds. varying in colour from a "hy[illegible]
red" to a "Chestnut brown". — As seen under
a lens. these clouds consisted of minute
points. apparently injected with a coloured fluid. The whole animal presented a most extraordinary mottled appearance. & much sur[illegible]

when in shallow place

A typical page from Charles's Zoology Notes

consisted of minute points apparently injected with a coloured fluid. The whole animal presented a most extraordinary mottled appearance, & much surprised every body who saw it. The edges of the sheath were orange, this likewise varied its tint. The animal seemed susceptible to small shocks of galvanism: contracting itself & the parts between the point of contact of wires, became almost black. This in a

> lesser degree followed from scratching the animal with a needle. The cups were in double rows on the arms & coloured reddish. The eye could be entirely closed by a circular eyelid, the pupil was of a dark blue. The animal was slightly phosphorescent at night.[52]

Charles was greatly excited by what he thought was a new discovery, and described it enthusiastically in his first letter to Henslow. But Henslow replied that he too had seen the colour changes of an octopus he had caught at Weymouth, and that the phenomenon had also been reported by others. Cuvier had indeed mentioned the ability of an octopus to outdo a chameleon in this respect, but Charles was nevertheless the first to give an accurate description of the properties of its chromatophores, the pigment cells in the skin whose rapid contraction and expansion under nervous control are responsible for the vivid colour changes in octopus and other cephalopods such as cuttlefish and squid. Their function is not only to camouflage the animals when they move to new surroundings, but as has only been appreciated very recently, to provide a means of communication between them.

In his *Autobiography*, Charles wrote:

> The investigation of the geology of all the places visited was far more important than natural history, as reasoning here comes into play. On first examining a new district nothing can appear more hopeless than the chaos of rocks; but by recording the stratification and nature of the rocks and fossils at many points, always reasoning and predicting what will be found elsewhere, light soon begins to dawn on the district, and the structure of the whole becomes more or less intelligible. I had brought with me the first volume of Lyell's *Principles of Geology*, which I studied attentively; and this book was of the highest service to me in many ways. The very first place which I examined, namely St. Jago in the Cape Verde Islands, showed me clearly the wonderful superiority of Lyell's manner of treating geology, compared with that of any other author whose works I had with me or ever afterwards read.

January 1832

His copy of Volume 1 of Lyell's *Principles* was inscribed 'Given me by Capt. F.R. C.Darwin', and he had been advised by Henslow to read it 'but on no account to accept the views therein advocated'.

Charles's first geological project was to examine the structure of Quail Island, which as it happened was painted by the artist Conrad Martens about eighteen months later, on his way out to join the *Beagle* at Monte Video (see Plate 1). It was a 'miserable desolate spot less than a mile in circumference' in the harbour of Porto Praya, but served Charles usefully as a key to the structure of the main island of St Jago. He wrote in his *Autobiography*:

> The geology of St Jago is very striking yet simple: a stream of lava formerly flowed over the bed of the sea, formed of triturated recent shells and corals, which it has baked into a hard white rock. Since then the whole island has been upheaved. But the line of white rock revealed to me a new and important fact, namely that there had been afterwards subsidence round the craters, which had since been in action, and had poured forth lava. It then first dawned on me that I might perhaps write a book on the geology of the various countries visited, and this made me thrill with delight. That was a memorable hour to me, and how distinctly I can call to mind the low cliff of lava beneath which I rested, with the sun glowing hot, a few strange desert plants growing near, and with living corals in the tidal pools at my feet.

In the notes made at the time Charles's interpretation was that both islands were volcanic, and had at some not too distant time been submerged beneath the sea, where they quietly collected beds of marine material, followed by another layer of molten lava.[53] 'The whole mass was then raised, since which or at the time there has been a *partial sinking*. I judge of this from the appearance of distortion, & indeed the distant line of coast seen to the East, which is considerably higher, bears me out.'

This was accompanied by a section drawing of Quail Island showing the successive layers. Tests on specimens from the white layer

January 1832

D showed that it 'Effervesces readily with Mur: Acid, gives precipitate with Oxalate of Ammonia. – Under Blowpipe becomes slowly caustic, & with heat Cobalt remains of a Violet colour. – Carbonate of Magnesia. (?) Carb. of Lime.' (The white line may be seen in Plate 1.)

In his first independent geological project, Charles's careful analysis of the sequence of rocks in Quail Island showed with what great effect he had followed the teaching of Sedgwick and Henslow. His notes also reveal how geology allowed him from the start to exercise to the full his latent passion for argument and theorisation. On completing his notes on Quail Island, he immediately reread them and wrote, 'I have drawn my pen through those parts which appear absurd,' and a year later he added a long list of further comments and theories. At the same time he immediately fell in wholeheartedly with Lyell's gradualist and not yet generally accepted approach that geological changes resulted from slow processes operating over a long period of time. Thus his evidence clearly supported the view that

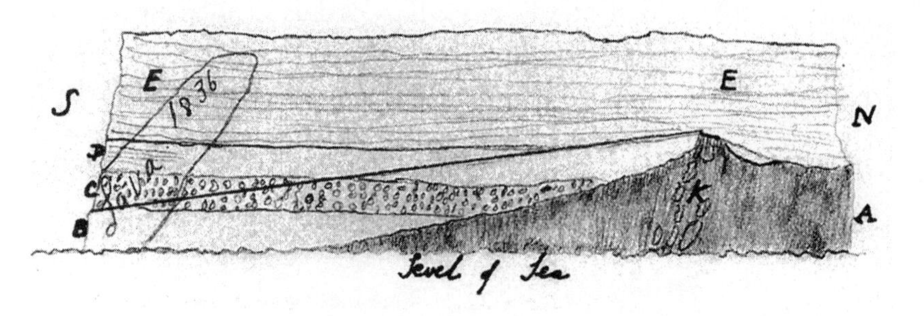

Section drawing at Quail Island from Charles's Geology Notes. The lowest rocks, labelled A, were black basalt containing crystals of olivine and other minerals, often decomposed into a clay or harder rock. Lower bed B was more soily, and contained many snail shells. Bed C was a mixture of gravel, large rounded stones and sand, with shells of present-day tropical species. The upper parts of bed D were mostly white or grey sand formed of minute shells and corals, while lower down it resembled a wall built of a fused mass of pea-sized globules of a white mineral. Where it was in close contact with layer E, it had become very hard and was white spotted with yellow. The upper cap E looked like basalt, and varied in depth between six and twenty feet. The note in pencil 'Lava 1836' was added when the *Beagle* briefly called at St Jago on the way home.

both subsidence and elevation of the land must have taken place over an appreciable area in the not too distant past, and he was led to agree with Lyell that the forces involved might act slowly and evenly so as to leave superficial features of the landscape and buildings undisturbed. Another relevant factor in the story was Charles's identification of the dust that thickly coated the ship throughout their visit to St Jago, 'to the great injury of fine astronomical instruments', as volcanic in origin.

Charles reacted to his first day on St Jago with the enthusiasm that never deserted him:

> I returned to the shore, treading on Volcanic rock, hearing the notes of unknown birds, & seeing new insects fluttering about still newer flowers. It has been for me a glorious day, like giving to a blind man eyes – he is overwhelmed with what he sees & cannot justly comprehend it. Such are my feelings, & such may they remain.

CHAPTER 5

Across the Equator to Bahia

On the day originally fixed for sailing on across the Atlantic to Brazil, FitzRoy was busy on shore complying with Captain Beaufort's strict instructions that no port should be quitted before not only the magnetic angle, but also the dip and daily variation had been ascertained. On 8 February the instruments were re-embarked, and after swinging the ship and determining less than twenty minutes' difference in any position of the bearing of the peak eleven miles away, the *Beagle* weighed anchor and sailed. On 10 February they came alongside the packet *Lyra*, on passage from London to Rio de Janeiro, and were pleased to find that she was carrying a box of six sounding-leads for them, modified by their designer to operate satisfactorily at depths well below a hundred fathoms. Charles posted a brief letter to his father, in case it might arrive sooner than a long one due to be dispatched from Bahia, in which he said:

> I think, if I can so soon judge, I shall be able to do some original work in Natural History – I find there is so little known about many of the Tropical animals.

At sunset on 15 February the St Paul Rocks were seen on the horizon, these being the summit of a sunken mountain, and further from land than FitzRoy had ever seen a group of such small rocks. At daylight next morning the sea was smooth, and while the *Beagle* sailed round so that Stokes could take angles and make soundings, two boats were sent out to enable FitzRoy, Charles and a party to land on the rocks and examine them. As FitzRoy described it:

February 1832

> When our party had effected a landing through the surf, and had a moment's leisure to look about them, they were astonished at the multitudes of birds which covered the rocks, and absolutely darkened the sky. Mr Darwin afterwards said that till then he had never believed the stories of men knocking down birds with sticks; but there they might be kicked before they would move out of the way. The first impulse of our invaders of this bird-covered rock was to lay about them like schoolboys; even the geological hammer at last became a missile. 'Lend me a hammer?' asked one. 'No, no' replied the owner, 'you'll break the handle'; but hardly had he said so, when, overcome by the novelty of the scene, and the example of those around him, away went the hammer, with all the force of his own right arm.[54]

In his own account Charles did not deny that he had been somewhat carried away. So he participated in the slaughter of birds on land while a similar struggle to obtain fish for the cooking pot was also taking place in the surrounding waters, both birds and fish being welcome to men who had been living too long on salt provisions. But he nevertheless found time to note that unlike almost all other isolated rocks in mid-ocean, St Paul was exceptionally not volcanic in origin, but was a mineral unfamiliar to him that incorporated streaks of serpentine.* The surrounding waters were very deep, so that it was the tip of a very large and steeply sided mountain. His conclusion was correct, and the modern view is that St Paul is an important example of the primordial material of the earth's mantle modified to become the basalt layer of the oceanic crust. The only birds to be seen were boobies, a species of gannet, and noddies, a species of tern; and the only other animals of any size were large tropical crabs of the genus *Grapsus*. Not a single plant, nor even a lichen, could Charles find growing on the rocks. There were some ticks and mites, and a small brown moth feeding on feathers that could have arrived with the birds; a rove beetle and a woodlouse

* An ancient greenish-coloured mineral consisting of magnesium silicate with chemically bound water.

from beneath the dung; and a large number of spiders that presumably preyed on the other insects. He reflected that since the first colonists of the coral islets in the South Seas were probably similar, 'it destroys the poetry of the story to find that these little vile insects should thus take possession before the cocoanut tree and other noble plants have appeared'.

The *Beagle* sailed on. They were now close to the Equator, and preparations were set in hand for the traditional naval ceremonies that accompanied 'crossing the line'. Soon after dark they were hailed by the gruff voice of a pseudo-Neptune. The Captain held a conversation with him through a speaking-trumpet, and it was arranged that in the morning he would visit the ship.

The proceedings next day were vividly described from memory nearly sixty years later by the then fourteen-year-old Midshipman Philip Gidley King:

> The effect produced on the young naturalist's mind was unmistakably remarkable. His first impression was that the ship's crew from Captain downwards had gone off their heads. 'What fools these sailors make of themselves', he said as he descended the companion ladder to wait below till he was admitted. The Captain received his godship and Amphitrite his wife with becoming solemnity; Neptune was surrounded by a set of the most ultra-demoniacal looking beings that could be well imagined, stripped to the waist, their naked arms and legs bedaubed with every conceivable colour which the ship's stores could turn out, the orbits of their eyes exaggerated with broad circles of red and yellow pigments. Those demons danced a sort of nautical war dance exulting on the fate awaiting their victims below. Putting his head down the after companion the captain called out 'Darwin, look up here!' Up came the young naturalist in wonderment but yet prepared for any extravagance in the world that seamen could produce. A gaze for a moment at the scene on deck was sufficient, he was convinced he was amongst madmen, and giving one yell, disappeared again down the ladder. He was of course the first to be called by the official secretary, and Neptune received him with grace and

Crossing the Line, by Augustus Earle

courtesy, observing that in deference to his high standing on board as a friend and messmate of the Captain his person would be held sacred from the ordinary rites observed in the locality. Of course Mr Darwin readily entered into the fun and submitted to a few buckets of water thrown over him and the Captain as they sat together by one of the youngsters as if by accident.[55]

From Charles's own account, he was treated with rather less courtesy than King remembered:

Before coming up, the constable blindfolded me & thus lead along, buckets of water were thundered all around. I was then placed on a plank, which could be easily tilted up into a large bath of water. They then lathered my face & mouth with pitch and paint, & scraped some of it off with a piece of roughened iron hoop. A signal being given I was tilted head over heels into the water, where two men received me

February 1832

> & ducked me. At last, glad enough, I escaped. Most of the others were treated much worse, dirty mixtures being put in their mouths & rubbed on their faces. The whole ship was a shower bath, & water was flying about in every direction, of course not one person, even the Captain, got clear of being wet through.

Although FitzRoy condemned the practice as an absurd and dangerous piece of folly, he also defended its survival on the grounds that 'its effects on the minds of those engaged in preparing for its mummeries, who enjoy it at the time, and talk of it long afterwards, cannot easily be judged of without being an eyewitness'.

The *Beagle*'s next port of call on 20 February was at Fernando Noronha, another isolated group of small islands, where the most prominent feature was a conical hill on the principal island rising very steeply to a peak a thousand feet high, and seemingly overhanging the shore on one side. Near its summit a permanently manned lookout station was maintained by the Brazilian government. According to Beaufort's programme, FitzRoy was required to verify some measurements of longitude made a few years earlier by another survey ship in pendulum experiments conducted in the Governor's house.

Fernando Noronha, by Conrad Martens

February 1832

With the *Beagle* lying offshore that evening before anchoring in the harbour, Lieutenant Sulivan skilfully harpooned a large porpoise, and moments later 'a dozen knives were skinning him for supper'. In the morning, landing despite the high surf as near as possible to the house where the previous observations had probably been made, FitzRoy took his shots of the sun and compared his chronometers with those used on shore, while Charles spent 'a most delightful day in wandering about the woods'. He concluded that unlike the St Paul Rocks, Fernando Noronha consisted of a volcanic rock called phonolite,* which had probably been injected in a molten state among yielding strata, but was not of very recent origin. The island was thickly covered with trees, often coated with delicate blossoms, though because of the low rainfall their growth was not luxurious, and FitzRoy noted that firewood collected by the crew was full of centipedes and other noxious insects. There were no gaudy birds, no humming birds, and no flowers, so Charles felt that he had not yet seen the full grandeur of the Tropics.

At noon on 28 February the *Beagle* anchored in the great Bay of All Saints (Baia de Todos os Santos) on the mainland of Brazil, on the north side of which the fine old town of Bahia, now known as Salvador, was situated. The view of the town itself was magnificent, and when next morning Charles had ventured ashore, he wrote in his journal of what he saw with a characteristic aesthetic appreciation, coupled with a strictly practical conclusion:

> The day has passed delightfully: delight is however a weak term for such transports of pleasure – I have been wandering by myself in a Brazilian forest. Amongst the multitude it is hard to say what set of objects is most striking. The general luxuriance of the vegetation bears the victory: the elegance of the grasses, the novelty of the parasitical plants, the beauty of the flowers, the glossy green of the foliage, all tend to this end. A most paradoxical mixture of sound & silence

* A type of volcanic rock so named at that time because it rings when struck.

pervades the shady parts of the wood. The noise from the insects is so loud that in the evening it can be heard even in a vessel anchored several hundred yards from the shore. Yet within the recesses of the forest when in the midst of it a universal stillness appears to reign. To a person fond of Natural History such a day as this brings with it pleasure more acute than he ever may again experience. After wandering about for some hours, I returned to the landing place. Before reaching it I was overtaken by a Tropical storm. I tried to find shelter under a tree so thick that it would never have been penetrated by common English rain, yet here in a couple of minutes, a little torrent flowed down the trunk. It is to this violence we must attribute the verdure in the bottom of the wood. If the showers were like those of a colder clime, the moisture would be absorbed or evaporated before reaching the ground.

He took many more walks with King or another companion, and after collecting numerous small beetles and some geological specimens, reflected that:

It is a new & pleasant thing for me to be conscious that naturalizing is doing my duty, & that if I neglected that duty I should at the same time neglect what has for some years given me so much pleasure.

Sometimes it was driver ants that caught his attention:

Some of the smaller species migrate in large bodies. One day my attention was drawn by many spiders, Blattaæ [a species of cockroach] & other insects rushing in the greatest agitation across a bare bit of ground. Behind this every stalk & leaf was blackened by a small ant. They crossed the open space till they arrived at a piece of old wall on the side of the road. Here the swarm divided & descended on each side, by this many insects were fairly enclosed: & the efforts which the poor little creatures made to extricate themselves from such a death were surprising. When the ants came to the road they changed their course & in narrow files reascended the wall & proceeding along one

View of Bahia, by A. Earle

> side in the course of a few hours (when I returned) they all had disappeared. When a small stone was placed in the track of one of their files, the whole of them first attacked it & then immediately retired: it would not on the open space have been one inch out of their way to have gone round the obstacle, & doubtless if it had previously been there, they would have done so. In a few seconds another larger body returned to the attack, but they not succeeding in moving the stone, this line of direction was entirely given up.

On another day he shot a most beautiful large lizard, but he complained that both here and at Rio de Janeiro, birds seemed to be unexpectedly scarce in the tropical jungle. Had he, however, set up a modern mist net in a clearing, and left it unobserved for an hour, he would have been better impressed by the large number of small birds that would have been caught in it.

Confined on board the *Beagle* by a badly swollen knee for a couple of weeks, Charles captured a puffer fish *Diodon* swimming in its unexpanded form alongside the ship, and since he was always interested in the mechanics of animal movements, wrote a closely analysed account of its behaviour, as usual unafraid to contradict the authorities if necessary:

March 1832

On head four soft projections; the upper ones longer like the feelers of a snail. Eye with pupil dark blue; iris yellow mottled with black. The dorsal, caudal & anal fins are so close together that they act as one. These, as well as the Pectorals which are placed just before branchial apertures, are in a continued state of tremulous motion even when the animal remains still. The animal propels its body by using these posterior fins in same manner as a boat is sculled, that is by moving them rapidly from side to side with an oblique surface exposed to the water. The pectoral fins have great play, which is necessary to enable the animal to swim with its back downwards. When handled, a considerable quantity of a fine "Carmine red" fibrous secretion was emitted from the abdomen & stained paper, ivory &c of a high colour. The fish has several means of defence, it can bite hard & can squirt water to some distance from its Mouth, making at the same time a curious noise with its jaws. After being taken out of water for a short time & then placed in again, it absorbed by the mouth (perhaps likewise by the branchial apertures) a considerable quantity of *water* & air, sufficient to distend its body into a perfect globe. This process is effected by two methods: chiefly by swallowing & then forcing it into the cavity of the body, its return being prevented by a muscular contraction which is externally visible; and by the dilatation of the animal producing suction. The water however I observed entered in a stream through the mouth, which was distended wide open & motionless; hence this latter action must have been caused by some kind of suction. When the body is thus distended, the papillæ with which it is covered become stiff, the above mentioned tentacula on the head being excepted. The animal being so much buoyed up, the branchial openings are out of water, but a *stream* regularly flowed out of them which was as constantly replenished by the mouth. After having remained in this state for a short time, the air & water would be expelled with considerable force from the branchial apertures & the mouth. The animal at its pleasure could emit a certain portion of the water & I think it is clear that this is taken in partly for the sake of regulating the specific gravity of its body. The skin about the abdomen is much looser than that on the back & in consequence is

> most distended; hence the animal swims with its back downwards. Cuvier doubts their being able to swim when in this position; but they clearly can not only swim forward, but also move round. This they effect, not like other fish by the action of their tails, but collapsing the caudal fins, they move only by their pectorals. When placed in fresh water seemed *singularly little inconvenienced.*

The prevailing rock in Bahia was gneiss-granite.* An interesting point was that in the immediate neighbourhood of Bahia, the foliations tended to be lined up with the coastline striking E 50°N, in agreement with the observations of Humboldt in Venezuela and Colombia.

It was at Bahia that one of Charles's most violent quarrels with FitzRoy arose. When he first landed there he was horrified to find himself in a country that was still a haven for 'that scandal to Christian Nations, Slavery' by legally importing slaves from Africa. This practice continued, thanks to the dependence of the Brazilian coffee-growers on slave labour, until it was abolished a quarter of a century later in response to sustained pressure from the British government. Slavery was an issue that always aroused Charles's strongest emotions, brought up as he had been in a family where both of his grandfathers had played prominent parts in the anti-slavery movement during the last twenty years of the eighteenth century, and which numbered influential Whig campaigners for the abolition of slavery among their friends. Two weeks later, Captain Paget of HMS *Samarang*, when dining with FitzRoy on the *Beagle*, regaled the company with horrific facts about the practice of slave owners in Brazil. As Charles recorded in his journal, Paget also proved the utter

* Granite is the best-known and most abundant of high-grade igneous rocks, containing a large proportion of silica or quartz. After metamorphosis by high pressure and high temperature over a large region, it may become a gneiss with coarse foliations or bands of segregated light and dark minerals. It may also pass by the disappearance of its quartz and mica, and by the feldspar losing its red colour, into a brilliantly grey primitive greenstone.

falseness of the view that even the best-treated of the slaves did not wish to return home to their countries. What Charles did not record at the time, but only revealed much later, was the sequel:

> Early in the voyage at Bahia in Brazil, FitzRoy defended and praised slavery, which I abominated, and told me that he had just visited a great slave-owner, who had called up many of his slaves and asked them whether they were happy, and whether they wished to be free, and all answered 'No'. I then asked him, perhaps with a sneer, whether he thought that the answers of slaves in the presence of their master was worth anything. This made him excessively angry, and he said that as I doubted his word, we could not live any longer together. I thought that I should have been compelled to leave the ship; but as soon as the news spread, which it did quickly, as the captain sent for the first lieutenant to assuage his anger by abusing me, I was deeply gratified by receiving an invitation from all the gun-room officers to mess with them. But after a few hours FitzRoy showed his usual magnanimity by sending an officer to me with an apology and a request that I would continue to live with him.[56]

As on other occasions, FitzRoy's anger was short-lived. Moreover, as he had already shown by his actions, he was always very sympathetic to natives, slaves and underdogs of all kinds, so that his outburst was perhaps more a reflection of his Tory political views than of his true feelings for humanity. Charles's point was well taken, and when writing from Monte Video to Beaufort in July 1833, FitzRoy said, 'If other trades fail, when I return to old England (if that day ever arrives) I am thinking of raising a crusade against the slavers! Think of *Monte Video* having sent out *four slavers*!!! . . . The *Adventure* will make a good privateer!!' And by the end of the voyage his views on the evil of slavery in Brazil were fully in agreement with those held by Charles.[57]

CHAPTER 6

Rio de Janeiro

On 18 March, after taking further soundings for the chart of the Bay of All Saints, the *Beagle* sailed slowly out in a light wind, and headed for the Abrolhos, a group of uninhabited islets off the coast of Brazil some 350 miles south of Bahia. Five days later the wind was still light, but there was a sufficient swell to make Charles uncomfortable. Occupation was always the best cure, so he settled down at his microscope to examine a mould called mucor growing on ginger from the steward's cupboard. He wrote in his notes:

> Mucor growing on green ginger: colour yellow, length from 1/20 to 1/15 of an inch. Diameter of stalk .001, of ball at extremity .006. Stalk transparent, cylindrical for about 1/10 of length, near to ball it is flattened, angular & rather broarder:* Terminal spherule full of grains, .0001 in diameter & sticking together in planes: When placed in water the ball partially burst & sent forth with granules large bubbles of air. A rush of fluid was visible in the stalk or cylinder. If merely breathed on, the spherule expanded itself & three conical semitransparent projections were formed on surface. (Much in the same manner as is seen in Pollen) These cones in a short time visibly were contracted & drawn within the spherule.

Unfortunately the specimen of the mould *Mucor* (Mucoraceae) was not well preserved, and Henslow wrote to Charles in January 1833 after receiving the first consignment from the *Beagle*, 'For goodness

* The aberrant spelling 'broard' was one of Charles's habitual idiosyncrasies.

sake what is N°. 223; it looks like the remains of an electric explosion, a mere mass of soot – something very curious I dare say.'

Around the Abrolhos there were shallow rocky shoals stretching far out to the east. One of the tasks allotted to the *Beagle* by Beaufort was to determine the precise extent of these shoals. FitzRoy therefore steered south-east to the latitude of the Abrolhos, and then turned west, sounding all the time, until a well-defined rocky bank was reached at a roughly constant depth of thirty fathoms. After spending two days surveying parts of the Abrolhos that had not been properly covered by a French expedition under Baron Roussin in 1818–21, perhaps because of the disconcertingly sudden changes in depth called by the French '*coups de sonde*', two parties landed on 29 March. Charles launched an attack on the rocks and insects and plants, while members of the crew began a much more bloody one on the birds, of which an enormous number were slaughtered. Charles reported to FitzRoy that the rocks, rising to about a hundred feet above the sea in horizontal strata, were of gneiss and sandstone. The general description of the islands entered in his notes was:

> The Abrolhos Islands seen from a short distance are of a bright green colour. The vegetation consists of succulent plants & Gramina [grasses], interspersed with a few bushes & Cactuses. Small as my collection of plants is from the Abrolhos I think it contains nearly every species then flowering. Birds of the family of Totipalmes [an old group name for some web-footed sea birds] are exceedingly abundant, such as Gannets, Tropic birds & Frigates. The number of Saurians is perhaps the most surprising thing, almost every stone has its accompanying lizard: Spiders are in great numbers: likewise rats: The bottom of the adjoining sea is thickly covered by enormous brain stones [solitary stony corals similar in appearance to a brain]; many of them could not be less than a yard in diameter.

The *Beagle* sailed on towards Rio, and on 1 April all hands were busy making fools of one another. The hook was easily baited, and when Lieutenant Sulivan cried out, 'Darwin, did you ever see a

Grampus: Bear a hand then,' Charles rushed out in a transport of enthusiasm, and was received by a roar of laughter from the whole watch.

Eighty miles from Rio they passed close to the promontory of Cape Frio, where not many years ago gleaming white sand still covered the shore, but today there is a line of skyscrapers. FitzRoy was anxious to revisit the scene where, on the evening of 5 December 1830, the frigate HMS *Thetis*, bound urgently for England with a cargo of treasure, had been battling desperately against contrary winds and was carried far off course by an unsuspected current, until in strong rain and very poor visibility she had sailed at nine knots directly on to the cliffs at Cape Frio, bringing down all three of her masts and injuring many men. In the subsequent struggles, with waves breaking heavily on the hull, twenty-five members of the crew were lost, and the ship quickly sank. FitzRoy had at one time served as a lieutenant on the *Thetis*, and concluded his deeply felt account of this tragic accident with the words:

> Those who never run any risk; who sail only when the wind is fair; who heave to when approaching land, though perhaps a day's sail distant; and who even delay the performance of urgent duties until they can be done easily and quite safely; are, doubtless, extremely prudent persons: but rather unlike those officers whose names will never be forgotten while England has a navy.[58]

Arriving at Rio de Janeiro on the evening of 4 April, Charles proudly noted that 'In most glorious style did the little *Beagle* enter the port and lower her sails alongside the Flagship . . . Whilst the Captain was away with the commanding officer, we tacked about the harbor & gained great credit from the manner in which the Beagle was manned & directed.' As Philip Gidley King remembered it:

> Though Mr Darwin knew little or nothing of nautical matters, on one day he volunteered his services to the First Lieutenant. The occasion was when the ship first entered Rio Janeiro. It was decided to make a

display of smartness in shortening sail before the numerous men-of-war at the anchorage under the flags of all nations. The ship entered the harbour under every yard of canvas which could be spread upon her yards including studding sails aloft on both sides, the lively sea breeze which brought her in being right aft. Mr Darwin was told to hold to a main royal sheet in each hand and a top mast studding sail tack in his teeth. At the order 'Shorten Sail' he was to let go and clap on to any rope he saw was short-handed – this he did and enjoyed the fun of it, afterwards remarking 'the feat could not have been performed without him'.[55]

In view of the political instability at that period of Brazil and the newly liberated countries on its southern borders, the Royal Navy maintained a squadron of ships at the magnificent harbour of Rio de Janeiro for the general protection of British interests in South America. It was commanded by Admiral Sir Thomas Baker. While Charles was assisting the *Beagle* so skilfully to shorten sail, the Captain was receiving orders from the Commander-in-Chief for the exact position to be taken up by the *Beagle* and other ships of the squadron in case marines had to be landed to assist in quelling a mutiny that had broken out among the troops in the town. Fortunately the need did not arise, and all on the *Beagle* settled down happily to read their accumulated mail from home.

The next morning, Charles landed with the ship's first official artist Augustus Earle at the Palace steps. Earle had once lived in Rio for some while, and after he had introduced Charles to the centre of the city, they found themselves 'a most delightful house' at Botafogo which would provide them with excellent lodgings. Its situation, as painted by Conrad Martens when he was passing through Rio a few months later (Plate 2), was an attractively rural one, but nowadays the shore of Botafogo is regrettably occupied by a sprawling network of multi-lane superhighways. The house was in due course also shared with 'Miss Fuegia Basket, who', remarked Charles, 'daily increases in every direction except height', with the Sergeant of the ship's marines, and with young Philip Gidley King, who wrote of it with affection:

April 1832

The mole, palace and cathedral at Rio, by Augustus Earle

At Rio Janeiro Mr Darwin thoroughly enjoyed the new life in a tropical climate. Hiring a cottage at Botafogo, a lovely land-locked bay with a sandy beach of a dazzling whiteness, Mr Darwin took for one of his shore companions the writer, who from having been in the former voyage with his father although then of tender years was able to remember and to recount to the so far inexperienced philosopher his own adventures. "Come King" he would say "you have been round Cape Horn and I have not yet done so, but do not come your traveller's yarns on me". One of these was that he had seen whales jump out of the water all but their tails, another that he had seen ostriches swimming in salt water. For disbelieving these statements however, Mr Darwin afterwards made ample reparation. The first was verified one fine afternoon on the East coast of Tierra del Fuego. A large number of whales were around the ship, the Captain, the "Philosopher" and the Surveyors were on the poop, presently Mr Darwin's arm was seized as a gigantic beast rose three fourths of his huge body out of the water. "Look Sir look! Will you believe me now?" was the exclamation of the hitherto discredited youth. "Yes! anything you tell me in future" was the quick reply of the kind-hearted naturalist.[55]

April 1832

It was in the Beagle Channel on 28 January 1833 that Charles was thus enlightened:

> the day was overpowringly [*sic*] hot, so much so that our skin was burnt; this is quite a novelty in Tierra del F. The Beagle Channel is here very striking, the view both ways is not intercepted, & to the West extends to the Pacific. So narrow and straight a channell & in length nearly 120 miles, must be a rare phenomenon. We were reminded that it was an arm of the sea by the number of Whales, which were spouting in different directions: the water is so deep that one morning two monstrous whales were swimming within stone throw of the shore.[59]

Charles at once set about organising an expedition on horseback to the Rio Macaé, some one hundred miles to the north-east of Rio. His 'extraordinary & quixotic set of adventurers' consisted firstly of an Irish businessman, Patrick Lennon, who had lived in Rio for twenty years and owned an estate near the mouth of the Macaé that he had not previously visited; he was accompanied by a nephew. Then there was Mr Lawrie, 'a well informed clever Scotchman, *selfish unprincipled* man, by trade partly slave merchant partly Swindler', with a friend who was apprentice to a druggist, and whose elder brother's Brazilian father-in-law Senhor Manuel Figuireda owned a large estate on the Macaé at Socégo. As a guide for the party Charles took along a black boy. The first obstacle was to obtain passports for an excursion to the interior. The local officials were somewhat less than helpful, 'but the prospect of wild forests tenanted by beautiful birds, Monkeys & Sloths, & Lakes by Cavies & Alligators, will make any naturalist lick the dust even from the foot of a Brazilian'.

The exotic cavalcade set out on 8 April, and Charles was entranced by the stillness of the woods – except for the large and brilliant butterflies which lazily fluttered about, with blue the prevailing tint – and by the infinite numbers of lianas and parasitical plants, whose beautiful flowers struck him as the most novel object to be seen in a tropical forest. In the evening the scene by the dimmed light of the moon was most desolate, with fireflies flitting by and the solitary snipe uttering its plaintive cry while the distant and sullen roar of the sea scarcely broke the quiet of the night. The inn at which they spent their first night sleeping on straw mats was a miserable one, though at others they fared sumptuously with wine and spirits at dinner, coffee in the evening, and fish for breakfast. The five days needed for the journey to the mouth of the Macaé were often strenuous, and the amount of labour that their horses could perform was impressive, even on the occasion when the riders had to swim alongside them to cross the Barro de St João.

On 13 April they rested at Senhor Figuireda's luxurious *fazenda* at Socégo, where Charles was relieved to see how kindly the slaves were treated, and how happy they seemed. Two days later he had a very different impression of slavery when Mr Lennon threatened to sell at a public auction an illegitimate mulatto child to whom his agent was much attached, and even to take all the women and children from their husbands to sell them separately at the market in Rio. Despite his feeling that Mr Lennon was not at heart an inhumane person, Charles reflected ruefully on the strange and inexplicable effect that prevailing custom and self-interest might have on a man's behaviour. It was agreed that Senhor Manuel should be asked to arbitrate in the quarrel, which he presumably did in favour of the slaves, although Charles did not report on the outcome. Charles returned to Socégo, where he spent the most enjoyable part of the whole expedition collecting insects and reptiles in the woods, and admiring the trees:

> The forests here are ornamented by one of the most elegant, the Cabbage-Palm, with a stem so narrow that with the two hands it may be clasped, it waves its most elegant head from 30 to 50 feet above the ground. The soft part, from which the leaves spring, affords a most excellent vegetable. The woody creepers, themselves covered by creepers, are of great thickness, varying from 1 to nearly 2 feet in circumference. Many of the older trees present a most curious spectacle, being covered with tresses of a liana, which much resembles bundles of hay. If the eye is turned from the world of foliage above, to the ground, it is attracted by the extreme elegance of the leaves of numberless species

> of Ferns & Mimosas. Thus it is easy to specify individual objects of admiration; but it is nearly impossible to give an adequate idea of the higher feelings which are excited; wonder, astonishment & sublime devotion fill & elevate the mind.

For the journey home, when Charles was accompanied only by Mr Lennon, the same route was followed, though back in Rio, having carelessly lost their passports, they had some difficulty in proving that their horses were not stolen. Charles returned to the *Beagle*, where he learnt that the surgeon Robert McCormick had been 'invalided', that is to say had quarrelled with the Captain and the First Lieutenant, and was about to go back to England on HMS *Tyne*. The news did not greatly distress Charles, for he had decided even before leaving Devonport that 'my friend the Doctor is an ass . . . at present he is in great tribulation, whether his cabin shall be painted French Grey or a dead white – I hear little excepting this subject from him'. And at St Jago McCormick had revealed himself as 'a philosopher of rather an antient date; at St Jago by his own account he made *general* remarks during the first fortnight & collected particular facts during the last'. Robert McCormick was an ambitious Scot, determined to make a career for himself as a naval surgeon, who had sailed to the Arctic in 1827 with William Edward Parry as assistant surgeon on the *Hecla*. His nose was put thoroughly out of joint on the *Beagle* by finding that Charles had been introduced by the Captain to look after natural history, one of the traditional responsibilities of the ship's surgeon. He subsequently sailed to the Antarctic as surgeon on the *Erebus*, and took part in the search for Franklin in the Arctic in 1852–53. But when he finally retired in 1865, the professional recognition that he had sought for so long still eluded him. He was succeeded as acting surgeon on the *Beagle* by Benjamin Bynoe, with whom Charles remained on the best of terms for the rest of the voyage.

On 25 April Charles suffered on a small scale what he described as some of the horrors of a shipwreck, when two or three large waves swamped the boat from which he was landing his possessions to transfer them to Botafogo, though nothing was completely spoiled. The following day he wrote an account of the disaster to his sister Caroline, also reporting to her:

> I send in a packet, my commonplace Journal. I have taken a fit of disgust with it & want to get it out of my sight. Any of you that like may read it, a great deal is absolutely childish. Remember however this, that it is written solely to make me remember this voyage, & that it is not a record of facts but of my thoughts, & in excuse recollect how tired I generally am when writing it . . . Be sure you mention the receiving of my journal, as anyhow to me it will be of considerable future interest as an exact record of all my first impressions, & such a set of vivid ones they have been must make this period of my life always one of interest to myself. If you will speak quite sincerely, I should be glad to have your criticisms. Only recollect the above mentioned apologies.[60]

During the next few days Charles was taken by FitzRoy to dine more than once with Mr Aston, representative of the English government, at meals which to his surprise 'from the absence of all form almost resembled a Cambridge party'. He also dined with the Admiral, Sir Thomas Baker, no doubt with the greater formality of the Navy, and was taken to watch the impressive spectacle of an official inspection of the seventy-four-gun battleship *Warspite*.

A week later the *Beagle* sailed back to Bahia to find an explanation for the discrepancy of four miles in the meridian distance between the Abrolhos Islands and Rio de Janeiro shown in Baron Roussin's chart as compared with the *Beagle*'s measurements. In a private letter to FitzRoy, Beaufort later commended his 'daring' for thus having turned back without prior instruction from the Admiralty.[61] It turned out that Baron Roussin's placing of the Abrolhos was correct, but not that of Rio, confirming that FitzRoy's twenty-two chronometers and his dependence on a connected chain of meridian distances was the most reliable method of finding the precise longitude. This information was duly conveyed to the French commander-in-chief at Rio.

June 1832

A less happy piece of news was that three members of a party who had sailed in the ship's cutter to the river Macacu shortly before the *Beagle*'s departure – an extraordinarily powerful seaman called Morgan, Boy Jones who had just been promised promotion, and Charles's young friend Midshipman Musters – had been stricken with fatal attacks of malaria a few days later, and were buried at Bahia. FitzRoy considered that the danger of contracting the disease appeared to be greatest while sleeping, while Charles found it puzzling that the fever so often came on several days after the victim had returned to a seemingly pure atmosphere. The full details of the role of mosquitoes as the vector in the transmission of malaria were made clear by Sir Ronald Ross only in 1897.

For the next two months Charles assiduously explored Rio and the surrounding country, and on alternate days wrote up his notes and sorted out the specimens that he had collected, for he found that one hour's collecting often kept him busy for the rest of the day. He noted that whereas 'The naturalist in England enjoys in his walks a great advantage over others in frequently meeting something worthy of attention; here he suffers a pleasant nuisance in not being able to walk a hundred yards without being fairly tied to the spot by some new & wondrous creature.' A discovery that particularly thrilled him was to find in the forest what was evidently a species of flatworm related to Cuvier's Planaria, but which he thought was generally regarded as a strictly marine animal. He wrote:

> June 17th. This very extraordinary animal was found, under the bark of a decaying tree, in the forest at a considerable elevation. The place was quite dry & no water at all near. Body soft, parenchymatous,* covered

* 'Parenchymatous' denotes an invertebrate animal consisting largely of soft tissues other than muscle and connective tissue. The animal in question would now be classified as a turbellarian flatworm in order Tricladida, whose members are found mainly in marine or fresh water, though some like this one, now known as *Geoplana vaginuloides* Darwin, are terrestrial. Charles later collected other new species of marine flatworms in order Polycladida, and published a short paper about them in 1844.

with *slime* (like snails & leaving a track), not much flattened. When fully extended, 2 & ¼ inches long: in broardest parts only .13 wide. Back arched, top rather flat; beneath, a level crawling surface (precisely resembles a gasteropode [snail], only not separated from the body), with a slightly projecting membranous edge. Anterior end *extremely* extensible, *pointed* lengthened; posterior half of body broardest, tail bluntly pointed.

Colours: back with glossy black stripe; on each side of this a primrose white one edged externally with black; these stripes reach to extremities, & become uniformly narrower. sides & foot dirty "orpiment orange". From the elegance of shape & great beauty of colours, the animal had a very striking appearance.

The anterior extremity of foot rather grooved or arched. On its edge is a regular row of round black dots (as in marine Planariæ) which are continued round the foot, but not regularly; foot thickly covered with very minute angular white marks or specks. On the foot in centre, about 1/3 of length from the tail, is an irregular circular white space, free from the specks. Extending through the whole width of this, is a transverse slit, sides straight parallel, extremities rounded, 1/60th of inch long, tolerably apparent (i.e. with my very weak lens).[62]

The colours in inverted commas quoted by Charles in this and other descriptions of his specimens were taken from a neat little colour atlas by Patrick Syme in the *Beagle*'s library, of which Charles made frequent use. The copy of this atlas that survived among his books at Down House in Kent is, however, spotless, so that the *Beagle*'s hard-worked copy evidently had to be replaced after his return to England. In a letter to Henslow begun on 23 July 1832, Charles said: 'Amongst the lower animals, nothing has so much interested me as finding 2 species of elegantly coloured true Planariæ, inhabiting the dry forest! The false relation they bear to snails is the most extraordinary thing of the kind I have ever seen. In the same genus (or more truly family) some of the marine species possess an organization so marvellous that I can scarcely credit my eyesight.'[63] Henslow was unconvinced, and on page 5 of the edition of Charles's

Corcovado Mountain at Rio de Janeiro, by Conrad Martens

letters to him printed for private distribution by the Cambridge Philosophical Society in 1835, the word 'true' was omitted, and '(?)' was added after 'Planariæ'. Charles's observations on the anatomy and behaviour of these flatworms were nevertheless mainly correct, except that he thought they fed on decayed wood, whereas in fact they are carnivorous.

Charles was taken hunting one day by a wealthy priest who had a pack of five exotically-named dogs that were released into a forest of huge trees and left to pursue their own small deer and other game. In the intervals, the hunters with guns shot toucans and beautiful little green parrots in a rather aimless fashion. Charles was taken to see a bearded monkey shot the previous day, but did not record having seen a live one.

Once again he was disappointed in the Brazilian birds, which made surprisingly little show in their native country. One of the most characteristic sounds in Rio today is the repeated call of the tyrant-flycatchers, but they do not possess the harmonious voice of the crotophaga, related to the parrots, of which Charles brought back a specimen with a stomachful of insects.

May 1832

The Sugar Loaf at Rio de Janeiro, by Conrad Martens

Better vocalists were found elsewhere, for in torrents of rain that soaked the fields he found a toad that sang through its nose at a high pitch, and then an equally musical frog:

> On the back, a band of "yellowish brown" width of head, sides copper yellow; abdomen silvery yellowish white slightly tuberculated: beneath the mouth, smooth dark yellow; under sides of legs leaden flesh colour. Can adhere to perpendicular surface of glass. The fields resound with the noise which this little animal, as it sits on a blade of grass about an inch from the water, emits. The note is very musical. I at first thought it must be a bird. When several are together they chirp in harmony; each beginning a lower note than the other, & then continuing upon two (I think these notes are thirds to each other).

In addition to its ability to climb up a sheet of glass, the musician had some interesting parasites on its skin, and these too were preserved for identification.

May 1832

A favourite excursion made by Charles several times with friends from the *Beagle* was to climb to the summit of the Corcovado mountain, a huge mass of naked granite looking down on Rio, where a century later the huge statue of Christ would be erected. On 30 May, Charles took his mountain barometer with him, and determined the height of the mountain to be 2225 feet above sea level, though possibly the figure of 2330 feet obtained on another occasion by Captain P.P. King was more reliable.

It was while he was in Rio that Charles wrote to Henslow: 'I am at present red-hot with spiders, they are very interesting, & if I am not mistaken, I have already taken some new genera.' He had indeed, and one of his captures was a crab spider of the family Thomisidae:

> Evidently by its four front strong equal legs being much longer than posterior; by its habits on a leaf of a tree, is a Laterigrade: It differs however most singularly from that tribe & is I think a new genus. Eyes 10 in number, (!?) anterior ones red, situated on two curved *longitudinal* lines, thus the central triangular ones on an eminence: Machoires rounded inclined: languettes bluntly arrow shaped: Cheliceres powerful with large aperture for poison. Abdomen encrusted & with 5 conical peaks: Thorax with one small one: Crotchets to Tarsi, very strong (& with 2 small corresponding ones beneath?) Colour snow white, except tarsi & half of leg bright yellow. also tops of abdominal points & line of eyes black. It must I think be new. Lithetron paradoxicus Darwin!!! Taken in the forest.[65]

Charles's occasional lapses into French in his notes were the consequence of his dependence on books by the *encyclopédistes* Cuvier, Lamarck, Lamouroux and others in the *Beagle*'s library, his favourite being the seventeen volumes of the *Dictionnaire classique d'histoire naturelle*, edited by Jean Baptiste Genevieve Marcellin Bory de Saint-Vincent.

Although spiders are important insect predators, Charles found that sometimes the tables were turned, for he came upon wasps known as mud daubers of the family Sphecidae that hunt spiders as food for their larvae. He wrote:

> I have frequently observed these insects carrying dead spiders, even the powerful genus Mygalus, & have found the clay cells made for their larvæ, filled with dying & dead small spiders: to day (June 2^{d}) I watched a contest between one of them & a large Lycosa. The insect dashed against the spider & then flew away; it had evidently mortially [*sic*] wounded its enemy with its sting; for the spider crawled a little way & then rolled down the hill & scrambled into a tuft of grass. The Hymenoptera [wasp] most assuredly again found out the spider by the power of smell; regularly making small circuits (like a dog) & rapidly vibrating its wings & antennæ: It was a most curious spectacle: the Spider had yet some life, & the Hymenop was most cautious to keep clear of the jaws; at last being stung twice more on under side of the thorax it became motionless. The hymenop. apparently ascertained this by repeatedly putting its head close to the spider, & then dragged away the heavy Lycosa with its mandibles. I then took them both.[65]

'Whilst on board the *Beagle*,' wrote Charles in his *Autobiography*, 'I was quite orthodox, and I remember being heartily laughed at by several of the officers (though themselves orthodox) for quoting the Bible as an unanswerable authority on some point of morality.' So at this time he had not yet begun to think seriously about the manner in which new species of animals might come into being, and his orthodoxy included a belief in a world tenanted by constant species that had originated at specific centres of creation. Since he was well-versed by now in the first two volumes of Charles Lyell's *Principles of Geology*, this does not of course mean that he subscribed to the absolute truth of the first book of Genesis, nor to the accuracy of Bishop Usher's calculations of the age of the earth. But he had been impressed at Cambridge by William Paley's argument in his *Natural Theology* that in looking at the living world 'The marks of *design* are too strong to be gotten over. Design must have had a designer. That designer must have been a person. That person is God.' In due course his faith in Paley waned, but as will be seen he continued to speak of a Creator in his notes until 1836, so that specifically on the evolutionary front his thoughts had not yet moved far when he was in Brazil.

All the same, he had already made significant advances in two important biological fields of which he was one of the founding fathers. Thus from the very beginning of the voyage he regarded the behaviour of the animals he observed as equal in importance to the anatomical differences between them in distinguishing between species. A good example was provided by his comments on the butterfly *Papilio feronia*:

> This insect is not uncommon & generally frequents the Orange groves; it is remarkable in several respects. It flies high & continually settles on the trunks of trees; invariably with its head downwards & with its wings expanded or opened to beyond the horizontal plane. It is the only butterfly I ever saw make use of its legs in running, this one will avoid being caught by shuffling to one side. Some time ago I saw several pairs, I presume males & females, of butterflies chasing each other, & which from appearance & *habits* were I am sure the same species as this. Strange as it may sound, they when fluttering about emitted a noise somewhat similar to cocking a small pistol; a sort of a click. I observed it repeatedly. June 28th. In same place I observed one of these butterflies resting as described on a trunk of tree; another happening to fly past, immediately they chased each other, emitting (& there could be no mistake the space being open) the peculiar noise: this is continued for some time & is more like a small toothed wheel passing under a spring pawl. – The noise would be heard about 20 yards distant. This fact would appear to be new.[66]

A preliminary examination of a specimen of the butterfly in 1837 by G.R. Waterhouse at the London Zoological Society provided no explanation for the source of the peculiar noise, but a few years later it was found by another entomologist that it was produced by a sort of drum at the base of the fore wings, together with a screw-like diaphragm in the interior.

June 1832

Another important branch of biology in which Charles was a leading pioneer, along with Linnaeus, Buffon and Humboldt, is the study of mutual relations between animals and their environment, for which the term 'ecology' was introduced in 1873. Here too, Charles's basically new way of thinking was apparent from the first in his notes. Summarising his general observations on what he had seen in Rio, he wrote:

> I could not help noticing how exactly the animals & plants in each region are adapted to each other. Every one must have noticed how Lettuces & Cabbages suffer from the attacks of Caterpillars & Snails. But when transplanted here in a foreign clime, the leaves remain as entire as if they contained poison. Nature, when she formed these animals & these plants, knew they must reside together.

Referring to collections of insects he had made on the shore behind the Sugar Loaf in Rio, he said that since the situation was much the same as that of Barmouth when he was collecting there in August 1828, many of the species would be closely allied. On another occasion he wrote:

> In my geological notes I have mentioned the lagoons on the coast which contain either salt or fresh water. The Lagoa near the Botanic Garden is one of this class. The water is not so salt as the sea, for only once in the year a passage is cut for sake of the fishes. The beach is composed of large grains of quartz & very clean. If cemented into a breccia or sandstone it would precisely resemble a rock at Bahia containing marine shells. A small Turbo [a turban snail] appeared the only proper inhabitant, & thus differed from the lagoons on the Northern coast in the absence of those large bodies of Bivalves. I was surprised on the borders to see a few Hydrophili [water beetles] inhabiting this salt water, & some Dolimedes [a nursery web spider] running on the surface.

CHAPTER 7

An Unquiet Trip from Monte Video to Buenos Aires

At nine o'clock in the morning on 5 July the *Beagle* sailed out of the harbour at Rio on a gentle breeze, hailed by a salute of hearty cheers from the crews of HMS *Warspite* and HMS *Samarang*. FitzRoy noted with some satisfaction that 'Strict etiquette might have been offended at such a compliment to a little ten-gun brig, or indeed to any vessel unless she were going out to meet an enemy, or were returning into port victorious: but although not about to encounter a foe, our lonely vessel was going to undertake a task laborious, and often dangerous, to the zealous execution of which the encouragement of our brother-seamen was no trifling enducement.'

For the next three weeks the *Beagle* sailed on to the south, sometimes in light winds when progress was disappointingly slow, sometimes in gales when even the sight of a whale possessed little interest to Charles's jaundiced eyes, but best when the studding sails were 'alow & aloft – that is wind abaft the beam & favourable'. On the morning of 14 July Charles noted:

> I was much interested by watching a large herd of Grampuses, which followed the ship for some time. They were about 15 feet in length, & generally rose together, cutting & splashing the water with great violence. In the distance some whales were seen blowing. All these have been the black whale. The Spermaceti is the sort which the Southern Whalers pursue.

The grampuses, which on this occasion were genuine ones, were probably a group of juvenile pilot whales, totally black in colour, with bulging foreheads full of sperm oil.

Four days later, Charles wrote:

> A wonderful shoal of Porpoises, at least many hundreds in number, crossed the bows of our vessel. The whole sea in places was furrowed by them. They proceeded by jumps in which the whole body was exposed, & as hundreds thus cut the water it presented a most extraordinary spectacle. When the ship was running 9 knots these animals could with the greatest ease cross & recross our bows & then dash away right ahead, thus showing off to us their great strength & activity.

The *Beagle* sailed on in the variable weather characteristic of the entrance to the Rio Plata. Close to the mouth of the river on a particularly dirty night, the ship was surrounded by penguins and seals which made such curious noises that the Master reported to the First Lieutenant that he had heard cattle lowing on the shore. On the morning of 26 July, the *Beagle*'s anchoring at Monte Video was, according to Charles, quickly followed by the arrival alongside of six heavily-armed boats from the frigate HMS *Druid*, containing forty marines and a hundred sailors. The frigate's Captain Hamilton explained that the current military government had just seized four hundred horses belonging to a British subject, and that he aimed to provide sufficient visible support for the opposition party to bring about a restitution of the horses.* It seemed that such disputes were usually won without bloodshed by the side that succeeded in looking the stronger. This episode was an eye-opener for Charles on the vagaries of South American politics, but FitzRoy did not regard

* Under their first Emperor Dom Pedro I, the Brazilians had in 1821 annexed the eastern part of Uruguay. Insurgents led by Juan Antonio Lavalleja (1784–1853) fought back, until in 1828 Uruguay's independence was at last recognised. When the *Beagle* arrived in Monte Video on 26 July 1832 the fighting in progress was between rival political parties.

August 1832

The mole at Montevideo, by Augustus Earle

the incident as worthy of mention in his account of the day, merely recording that he was occupied with observations for his chronometers, and preparing for surveying the coasts south of the Rio Plata.

On the following morning, FitzRoy and Charles landed on Rat Island, where one of them took sights, while the other found, but did not preserve, a species of legless lizard known as a skink. On 28 July Charles visited the Mount, the little hill 450 feet high that dominated the district and gave Monte Video its name. He decided that the view from the summit was the most uninteresting that he had ever seen – like Cambridgeshire but without even any trees.

Two days later, FitzRoy got wind of the remains of some hydrographical information collected by Spain that was preserved in the archives of Buenos Aires, and on 2 August the *Beagle* sailed to the south bank of the Rio Plata in search of it. As explained by Charles, they had a disconcerting reception:

> We certainly are a most unquiet ship; peace flies before our steps. On entering the outer roadstead, we passed a Buenos Ayres guard-ship. When abreast of her she fired an empty gun, we not understanding this sailed on, & in a few minutes another discharge was accompanied by the whistling of a shot over our rigging. Before she could get another gun

ready we had passed her range. When we arrived at our anchorage, which is more than three miles distant from the landing place, two boats were lowered, & a large party started in order to stay some days in the city. Wickham went with us, & intended immediately going to Mr Fox, the English minister, to inform him of the insult offered to the British flag. When close to the shore, we were met by a Quarantine boat which said we must all return on board, to have our bill of health inspected, from fears of the Cholera. Nothing which we could say about being a man of war, having left England 7 months & lying in an open roadstead, had any effect. They said we ought to have waited for a boat from the guard-ship & that we must pull the whole distance back to the vessel, with the wind dead on end against us & a strong tide running in. During our absence, a boat had come with an officer whom the Captain soon dispatched with a message to his Commander to say 'He was sorry he was not aware he was entering an uncivilized port, or he would have had his broardside ready for anwering his shot'. When our boats & the health one came alongside, the Captain immediately gave orders to get under weigh & return to M Video. At the same time sending to the Governor, through the Spanish officer, the same messuages [*sic*] which he had sent to the Guard-ship, adding that the case should be throughily [*sic*] investigated in other quarters. We then loaded & pointed all the guns on one broadside, & ran down close alongside the guard-ship. Hailed her & said that when we again entered the port, we would be prepared as at present & if she dared to fire a shot we would send our whole broardside into her rotten hulk.* We are now sailing

* The Beagle was nominally a ten-gun brig, but for this surveying voyage the Admiralty firmly refused to fit more than seven guns, comprising a six-pound boat-carronade on the forecastle, and before the chesstree two six-pound guns. Abaft the mainmast there were two nine-pound and two six-pound guns. At FitzRoy's insistence all the guns were brass, to avoid interference with the compasses. In Rio he further increased his armament by buying two more brass nine-pounder long guns, bringing the total up to nine. He was all the same very reluctant ever to fire them, for fear of disturbing his twenty-two chronometers. See Keith S. Thomson, *HMS Beagle: The Story of Darwin's Ship*. W.W. Norton, New York and London, 1995.

quietly down the river. From M Video the Captain intends writing to Mr Fox & to the Admiral, so that they may take effective steps to prevent our Flag being again insulted in so unprovoked a manner.

The following day, after another tricky passage along the muddy and winding channel of the Rio Plata, with banks often marked by old wrecks – 'it is an ill wind which blows nobody any good' said Charles – the *Beagle* arrived at Monte Video after sunset, and the Captain immediately went on board the *Druid*. He returned with the news that the *Druid* would next morning sail for Buenos Aires, and demand an apology for the guard-ship's conduct. Charles noted belligerently, 'Oh I hope the Guard-ship will fire a gun at the frigate; if she does, it will be her last day above water.'

A fortnight later the *Druid* returned from Buenos Aires with a long apology from the government for the insult offered to the *Beagle*. The captain of the guard-ship had immediately been arrested, and it was left to the British Consul whether he should any longer retain his commission. It seemed nevertheless that the Argentinians had voiced some complaint against FitzRoy's undiplomatic language, for reporting later to the Hydrographer in London on his conduct of the affair, FitzRoy wrote:

> With reference to the expressions which have offended the Buenos Airean Government, I beg to inform you, and I request that you will make it known, if necessary, that I did *not* say, that 'I should go to some other country where the government was more civilized', but that my expression to the health officer *was*, 'Say to your government that I shall return to a more civilized country where boats are sent more frequently than balls.' In hailing the guard vessel I did *not* in *any* way allude to the *government* and my words to *her commander* were 'If you dare to fire another shot at a British man-of-war you may expect to have your hulk sunk, and if you fire at *this* vessel, I will return a broadside for every shot!'.[67]

In the meantime, further trouble of a different kind had arisen in Monte Video. On 5 August, Charles wrote in his journal:

August 1832

The custom house at Monte Video, by Augustus Earle

This has been an eventful day in the history of the *Beagle*. At 10 oclock in the morning the Minister for the present military government came on board & begged for assistance against a serious insurrection of some black troops. Cap FitzRoy immediately went ashore to ascertain whether it was a party affair, or that the inhabitants were really in danger of having their houses ransacked. The head of the Police (Damas) has continued to power through both governments, & is considered as entirely neutral; being applied to, he gave it as his opinion that it would be doing a service to the state to land our force. Whilst this was going on ashore, the Americans landed their boats & occupied the Custom House. Immediately the Captain arrived at the mole, he made us the signal to hoist out & man our boats. In a very few minutes, the Yawl, Cutter, Whaleboat & Gig were ready with 52 men heavily armed with Muskets, Cutlasses & Pistols. After waiting some time on the pier Signor Dumas arrived & we marched to a central fort, the seat of government. During this time the insurgents had planted artillery to command some of the streets, but otherwise remained quiet. They had previously broken open the Prison & armed the prisoners. The chief cause of apprehension was owing to their being in possession of the citadel which contains all the ammunition. It is suspected that all this disturbance is owing to the mæneuvring of the former constitutional government. But the politicks of the place are quite unintelligible: it has always been said that the interests of the soldiers & the present government are identical, & now it would seem to be the reverse. Capt. FitzRoy would have nothing to do with all this: he would only remain to see that private property was not attacked. If the National band were not rank cowards, they might at once seize the citadel & finish the business; instead of this, they prefer protecting themselves in the fortress of St. Lucia. Whilst the different parties were trying to negociate matters, we remained at our station & amused ourselves by cooking beefsteaks in the Courtyard. At sun-set the boats were sent on board & one returned with warm clothing for the men to bivouac during the night. As I had a bad headache, I also came & remained on board. The few left in the Ship under the command of Mr Chaffers [the Master] have been the most busily engaged of the whole crew. They have triced up the Boarding netting, loaded & pointed the guns, & cleared for action. We are now at night in a high state of preparation so as to make the best defence possible, if the *Beagle* should be attacked. To obtain ammunition could be the only possible motive. 6th. The boats have returned. Affairs in the city now more decidedly show a party spirit, & as the black troops are enclosed in the citadel by double the number of armed citizens, Capt FitzRoy deemed it advisable to withdraw his force. It is probable in a very short time the two adverse sides will come to an encounter under such circumstances. Capt FitzRoy being in possession of the central fort would have found it very difficult to have preserved his character of neutrality. There certainly is a great deal of pleasure in the excitement of this sort of work – quite sufficient to explain the reckless gayety with which sailors undertake even the most hazardous attacks. Yet as time flies, it is evil to waste so much in empty parade.

FitzRoy's withdrawal was followed by a fair amount of skirmishing and continued disorder. The military governor, Juan Antonio Lavalleja, was the man who had first established the independence of Uruguay in 1828. When he entered the town of Monte Video, he was said by Charles to have been well received by everybody except for his own black troops. He threatened to expel them from the citadel, and planted some guns to command the gate. During the night the blacks then made a sally, volleys of musketry were heard in the city, and it seemed on the *Beagle* that there might have been heavy fighting. But in fact not a single person was wounded, because according to Charles both parties were determined not to come within musket range either of one another or of the black troops. The next day, support for Lavalleja quickly evaporated, and he made a strategic retreat from the scene, leaving the field to his rival the former president, Don Fructuoso Rivera. Fierce party quarrels continued to take place in the town, and until 12 August the shops were all shut and the inhabitants were obliged to keep within their houses. Don Fructuoso then reappeared, and restoration of the constitutional government was proclaimed. Two days later the President made his formal re-entry into the town, and his government was considered to be in office once again. It was reported to Charles, perhaps by the merchant Mr Parry with whom he had earlier dined, that the spectacle was a magnificent one, with 1800 wild gaucho (Argentinian cowboy) cavalry in support, many of whom were curiously-dressed Indians with splendid horses.

FitzRoy was pleased to be told by the principal persons whose lives and property were threatened that the presence of the *Beagle*'s crewmen had certainly prevented bloodshed. Charles concluded that 'One is shocked at the bloody revolutions in Europe, but after seeing to what an extent such imbecile changes can proceed, it is hard to determine which of the two is most to be dreaded.' Considering that like patriots in neighbouring countries, Lavalleja and his predecessors had had a severe struggle against the Spanish overlords, followed by fights against both Portuguese and Brazilian forces trying to take advantage of the weakness of the small Republica Oriental del Uruguay, Charles was perhaps being rather severe. And Uruguay remained for some years to come in a state of intermittent civil war between Lavalleja's supporters, named the Blancos because they carried white flags, and the Colorados once led by Don Fructuoso, with red ones.

August 1832

In his general notes on what wildlife he had seen in Monte Video, Charles recorded that:

> Birds are abundant on the plains & are brilliantly coloured. Starlings, thrushes, shrikes, larks & partridges are the commonest. Snipes here frequently rise & fly up in great circles; in their flight, as they descend, they make that peculiar buzzing noise, which few which breed in England are known to do. On the sand-banks on the coast are large flocks of Rhynchops [scissor-beaks]; these birds are generally supposed to be the inhabitants of the Tropics. Every evening they fly out in flocks to the sea & return to the beach in the morning. I have seen them at night, especially at Bahia Blanca, flying round a boat in a wild rapid irregular manner, something in same manner as Caprimulgus [nightjar] does. I cannot imagine what animals they catch with their singular bills.

When next he encountered *Rhyncops* he had more to say about the function of their beaks.

During the last few days before departing on the *Beagle*'s first cruise, Charles's most notable achievement was, after a long chase among the rocks at the Mount, to shoot through the head a large female capybara, in structure a huge guinea-pig, in habits a water rat. She weighed ninety-eight pounds, and could not readily be preserved apart from a tick crawling on her skin. He also found two more new species of turbellarian worm under dry stones on the Mount, collected some elegant snakes and frogs, and as usual captured a host of spiders and beetles.

CHAPTER 8

Digging up Fossils in the Cliffs at Bahia Blanca

The Beagle was now ready to proceed on her first cruise to the south, surveying the coastal waters between Buenos Aires and Bahia Blanca. At first the steep waves in the shallow water at the mouth of the Rio Plata caused so much spray to break over the ship that Charles had seldom felt a more disagreeable sensation in his stomach. In more open water after rounding Point Piedras, matters improved, and on 24 August in ten fathoms of water slightly north of Cape Corrientes in latitude 37°26´S, Charles found 'incredible numbers' of the very simple animal that he had encountered earlier in the voyage, north of St Jago (see p.54) and off the Abrolhos Islands, and had been unable to identify. This time he was able to describe in some detail the anatomy of these arrow worms or chaetognaths as they came to be called, and twelve years later published a short paper about them.[68]

Two days later, in latitude 38°20´S, Charles found some of what he called corallines, the insignificant but exceedingly numerous little organisms of doubtful nature that encrusted rocks and fronds of seaweed like a moss growing on the surface, to which he had been introduced by Robert Grant at Edinburgh (see p.6). On examination of the specimens under his microscope, he immediately and correctly identified them as closely related to what would today be called 'bryozoans' of genus *Flustra* that he had seen in the Firth of Forth. But a vital question about them that had still not finally been settled was whether they should be classified as belonging to the plant or the animal kingdom. Were they plants capable of living wholly on

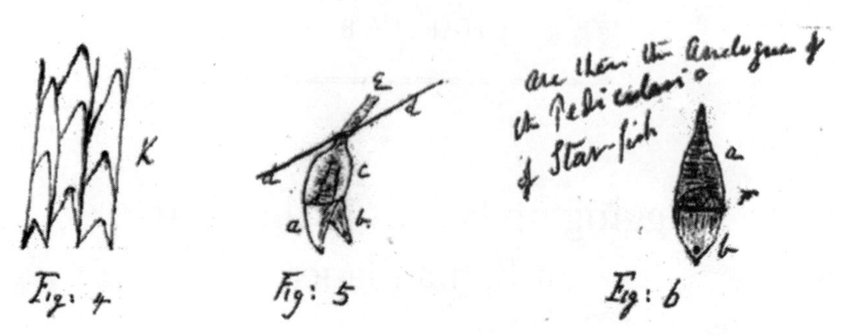

The bryozoan with a vulture's beak capsule

inorganic substances, deriving their energy from light by photosynthesis, and having neither powers of locomotion nor special organs of sensation or digestion? Or were they living animals endowed with identifiable digestive and sensory systems, and capable of undertaking voluntary motion?

What immediately excited Charles about these particular corallines were the pointed box-like individual cells shown in Fig. 4 of the drawing that he made under his microscope, which are now termed 'zooids', and contained not only the retractable polyps serving as feeding units, but were mostly equipped in addition with very curious capsules resembling in shape though not in size the beak and head of a vulture, as may be seen in Fig. 5.* The feature of the capsules that Charles found so remarkable was the continual movement that they made. In his notes he wrote:

* The stems of the coralline, up to two inches high in all, branched irregularly and had several rows of the tiny box-like zooids, which contained a retractable feeding unit or polyp, complete with a mouth and an anus, and mostly a vulture's head capsule as well. The capsules as drawn in Fig. 5 were 1/75th of an inch in length, with a head, labelled (c), an upper mandible (a), and a lower mandible (b), the dot representing a tooth. They were connected to the middle of the outer edge (dd) of the cell by a short stem or peduncle (e) extending internally to the base of the cell. Fig. 6 shows the mouth wide open seen from the front, F being a semi-circular opening at the base of the upper mandible.

September 1832

> When the Coralline is in water, whether the Polype is within or out of cell, the capsule generally is wide open (as in Fig. 6), & the whole head *on peduncle* turns backwards & forwards, vertically going through at least 90°. They perform the whole motion in about 5 seconds. Most of the capsules perform it isochronously. Occassionally [sic] they close for an instant the lower mandible. In a small branch, so many capsules moving caused in it a trembling. A point of needle being inserted within the jaws was always seized so fast as to be able to drag small branch. The motion in these became fainter as the Polype lost strength. Polype, although so irritable of motion, took no notice of the motion of Capsule. What office does this organ perform?[69]

The continual nodding of the capsules, which are known as 'pedunculate avicularia', is indeed an extremely striking phenomenon to watch under a dissecting microscope in those species of bryozoans that are equipped with them, which cannot be observed without the dismissal of any idea that such organisms might possibly be plants. Charles's Specimen No. 355 in Spirits of Wine was identified in 1901 by the curator of the Zoology Museum of Cambridge University, where it is safely stored in its bottle of alcohol, as belonging to genus *Bugula* in the same sub-order as *Flustra*. The question as to what exact function is served by the avicularia surprisingly still remains unanswered. As Charles himself pointed out, the polyps of all bryozoans have innumerable small tentacles for seizing their prey, so that the vultures' beaks would be superfluous for feeding purposes. And he scornfully dismissed the possibility of their being an organ of generation, which he considered to be 'the last resource in all puzzling cases'. A defensive role has been suggested, though there is no decisive evidence in its support, and perhaps protection from the accumulation of rubbish on the colony is the best answer.

On 6 September the *Beagle* arrived in Blanco Bay, where they met a small schooner sailing southwards from the new settlement at Bahia Blanca to the Rio Negro and beyond to catch some seals. On board was its owner, James Harris, a British trader who offered to guide them to a good anchorage at Port Belgrano at the head of the bay and thence to the settlement about twenty miles away. Next day, after a difficult trip in one of the ship's boats through a maze of muddy creeks, they reached the landing place, where what FitzRoy described as an assemblage of grotesque figures had come to meet them, in the mistaken expectation that they were bringing supplies from Buenos Aires for the needy colony. The welcoming party was led by the *Comandante* of the settlement, accompanied by a number of gauchos whose exotic costumes suggested to Charles that he had landed in the middle of Turkey; some irregular soldiers variously armed and clad, but very well mounted; and a group of almost naked Indian prisoners, who had been brought along in order to transport the supplies, but were currently occupied in devouring the remains of a half-roasted horse.

The boat's crew were then left to bivouac for the night at the landing place, while FitzRoy and the *Beagle's* purser on a loaned horse, and Harris and Charles mounted similarly behind two gaucho soldiers, rode the last four miles to the settlement. This was a fortress built by a French engineer four years previously, and grandly named 'La fortaleza protectora Argentina'. It occupied a polygon 282 yards in diameter, surrounded by a narrow ditch and mud walls that in places were nearly twenty feet high. It housed some four hundred people, most of them soldiers, and had been successfully defended several times against large bodies of Indians, whom the Spaniards were attempting to wipe out in a thoroughly barbarous fashion.

Charles and his gaucho arrived at the fortress slightly before the others, giving the second-in-command, who was an elderly major evidently reputed for his wisdom, time to cross-question him about the purpose of his visit. Unfortunately Harris had explained that Mr Darwin was '*un naturalista*', a term unknown to anyone at the settlement. The further explanation that this meant 'a man that knows every thing' merely increased the suspicions of the Major, who as an old Spaniard with the old feelings of jealousy, suspected that FitzRoy's party were spies sent to reconnoitre the position in preparation for an attack. When the Captain, wishing only to praise the bay, said that he could bring up into it even a line of battleships, Charles felt that 'the old gentleman was appalled, and in his mind's eye saw the British marines taking his fort'.

Although the party was treated with hospitality, and the *Comandante* himself was reasonably polite, they were kept throughout under uncomfortably close observation, and FitzRoy determined to return to the *Beagle* as early as possible next morning. When they got back to the ship, they found a troop of gaucho cavalry who had evidently been sent to keep an eye on them, but who were nevertheless happy to engage in a brisk trade for dollars of ostriches and their eggs, and of deer, guinea-pigs and armadillos. They also gave Charles his first lesson in their method of capturing ostriches with a lasso consisting of two heavy balls, or *bolas*, fastened to the ends of a long leather thong. Later he admitted, 'I never enjoyed anything so much as Ostrich hunting with the wild Soldiers.'

During his travels, Charles always kept with him a small volume of Milton's *Paradise Lost*. It was near Bahia Blanca on 11 September that he found a remarkable toad which looked as if it had first 'been steeped in the blackest ink, and then when dry, allowed to crawl over a board freshly painted the brightest vermilion, so as to colour the soles of its feet and parts of its stomach'. He felt that if it was new it should be christened 'diabolicus', for Milton must have alluded to this individual as being 'squat like a toad' and fit to 'preach in the ear of Eve'. Having found another of these toads at Maldonado in surroundings that seemed uncomfortably dry, he had carried it to a pool of water as a treat, but to his dismay found it unable to swim. Three years after Charles's death, his widow Emma wrote to their son William: 'I am reading your father's Journal after a long gap — it makes me feel so happy as if I was going with him; only I want to ask him so many questions. The real man comes out constantly, e.g. thinking to give a toad in a dry place in Rio Plata "quite a treat" by taking it to a pond & nearly drowning it.'

Later in September, Milton was still on Charles's mind when he wrote:

September 1832

> The sea from its extreme luminousness presented a wonderful & most beautiful appearance; every part of the water, which by day is seen as foam, glowed with a pale light. The vessel drove before her bows two billows of liquid phosphorus, & in her wake was a milky train. As far as the eye reached, the crest of every wave was bright, & from the reflected light, the sky just above the horizon was not so utterly dark as the rest of the Heavens. It was impossible to behold this plain of matter, as it were melted & consuming by heat, without being reminded of Milton's description of the regions of Chaos & Anarchy.

On 13 September the *Beagle's* anchorage was moved a few miles up the harbour in order to be nearer a newly discovered watering place. Charles was happy to find that he could spend September in Patagonia in much the same way as he would have at home, shooting all the game in sight. There were, all the same, some differences from shooting parties in England. The only birds in the game bag were ostriches, and one of the deer that he shot had been disposed of by the time some sailors came to collect it, for 'I left it on the ground a substantial beast, but in the evening the vultures & hawks had picked even the bones clean'. Most of Charles's fellow sportsmen were pure Indians, though he realised at dinner, watching 'their swarthy but expressive countenances', that they were not all that they seemed to be, for one of them began to flirt with him, and 'she pretended to be frightened of my gun & screamed out "no est cargado?" [isn't it charged?]'. Also he wrote that 'What we had for dinner today would sound very odd in England. Ostrich dumpling & Armadilloes; the former would never be recognised as a bird but rather as beef. The Armadilloes when unlike to the Gaucho's fashion, cooked without their cases, taste & look like a duck. Both of them are very good.'

22 September was a red-letter day for biology, when Charles went for a pleasant cruise around the bay with the Captain and Sulivan, and at Punta Alta about ten miles from the ship made a portentous discovery. In his pocket-book he wrote:

September 1832

> Sept 22d. Entrance of creek, dark blue sandy clay much stratified dipping to NNW or N by W at about 6°∠. On the beach a succession of thin strata dipping at 15° to W by S, conglomerate quartz & jasper pebbles with shells – vide specimens. On the coast about 12 feet high, in the conglom teeth & the thigh bone. Proceeding to NW there is a horizontal bed of earth containing much fewer shells, but armadillo – this is horizontal but widens gradually, hence I think conglomerate with broken shells was deposited by the action of tides, earth quietly. Is this above the clay which was seen a short time previously, covered by diluvium & sand hillocks as earthy bank? Thickened & cropped out in direction NNW it probably overlies the clay.[70]

It may be noted how commendably careful Charles was, on this his very first sight of a fossil vertebrate, to determine the precise strata in which the fossils were located. The next day he returned to Punta Alta better equipped for some digging, and after nearly three hours of hard work managed to extract from the soft rock the head of a large animal which he thought at first sight might be allied to a rhinoceros.* The untidy piles of fossils dumped by Charles on the spotless decks of the *Beagle* were wholly contrary to naval tradition, and as recalled long afterwards by his daughter Henrietta:

> My Father used to describe how Wickham, the first Lieutenant – a very tidy man who liked to keep the decks so that you could eat your dinner off them – used to say 'If I had my way, all your d—d mess would be chucked overboard, & you after it old Flycatcher.'[71]

But 'although Wickham always was growling at my bringing more dirt on board than any ten men, he is far the most conversible being

* It was in due course classified by the anatomist Richard Owen in London as *Scelidotherium*, belonging to the sub-order Xenarthra, containing the New World edentates that are placental mammals such as the extinct species *Megatherium*, *Mylodon* and *Glyptodon*, whose bones and teeth were also found by Charles at Punta Alta, together with modern anteaters, tree sloths and armadillos.

on board. I do not mean talks the most, for in that respect Sulivan quite bears away the palm.' So Charles remained on the best of terms with the First Lieutenant, 'a glorious fine fellow', for the whole of the voyage.

Then on 2 October FitzRoy anchored the *Beagle* off the cliffs at Monte Hermoso, close to the entrance to Blanco Bay, where he proposed to erect a sea-mark for the guidance of ships bound for Puerto Belgrano. While members of the crew got on with their building operation, Charles and Philip Gidley King set off in one direction to geologise, while the surgeon Bynoe went in another to shoot. King later described what happened next:

> The Beagle anchored some distance from that bar-bound Bay on the open coast line about a mile from the shore. Three of its crew were sent away and effected a landing through the surf. A day's provision had been supplied to them, supposing they would return in the evening. A gale from South East however sprang suddenly up, making a dead lee shore from which their little vessel could for two days be seen holding on to her two anchors pitching bows under. Mr Darwin and the writer were of the shore party which turned the boats bottom up on the beach for shelter for the nights they passed without more food than they had saved from their first supper, with the addition of a dead hawk picked up on the shore. The scene to Mr Darwin and indeed to others was a novel one, and the probability that the ship might drag her anchors or part her cables was present to many minds. However the gale moderated and a boat was seen to leave the Beagle, the Captain himself in it with a cask of provisions which was thrown into the surf just outside the breakers. Several splendid looking men soon stripped and went to land the cask, the danger being that it might be dashed to pieces on the rocks. The head of the cask was soon knocked in, and Mr Darwin and the party stood round it to receive each his share. Some record of this may be amongst Mr Darwin's notes, at any rate the foregoing is one amongst many others of his experiences. On shore we all admired the peace and thorough confidence the Captain had in his well-proved "holding gear", as shown by

his remaining in our sight instead of beating his ship off the lee shore as he might have done.[72]

Charles's account of the episode in his commonplace journal[73] closely confirmed those of King and FitzRoy, and he added that 'Nothing would break the wind, which was so cold that there was snow in the morning on the Sierra de Ventana. I never knew how painful cold could be. I was unable to sleep even for a minute from my body shivering so much.'

On 8 October 1832 Charles went back again to Punta Alta and found a jawbone containing a tooth that enabled him to identify it as belonging to a *Megatherium*. But when he noted in his pocket-book 'Megatherium like Armadillo case, teeth:' he had been misled by the statement in the *Dictionnaire classique d'histoire naturelle*, based on a mistake due to Cuvier, that it had recently been discovered that as in an armadillo, the skin of the huge fossil animal was reinforced by ossified polygons throughout its thickness. 'This is particularly interesting,' he noted in his journal, 'as the only specimens in Europe are in the Kings collection at Madrid, where for all purposes of science they are nearly as much hidden as if in their primæval rock.'

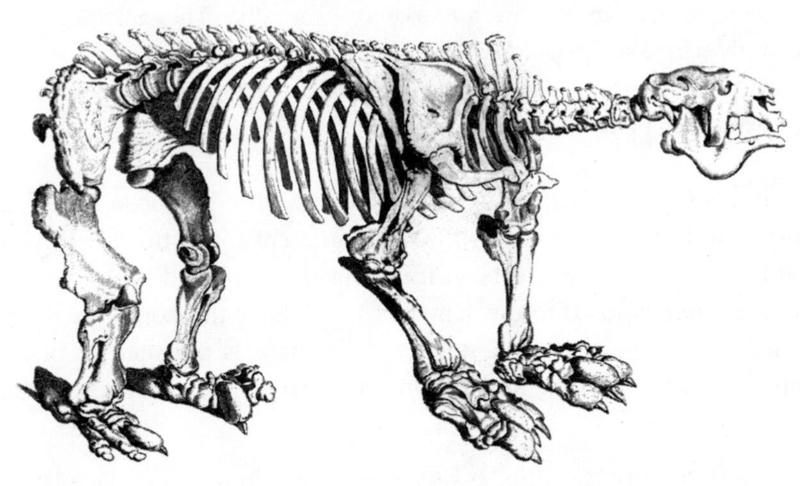

Reconstruction of *Megatherium* bones from Cuvier's *Recherches sur les ossements fossiles*, after he had corrected his mistake

October 1832

In a letter to Henslow, Charles wrote:

I have been very lucky with fossil bones; I have fragments of at least 6 distinct animals; as many of them are teeth I trust, shattered & rolled as they have been, they will be recognised. I have paid all the attention I am capable of to their geological site, but of course it is too long a story for here. – 1st. the Tarsi & Metatarsi very perfect of a Cavia: 2nd. the upper jaw & head of some very large animal, with 4 square hollow molars, & the head greatly produced in front. I at first thought it belonged either to the Megalonyx or Megatherium. In confirmation of this, in the same formation I found a large surface of the osseous polygonal plates, which "late observations" (what are they?) show belong to the Megatherium. Immediately I saw them I thought they must belong to an enormous Armadillo, living species of which genus are so abundant here: 3d. The lower jaw of some large animal, which from the molar teeth I should think belonged to the Edentata: 4th. some large molar teeth, which in some respects would seem to belong to an enormous Rodentia; 5th. also some smaller teeth belonging to the same order: &c &c. – If it interests you sufficiently to unpack them, I shall be very anxious to hear something about them: Care must be taken, in this case, not to confuse the tallies. They are mingled with marine shells, which appear to me identical with what now exist. But since they were deposited in their beds, several geological changes have taken place in the country.[73]

In his Geology Notes, Charles wrote of Monte Hermoso that 'The bones in their nature were singularly different from those at P. Alta; in the one case they had been immediately enveloped in the Tosca,* in the other exposed to the action of water. Here the bones were very hard & of a great specific gravity, their surfaces polished & blackened externally; in the smaller ones they from this cause resembled

* Tosca was the name given by the local inhabitants to the reddish soil mixed with clay and lime, and sometimes hardened into a rock, that was characteristic of the pampas.

1. St Jago, one of the Verds, bearing E by N distant 6 miles, 9 June 1833, by Conrad Martens.

2. Botofogo Bay, Rio de Janeiro, by Conrad Martens.

3. Slinging the monkey at Port Desire on Christmas Day 1833, by Conrad Martens.

4. *Rhea darwinii*, by John Gould.

5. Bivouac at Port Desire, 29 December 1833, by Conrad Martens.

6. Entrance to Port St Julian, by Conrad Martens.

October 1832

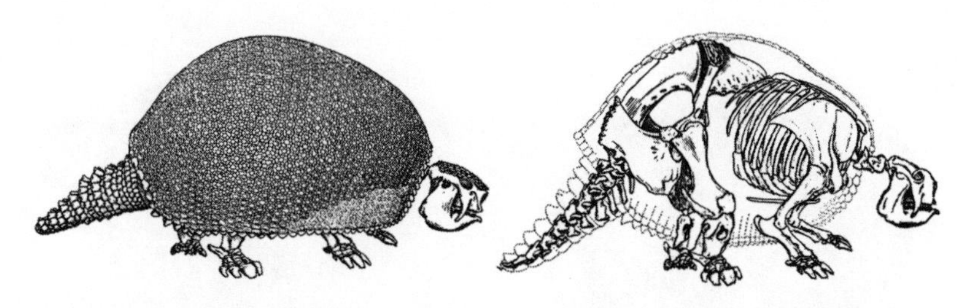

Reconstruction of *Glyptodon*, length about nine feet, with and without its carapace (Romer)

jet.' Among them were bones of some small rodents closely related to the existing agouti or guinea-pig (*Cavia patagonica*), others to the capybara and to the burrowing animal that he called the tucotuco (*Ctenomys braziliensis*). There were also pieces of the carapace composed of a mosaic of polygonal plates fused together of the *Glyptodon*, the giant extinct armadillo of the late Pleistocene period that was common in Argentina.* At Punta Alta there was an exceptionally well-conserved skeleton of the giant ground sloth *Scelidotherium*, with its limbs entombed in their correct relative positions. From the good condition of these fossils, and other features of the strata in which they were found, Charles concluded that they were in all probability coeval with one another, and with still-existing species of shells.

Although the fossil-laden cliffs at Punta Alta have today vanished entirely from sight beneath the principal Argentinian naval base of

* The earliest edentate mammals, so called because they had no incisor or canine teeth covered with hard enamel, or even were toothless, were primitive armadillos, similar in size to their modern descendants, that first appeared in South America at the end of the Paleocene epoch, some fifty million years ago. The more heavily armoured glyptodonts joined the fossil record in the Eocene, around thirteen million years later, accompanied in the Oligocene by ground sloths of modest stature, along with litopterns and notoungulates of steadily increasing size. In the Miocene and Pliocene the fossil beds of Patagonia were rich in the giant sloths and other herbivores that late in the Pleistocene were still flourishing at Pehuen-Co.

A piece of *Glyptodon* carapace from Pehuen-Co, near Monte Hermoso

Puerto Belgrano, the cliffs at Monte Hermoso thirty-five miles up the coast to the east still look much as they must have done in 1832, except that a thinning of the sand dunes that cap the harder sandstone may have slightly reduced their apparent height. One can still dig out of the cliffs jet-black fossils like those described by Charles in his notes. In the specimen illustrated above, the thickness of the carapace of the *Glyptodon* was just over an inch. A point of interest is the presence of several large holes penetrating into the carapace at the borders of the central polygon, whose function was to accommodate blood vessels, nerves and the bulb at the base of hairs, for like the primitive armadillo from which both the extinct giants and their modern cousins were descended, the animals must always have had a thin and hairy external skin outside the carapace.

What Charles did not see, and which would have delighted him immensely, were the lines of footprints of *Megatherium* which revealed themselves in a tidally eroded deposit on the beach at Pehuen-Co not far from Monte Hermoso, in October 1986.[76] Since their discovery, further footprints have come to light of *Megatherium*, of *Macrauchenia* with their distinctive three toes, of a *Glyptodon*, a paleollama and other extinct animals, together with a trackway of the existing bird *Rhea americana*; and the age of samples of the sediment around the footsteps has been determined by radioactive carbon dating and found to be 12,000 ± 110 years before the present.

October 1832

Two sets (above) of *Megatherium* footprints at Pehuen-Co, whose individual diameter was about 0.9 meters, of an animal walking bipedally, and (below) with one set crossing one another. From Aramayo and Manera[76]

The bones from Pehuen-Co have a radiocarbon age of sixteen thousand years, which agrees nicely with Charles's conclusion that they were relatively recent in date. It is not clear precisely what conditions prevailed when this particular group of animals came together for a short

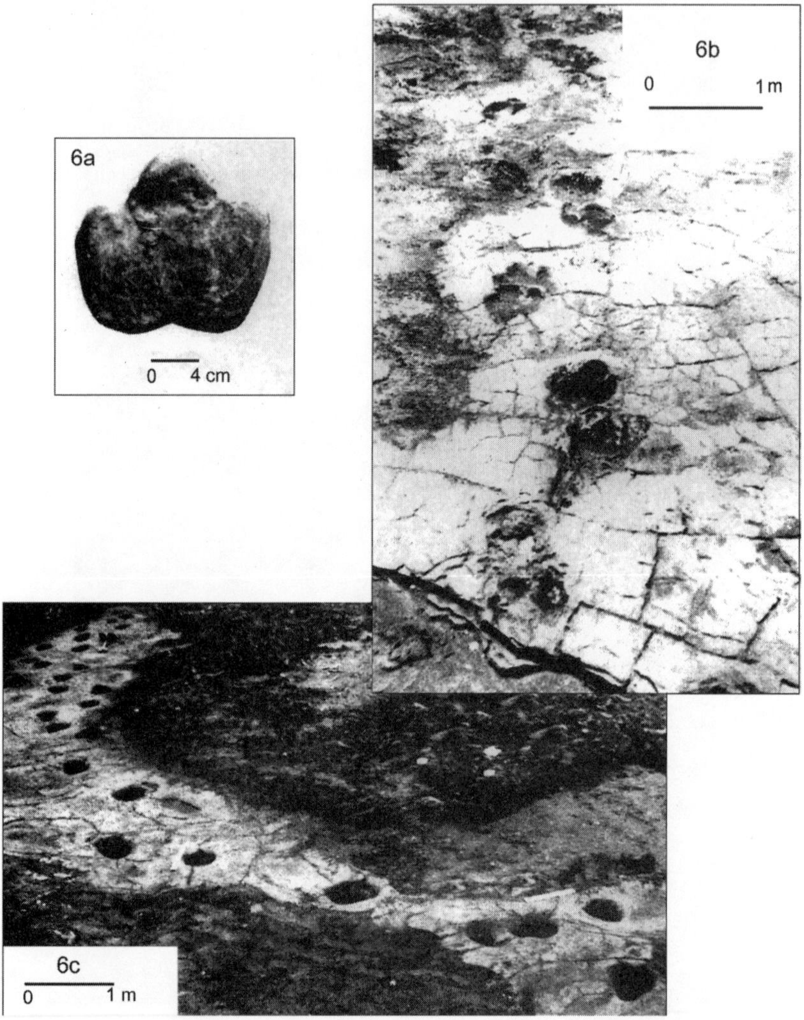

6a
0 4 cm

6b
0 1 m

6c
0 1 m

Footprints at Pehuen-Co with the characteristic three toes of *Macrauchenia* at Pehuen-Co. From Aramayo and Manera[75]

while twelve thousand years ago, with the ground on the beach in such a state as to take faithful impressions of their footsteps that were then covered with sand to be well preserved for rediscovery in the present day. One conclusion that follows from this fortunate accident is that the huge mammals that once flourished in South America

October 1832

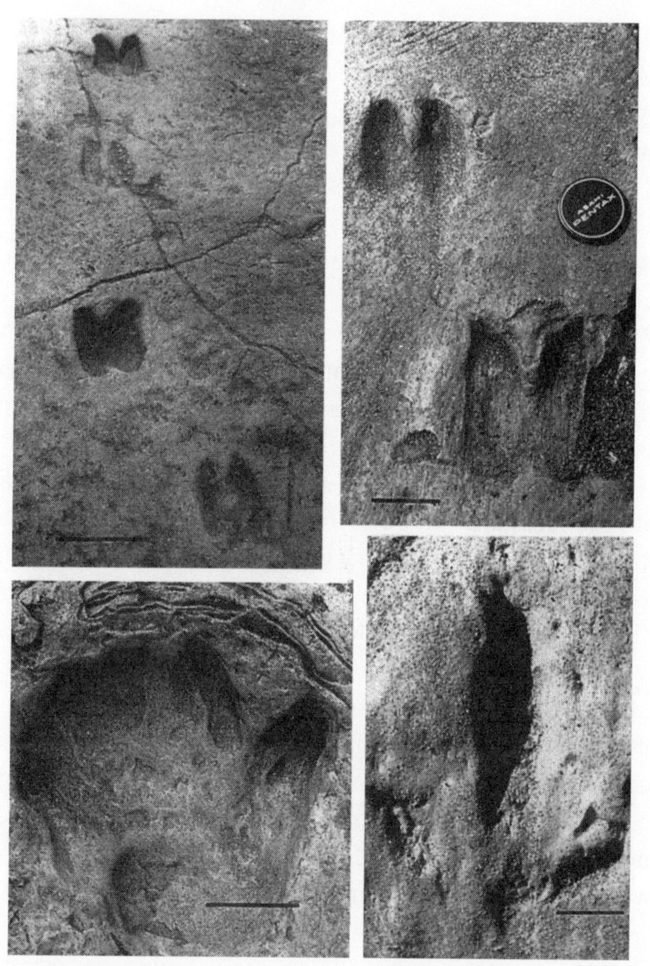

Footprints at Pehuen-Co of a paleollama (above, scales 4cm on left, 5cm on right), *Glyptodon* (below left, scale 5cm) and *Rhea americana* (below right, scale 2cm). From Aramayo and Manera[76]

survived somewhat longer than was previously thought, on this day they were parading themselves at Pehuen-Co, blissfully unconcious of their forthcoming extinction. Certainly they must have co-existed with the first humans to arrive in the area, and it is not improbable that the ancestors of the Araucanians contributed towards their extinction, though there is little direct evidence to prove it.

October 1832

At his first meeting with Harris a month earlier, FitzRoy had worked out a plan to hire by the month from him and his partner their two small schooners, the *Paz* of fifteen tons and the *Liebre* of nine, which were well suited to carry out, under the supervision of Wickham and Stokes, a detailed survey of the east coast of Patagonia between Blanco Bay and San Blas while the *Beagle* herself was otherwise engaged in Tierra del Fuego and the Falkland Islands. A serious difficulty recognised by FitzRoy was that he was not in fact authorised by the Admiralty to hire or purchase assistance in this way, but he was sanguine about the outcome, and was penalised most unfairly for his enthusiasm by having to meet much of the cost from his own pocket.

Both boats were initially covered thickly with dirt and soaked with seal and sea-elephant oil, but after Wickham and a small party had fitted them with altered spars, rigging and sails, and had given them a fresh coat of paint, they were transformed beyond recognition into smart little cock-boats, ready for their new task. FitzRoy was a trifle uneasy on seeing that the pilot of the *Liebre* was so large a man in comparison with his boat, but Wickham maintained that his weight was of great value for trimming the vessel, as was his strength for heaving her upright when she ran aground in the mud, though it had to be admitted that 'he did harm on one day, by going up to look-out, and breaking the mast'. However, it was evident that the two boats were well fitted for their new and important task.

CHAPTER 9

The Return of the Fuegians to Their Homeland

The *Beagle* sailed north to arrive at Monte Video on 25 October, anxious for letters from home and a fresh supply of bread. Six days later, Charles wrote a classical description of the gossamer spiders of family Linyphiidae:

> Sailing between M Video & B Ayres on Octob. 31st the rigging was coated with the Gossamer web: it had been a fine clear day with a fresh breeze. The next morning the ropes were equally fringed with these long streamers. On examining these webs I found great numbers of a small spider. On the second day (which was calmer) there must have been some thousands in the ship. When first coming in contact with the ropes, they were seated on the fine lines & not on the cottony mass. This latter appears to be only the separate lines collected by the wind. From the direction of the wind [they] must have travelled at least 60 miles from the Northern shore. They were some full grown & of both sexes & young ones; these latter, besides being of a smaller size, were more duskily coloured . . . These little spiders, after alighting on the ropes, were in their habits very active; They frequently let themselves fall from a small height & then reascend the attached line. Occasionally when thus suspended, the slightest breath of air would carry them out of sight on a rectangular course to the line of suspension. I never saw them rise at all: They formed an irregular net work amongst the ropes: Could run easily on water: Lifted up their front legs in attitude of attention. Seemed to have an inexhaustible stock of web: With their Maxillæ protruded, drank eagerly water . . . The above

mentioned facts in the occurrence of numerous (sufficient I think to account for the Gossamer) spiders of same species but different sexes & ages, on their webs, & at a great distance from the land & therefore liable to no mistakes demonstratively proved that the habit of sailing in the air as much belongs to a division in Spiders as diving in the water does to Argyroneta [the unique water spider of the northern hemisphere].[77]

On 2 November the *Beagle* anchored for the second time at Buenos Aires, where the guard-ship treated them with greater respect than on the previous occasion. Charles spent the next week shopping, and exploring the city and its immediate surroundings on horseback, accompanied by Lieutenant Hamond, who had been transferred to the ship from the *Druid*. Charles wrote to his sister Caroline:

> Our chief amusement was riding about & admiring the Spanish Ladies. After watching one of these angels gliding down the streets, involuntarily we groaned out, 'how foolish English women are, they can neither walk nor dress'. And then how ugly Miss sounds after Signorita; I am sorry for you all, it would do the whole tribe of you a great deal of good to come to Buenos Ayres.

A week later they moved north again to Monte Video, but unfavourable winds delayed their arrival there until 14 November. Although the dilatory method of doing things there did not hasten their progress, even the quarterdeck was soon crowded with provisions and stores to last for up to eight months during their cruise to the south, with an extra supply of iron and coal for the forge in case of any serious accident.

On 23 November there was a grand ball in the theatre in order to celebrate the restoration of the President, in which the *Beagle*'s men had played a part. Charles reported:

> It was a much gayer scene than I should have thought this place could have produced. The desire which the inhabitants have on such occasions

December 1832

of appearing splendidly dressed is excessive & to gratify it the ladies will spare no sacrifices. The music was in very slow time & the dancing, although most formal, possessed much gracefulness. The ball was given in the Theatre; nothing surprised me so much as the arrangements of the house; every part not actually occupied by the dancers was entirely open to the lowest classes of society, so that all the passages to the boxes, & back parts of the pitt, were filled by any people who liked to look on. And nobody ever seemed even to imagine the possibility of disorderly conduct on their parts. How different are the habits of Englishmen, on such Jubilee nights!

On the following evening Charles went to the theatre again for a performance of Rossini's opera *La Cenerentola*, but had nothing to say about the quality of the singing.

On 26 November the *Beagle* at last set off on her mission to repatriate FitzRoy's Fuegians, heading southwards along the coast of Patagonia, whose people had been named Patagónes – 'big feet of bears' – by Magellan at his first encounter at Port St Julian in 1520 with a native wearing huge fur boots.

Charles devoted himself mainly to collecting in his net the many species of shrimps and larval crustaceans that filled the water, and carefully examining their anatomy under his microscope. However, there were other animals about, and one day he wrote in his notebook:

December 4th.– About 10 miles off the Bay of San Blas, in the evening, the infinite numbers of Lepidoptera formed a most curious spectacle. They were of various species, but chiefly a yellow sort. With them were some moths & Hymenoptera, & even a Calosoma [a beetle] flew on board. The men all cried out "it is snowing butterflies"; at a distance it had this appearance. The butterflies were in bands or flocks of countless myriads, & as far as the telescope reached, they might be seen fluttering over the water. This took place in the evening; the morning had been calm & the day before very light variable winds. It is clear these insects had voluntarily come out to sea. It was the last

> day for most of them, for a strong breeze sprung up from the North, which must have destroyed the greater number. How are we to account for these flights, which others have also observed? Is it an instinct implanted in the animal to find new countries, its own one being overstocked by a particularly favourable year.[78]

FitzRoy was also impressed by the butterflies, whose swarm he estimated to have been not less than two hundred yards in height, a mile in width, and several miles in length. But he reckoned that they were white rather than yellow.

The *Beagle* called in at San Blas, to be greeted by Wickham with excellent progress to report on the surveys, though he and Stokes had been badly burnt by the sun despite their heavy beards, and had suffered much from sea-sickness caused by the violent movements of the little boats in the tide-races and eddies of the coastal waters. FitzRoy then laid out in detail the scope of the work to be done. Their instructions were to explore the coast even as far as Port Desire before rejoining the *Beagle* on her return from Tierra del Fuego and the Falkland Islands, either in March at the Rio Negro, or in Blanco Bay at the beginning of July.

The coast of Tierra del Fuego was reached on 16 December a little to the south of Cape St Sebastian, where the *Beagle*'s arrival was at once signalled by smoke from fires lit by the natives. In the same way, when three hundred years earlier Magellan's fleet had first sailed into the mouth of the Straits that came to bear his name, he saw many fires burning in the hills to the south, and called the country Tierra del Fuego, the Land of Fire. Near Cape Peñas a group of tall men on foot, nearly naked and accompanied by large dogs, could be seen with glasses. To FitzRoy this renewed but distant glimpse of the aborigines was 'deeply engaging', but York Minster and Jemmy Button asked him to fire at them, saying that they were 'Oens-men – very bad men'. According to E. Lucas Bridges, son of the missionary Thomas Bridges who in 1871 set up the Mission to the Fuegians at Ushuaia on the north shore of the Beagle Channel, who was brought up in Tierra del Fuego, the tall men living on the eastern coast of

Staten Land were members of the Eastern Ona or Aush tribe.[79] They were never on good terms with the Yaghan Indians of Jemmy Button's tribe, living on the south side of the Beagle Channel.

After anchoring offshore for several hours hoping to take some angles, the *Beagle* found herself rolling uncomfortably in a swell from a distant gale, and was obliged to sail on down the inhospitable coast of Staten Land to round Cape St Diego into the Strait of Le Maire and come to anchor in Good Success Bay. This sheltered harbour at the south-eastern tip of Tierra del Fuego was memorable for having been the anchorage used by Captain Cook before rounding Cape Horn on his first voyage to the Pacific in HMS *Endeavour* in 1768–71.[80] The botanists Joseph Banks and Daniel Solander who accompanied him went climbing at the Bay of Good Success on 16–17 January 1769, and were caught in a snowstorm in which two members of their group perished.

FitzRoy described the *Beagle*'s arrival on 17 December:

> As we sailed into Good Success Bay, a Fuegian yell echoed among the woody heights, and shout after shout succeeded from a party of natives, posted on a projecting woody eminence, at the north head of the bay, who were seen waving skins, and beckoning to us with extreme eagerness. Finding that we did not notice them, they lighted a fire, which instantly sent up a volume of thick white smoke. I have often been astonished at the rapidity with which the Fuegians produce this effect (meant by them as a signal) in their wet climate, where I have been, at times, more than two hours attempting to kindle a fire.

The next morning FitzRoy sent a boat with a large party of officers to communicate with the Fuegians, and Charles made his first acquaintance with them in the field. He wrote in his commonplace journal:

> As soon as the boat came within hail, one of the four men who advanced to receive us began to shout most vehemently, & at the same

A selection of Fuegians drawn by Robert FitzRoy

time pointed out a good landing place. The women & children had all disappeared. When we landed, the party looked rather alarmed, but continued talking & making gestures with great rapidity. It was without exception the most curious & interesting spectacle I ever beheld. I would not have believed how entire the difference between savage & civilized man is. It is greater than between a wild & domesticated animal, in as much as in man there is greater power of improvement. The chief spokesman was old & appeared to be head of the family; the three others were young powerful men & about 6 feet high. From their dress &c &c they resembled the representations of Devils on the Stage, for instance in [Weber's opera] Der Freischutz. The old man had a white feather cap, from under which black long hair hung round his face. The skin is dirty copper colour. Reaching from ear to ear & including the upper lip, there was a broard red coloured band of paint; & parallel & above this, there was a white one; so that the eyebrows & eyelids were even thus coloured. The only garment was a large guanaco skin, with the hair outside. This was merely thrown over their shoulders, one arm & leg being bare; for any exercise they must be absolutely naked.

Their very attitudes were abject, & the expression distrustful, surprised & startled. Having given them some red cloth, which they immediately placed round their necks, we became good friends. This was shown by the old man patting our breasts & making something like the same noise which people do when feeding chickens. I walked with the old man & this demonstration was repeated between us several times; at last he gave me three hard slaps on the breast & back at the same time, & making most curious noises. He then bared his bosom for me to return the compliment, which being done, he seemed highly pleased. Their language does not deserve to be called articulate: Capt Cook says it is like a man clearing his throat; to which may be added another very hoarse man trying to shout & a third encouraging a horse with that peculiar noise which is made in one side of the mouth. Imagine these sounds & a few gutturals mingled with them, & there will be as near an approximation to their language as any European may expect to obtain. Their chief anxiety was to obtain knives; this they showed by pretending to have blubber in their mouths, & cutting instead of tearing it from the body. They called them in a plaintive tone Cochilla, probably a corruption from a Spanish word.* They are excellent mimics, if you cough or yawn or make any odd motion they immediately imitate you. Some of the officers began to squint & make monkey like faces, but one of the young men, whose face was painted black with white band over his eyes was most successful in making still more hideous grimaces. When a song was struck up, I thought they would have fallen down with astonishment; & with equal delight they viewed our dancing, & immediately began themselves to waltz with one of the officers. They knew what guns were & much dreaded them, & nothing would tempt them to take one in their hands.

Jemmy Button came in the boat with us; it was interesting to watch their conduct to him. They immediately perceived the difference & held much conversation between themselves on the subject. The old man then began a long harangue to Jemmy, who said it was inviting him to stay with them, but the language is rather different & Jemmy could not talk to them. If their dress & appearance is miserable, their manner of living is still more so. Their food chiefly consists in limpets & muscles [*sic*], together with seals & a few birds; they must also catch occassionally [*sic*] a Guanaco. They seem to have no property excepting bows & arrows & spears: their present residence is under a few bushes by a ledge of rock: it is in no ways sufficient to keep out rain or wind, & now in the middle of summer it daily rains & as yet each day there has been some sleet. The almost impenetrable wood reaches down to high water mark, so that the habitable land is literally reduced to the large stones on the beach, & here at low water, whether it be night or day, these wretched looking beings pick up a livelihood. I believe if the world was searched, no lower grade of man could be found. The Southsea Islanders are civilized compared to them, & the Esquimaux in subterranean huts may enjoy some of the comforts of life.

---

* *Cuchilla* is the correct word for a knife.

December 1832

After dinner the Captain paid the Fuegians another visit. They received us with less distrust & brought with them their timid children. They noticed York Minster (who accompanied us) in the same manner as Jemmy, & told him he ought to shave, & yet he has not 20 hairs on his face, whilst we all wear our untrimmed beards. They examined the colour of his skin; & having done so, they looked at ours. An arm being bared, they expressed the liveliest surprise & admiration. Their whole conduct was such an odd mixture of astonishment & imitation, that nothing could be more laughable & interesting. The tallest man was pleased with being examined & compared with a tall sea-man, in doing this he tried his best to get on rather higher ground & to stand on tip-toes: He opened his mouth to show his teeth & turned his face en profil; for the rest of his days doubtless he will be the beau ideal of his tribe. Two or three of the officers, who are both fatter & shorter than the others (although possessed of large beards) were, we think, taken for ladies. I wish they would follow our supposed example & produce their "squaws". In the evening we parted very good friends, which I think was fortunate, for the dancing & "sky-larking" had occassionally bordered on a trial of strength.[81]

On the following day, Charles set out, accompanied by a seaman, to penetrate a little way into the country, but found the hills closely covered with a tangled mass of Antarctic beech trees not more than ten feet high through which walking was difficult. FitzRoy was certain that it was on the nearest hill that Banks and Solander had got themselves into such serious trouble, but Charles was determined to collect some alpine plants and insects, and tried again the next day. By sticking to a branch of a watercourse to avoid the possibility of losing himself, and taking advantage of the regular paths frequented by the guanacos that he saw from time to time, he reached the peak after climbing for two and a half hours. He was rewarded by a splendid view, some diminutive alpine flowers, quite a number of insects, and 'altogether most throughily enjoyed the walk'.

On 21 December the *Beagle* got under weigh at 4 a.m., doubled Cape Good Success, and with a fine easterly wind 'which is about as lucky &

rare an event as getting a prize ticket in a lottery' was anchored snugly on Christmas Day at Wigwam Cove on Hermit Island, slightly to the west of Cape Horn. All duty on board was suspended that day, and after breakfast Charles, Sulivan and Hamond made an energetic climb to the summit of Kater's Peak, towering 1700 feet above them.

For the rest of the month bad weather kept the *Beagle* at anchor, and although such conditions 'rendered boating rather disagreeable' Charles accompanied FitzRoy to reconnoitre the bays at the back of Hermit Island. In most of the coves there were wigwams that had evidently for many years been occupied spasmodically until the local supply of shellfish was exhausted. They were consequently perched on large hillocks of shells with bright green patches of wild celery, scurvy grass and other vegetation growing on them. There were also many curious seabirds, such as penguins that behaved more like fish than birds, and the logger-headed steamer duck renowned for its extraordinary manner of splashing and paddling along.

On 31 December, when there was a hope of not immediately being driven eastward, FitzRoy put the *Beagle* to sea again. For a fortnight the ship attempted vainly to work westward towards Christmas Sound in order to land York Minster and Fuegia Basket among their own people, and then to return eastward through the Beagle Channel to Jemmy Button's country. But when on 11 January 1833 they had fetched to within a mile of Christmas Sound, and momentarily caught sight of the mountain named York Minster by Captain Cook for its castellated form, a violent squall forced them to shorten sail and stand out to sea. The fury of the unbroken ocean could be judged from the fact that clouds of spray were carried over a precipice that Charles reckoned to have been two hundred feet high. Next day the gale freshened into a regular storm, and fearing that the close-reefed main-topsail might carry away, the Captain continued with merely the close-reefed storm-trysails and fore-staysail. Of the morning of 13 January, Charles noted:

> The worst part of the business is our not exactly knowing our position; it has an awkward sound to hear the officers repeatedly telling

the look out man to look well to leeward. Our horizon was limited to a small compass by the spray carried by the wind: the sea looked ominous; there was so much foam that it resembled a dreary plain covered by patches of drifted snow. Whilst we were heavily labouring, it was curious to see how the Albatross with its widely expanded wings glided right up the wind.

At noon the storm was at its height, & we began to suffer; a great sea struck us & came on board; the after tackle of the quarter boat gave way, & an axe being obtained they were instantly obliged to cut away one of the beautiful whale-boats. The same sea filled our decks so deep that if another had followed it is not difficult to guess the result. It is not easy to imagine what a state of confusion the decks were in from the great body of water. At last the ports were knocked open & she again rose buoyant to the sea. In the evening it moderated, & we made out Cape Spencer (near Wigwam Cove), & running in, anchored behind false Cape Horn.

Charles did not exaggerate, as is clear from FitzRoy's account of the critical moments:

Soon after one, the sea had risen to a great height, and I was anxiously watching the successive waves, when three huge rollers approached, whose size and steepness at once told me that our sea-boat, good as she was, would be sorely tried. Having steerage way, the vessel met and rose over the first unharmed, but of course her way was checked; the second deadened her way completely, throwing her off the wind; and the third great sea, taking her right abeam, turned her so far over that all the lee bulwark, from the cat-head to the stern davit, was two or three feet under water. For a moment, our position was critical; but, like a cask, she rolled back again, though with some feet of water over the whole deck. Had another sea then struck her, the little ship might have been numbered among the many of her class which have disappeared: but the crisis was past – she shook the sea off her through the ports, and was none the worse.

January 1833

In a footnote, FitzRoy added that the roller which almost hove the *Beagle* on her beam ends was the highest and most hollow that he had ever seen, excepting one in the Bay of Biscay and another in the South Atlantic. But for the last word on this dramatic episode, the memoirs of Lieutenant, later Admiral Sir Bartholomew Sulivan, must be recalled.[82] He said that the Captain always had the ports secured, but that he himself never liked this order, and told the carpenter always to have a handspike handy for eventualities. Shortly before the three huge waves struck the boat, Sulivan had relinquished the deck to FitzRoy, but on returning to the deck from below he found the carpenter up to his waist in water and standing on the bulwark, strenuously driving a handspike against the port, which he eventually burst open. Charles's recollection appears to confirm that the port did have to be knocked out in order to right the ship, so that it was evidently Sulivan's countermanding of FitzRoy's standing order that finally saved the day.*

In his List of Specimens not in Spirits, Charles wrote: 'All specimens from 888 to 900 much injured by the gale of Jany 13th – & Numbers 804 . . . 900 changed into 931 . . . 937.' In his commonplace journal he wrote: 'I find I have suffered an irreparable loss from yesterday's disaster, in my drying paper & plants being wetted with salt-water.' But there were compensations, and the Specimens in Spirits of Wine came to no harm, already stored comfortably in their bottles of alcohol.

Since it transpired that York Minster had no objection to being repatriated in company with Jemmy Button and Fuegia Basket, FitzRoy now gave up his efforts to return to the west of Tierra del Fuego, and sailed north-east to an excellent anchorage in Goree Sound, near the eastern entrance to the Beagle Channel. Here he proposed to stay for several weeks while the three Fuegians and Richard Matthews, who had been sent out by the Church Missionary Society in England to accompany them, were found somewhere to dwell.

* The *Beagle's* plight on 13 January 1833 has been vividly and accurately portrayed by the artist John Chancellor.

January 1833

Cape Horn and rocks off Cape Deceit, by Conrad Martens

There was no suitable land for a settlement at the eastern end of Navarin Island, so on 19 January a party of twenty-eight set off in three whale-boats, one towing the heavily laden ship's yawl, along the Beagle Channel towards Jemmy Button's country at the other end of Navarin Island. The yawl was loaded with a ludicrous assortment of goods – wine glasses, butter-bolts, soup tureens, a mahogany dressing-case, fine white linen and beaver hats – that the good people of Walthamstow mistakenly thought would be useful to Matthews for his mission to the Fuegians. One night was spent in uninhabited country, and two more in parts where naked Fuegians ran beside them on the shore uttering hideous yells, until on 22 January they turned into Ponsonby Sound and landed among the people called the 'Yapoo Tekeenica' by FitzRoy.

As was subsequently explained in the dictionary published by Thomas Bridges, this was something of a misnomer.[83] The word '*iapooh*' meant in fact an otter, while '*teke uneka*' meant 'I do not understand you'; so FitzRoy's original informant must have been trying to convey a lack of understanding rather than the name of the place. The canoe-using natives of the Beagle Channel called themselves 'Yamana' ('People'), while the particular group who lived around the Murray Narrow were the 'Yahgashagalumoala', or 'People from Mountain Valley Channel', which Thomas Bridges shortened to Yahgans. In this their first meeting with Europeans, Charles said that

January 1833

Fuegians at Woollya, by Robert FitzRoy

they only became friendly on being given small presents, such as red tape which was tied around their foreheads, and that both their first and last words were invariably 'Yammerschooner', meaning 'Give me'. His interpretation was correct, for in the Yamana–English dictionary 'yamašk-una' does indeed mean 'Do be liberal to me'.

The next morning, followed by a fleet of canoes paddled by Yahgans painted white, red and black like so many 'demoniacs', the boats proceeded down the Murray Narrow to Woollya, the cove where Jemmy Button lived. There was a cleared area of rich ground which seemed suitable for the cultivation of European vegetables, on which three houses were quickly built and two gardens were dug and planted. The proceedings were watched by more than a hundred Yahgans, who sat without showing great interest except when offered a present, or when grasping an opportunity to steal something. Jemmy's mother, brother and uncle arrived, but Charles noted that they greeted one another with no more excitement than two horses in a field. Despite their total lack of emotion, Jemmy was able to recognise his brother's voice at an astonishing distance, and the acuity of eyesight possessed by the Yahgans was quite remarkable. Three days later there was an alarm when the Yahgans suddenly all departed for no obvious reason, but the next morning they again settled down peaceably to watch, while the men in canoes were busy spearing fish. FitzRoy decided that despite the erratic and often inexplicable behaviour of the Fuegians it was safe to leave Matthews and his party on their own, and one whale-boat and the yawl were sent back to the *Beagle*, while the other two whale-boats, commanded by FitzRoy and Hamond, set out on a brief excursion to examine the north-west arm of the Beagle Channel.

The scenery where the channel divided was very grand, with lofty mountains on the right covered with a white mantle of perpetual snow, cascades of water pouring through the woods into the channel, and magnificent glaciers, beryl blue in colour and extending from the mountains to the water's edge. While the party was dining round a fire about half a mile from one such glacier, an incident occurred which was graphically described by FitzRoy:

> Our boats were hauled up out of the water upon the sandy point, and we were sitting round a fire about two hundred yards from them, when a thundering crash shook us: down came the whole front of the icy cliff, and the sea surged up in a vast heap of foam. Reverberating echoes sounded in every direction, from the lofty mountains which hemmed us in; but our whole attention was immediately called to great rolling waves which came so rapidly that there was scarcely time for the most active of our party to run and seize the boats before they were tossed along the beach like empty calabashes. By the exertions of those who grappled them or seized their ropes, they were hauled up again out of reach of a second and third roller; and indeed we had good reason to rejoice that they were just saved in time, for had not Mr Darwin, and two or three of the men, run to them instantly, they would have been swept away from us irrecoverably. Wind and tide would soon have drifted them beyond the distance a man could swim; and then, what prizes they would have been for the Fuegians, even if we had escaped by possessing ourselves of canoes. At the extremity of the sandy point on which we stood, there were many

> large blocks of stone, which seemed to have been transported from the adjacent mountains, either upon masses of ice, or by the force of waves such as those which we witnessed. Had our boats struck these blocks, instead of soft sand, our dilemma would not have been much less than if they had been at once swept away . . . The following day, the 30th, we passed into a large expanse of water, which I named Darwin Sound – after my messmate, who so willingly encountered the discomfort and risk of a long cruise in a small loaded boat.

In his own account, Charles made rather less of the episode, but nevertheless it was commemorated by naming after him the large mountain skirted by glaciers overlooking the sound, whose height was estimated by FitzRoy as 6800 feet. In a letter to his sister Catherine, Charles later described Sarmiento as 'the highest mountain in the South, excepting M.!!Darwin!!', but alas a modern atlas shows Sarmiento as 2300 metres, whereas Darwin is only 2135 metres.

FitzRoy pressed on to Whaleboat Sound and Stewart Island, of which he had all too many unhappy memories from his experiences there in 1830 when the whale-boat was stolen, and then turned back into the south-west arm of the Beagle Channel, where a few of the Alacaloofs from the western part of Tierra del Fuego were seen. At Point Divide the slate rock seemed to be of excellent quality and fit for roofing, but when would roofing slates ever be required in Tierra del Fuego? FitzRoy speculated that the accidental discovery of a valuable mine might effect great changes, but the whole area is still uninhabited and undeveloped, and likely to remain so. The evening before re-entering Ponsonby Sound, they met a large party of Yahgans from whom they purchased an excellent supper of fish in exchange for old buttons and some pieces of red cloth. But they were disturbed to see that some of the Indians were ornamented with rags of English clothing, and one woman was wearing a linen garment that had belonged to Fuegia Basket.

Arriving at the settlement on 6 February they were relieved when Matthews appeared dressed as usual. Jemmy and York were also dressed and looking well, but they said that Fuegia was in a wigwam. FitzRoy took Matthews into his boat to be questioned, and Jemmy was taken into the other boat. York waited on the beach, and nearly all the Fuegians squatted down on their hams to watch the proceedings, reminding Charles of a pack of hounds waiting for a fox to be unearthed. Matthews reported that from the moment of departure of the boats, both he, and Jemmy and York as well, had been closely watched by night and day and subjected, though without overt violence, to a regular system of plunder. Fortunately his own most valuable possessions had been hidden safely in a cave, and the large tools were concealed in the roof of his hut; but otherwise everything had been stolen. Only the men were responsible; the women had taken him into their wigwams, giving him a share of their food and asking for nothing in return. The garden had been trampled over repeatedly, although Jemmy had done his best to explain its object, and could only shake his head and say sorrowfully, 'My people very bad; great fool; know nothing at all; very great fool.'

Under these circumstances, FitzRoy decided that Matthews would have to be rescued as quickly as possible. Spreading his party about to create a degree of confidence, the wigwam was successfully cleared, the cave was emptied, and Matthews and the sailors were safely embarked. Useful articles such as axes, saws, gimlets, knives and nails were distributed among the natives. Bidding farewell to Jemmy and York, and promising to see them again in a few days, the *Beagle*'s men were able to their relief to depart peacefully from the wondering throng assembled on the beach. To save time, they turned south and sailed back outside Navarin Island to rejoin the ship in Goree Sound on the evening of 7 February. Their circuit had covered some three hundred miles and had provided an excellent geological section of the country.

Charles found it melancholy to leave the three Fuegians amongst their barbarous countrymen. He comforted himself with the thought that they had no personal fears, but although three years had sufficed on the face of it to change their habits from savage into European ones, he was afraid that whatever other ends their excursion to

England had served, it would not be conducive to their happiness. FitzRoy briefly revisited Woollya a week later, and found Jemmy complaining that strangers had been there with whom he and his people had 'very much jaw', and who had stolen two women, in exchange for which Jemmy's party had then stolen one of theirs. He was engaged in hollowing out the trunk of a large tree in order to make a canoe of the type that he had seen in Rio. Some of the vegetables in the garden were already sprouting. FitzRoy was left with 'rather sanguine hopes of their effecting among their countrymen some change for the better', but his optimism was not in the end justified.

After carrying out several further surveys in the southern part of Tierra del Fuego, the *Beagle* went back to Good Success Bay, and on 26 February set sail for the Falkland Islands.

CHAPTER 10

First Visit to the Falkland Islands

The *Beagle* had been instructed by Captain Beaufort that the time needed to conduct a complete survey of all parts of the Falkland Islands would not be justifiable, in view of the limited practical value of such an operation. But the Falklands were all the same a frequent resort of whalers, and it was of immense consequence to a vessel that had lost her masts, anchors, or a large part of her crew to have a precise knowledge of the port where she would be most likely to get help. Some sacrifice of time would therefore be in order for the approaches to the most advantageous harbours in the islands to be accurately surveyed, so that clear directions could be provided for recognising and entering these ports, accompanied by an account of the 'refreshments' that might be available at each of them. It might also be possible to trace out a line of deep soundings from the islands to the mainland of South America, which could be of great service to a ship finding itself in difficulty in the area to enable it to rectify its position.

When, at dawn on 1 March 1833, the *Beagle* reached Cape Pembroke at the eastern extremity of the Falkland Islands, FitzRoy did not know of the easiest access to the islands offered nearby by Port William, close to which the modern capital of the islands, Port Stanley, is located. He therefore worked to windward into Berkeley Sound, where he found on the north side of the sound a wrecked ship with her masts standing, and two other wrecks. The *Beagle* put down her anchor near the cove where in 1820 the French navigator Freycinet had beached his ship *L'Uranie* after striking a rock at the entrance to the sound. It was then learned from a boat which came alongside that the crew of the French whaler *Magellan* would be grateful for the *Beagle*'s assistance, having been driven from her anchors and totally wrecked by the tremendous storm of 12–13 January that the *Beagle* herself had experienced off Cape Horn. This all underlined the good sense of Beaufort's instructions.

The most surprising news was that because of a current dispute between the governments in Buenos Aires and Washington as to how a semblance of order might be maintained in the Falklands for the benefit of the shipping of all nations that used them, the Royal Navy had once more stepped in. As summarised by FitzRoy,[84] the legality of this act depended on the first sighting of the islands by a British navigator, John Davis, in 1592, and their rediscovery by Hawkins and Strong a century later. In the first half of the eighteenth century ships from St Malo regularly passed close to the East Falklands, which were named by the French Les Malouines, later changed by the Spaniards to Las Malvinas; a settlement was established by French colonists at Port Louis in 1764. In the following year the Union Jack was hoisted at Port Egmont in the West Falklands. In 1767 Spain laid claim to the settlement at Port Louis, and bought out the French. In 1774 the British withdrew from Port Egmont, which was poorly located and expensive to maintain, though the flag was left flying to indicate the continued right of possession. For some years the Spanish continued to keep a small garrison at Port Louis, but by 1810 there was no one to maintain any order in the Falklands over the many vessels engaged on whaling or sealing in Antarctic waters, who called at the islands and freely slaughtered the cattle, pigs and horses descended from those of the former colonists. Some use of the Malvinas nevertheless continued to be made from time to time by the government of Argentina for shipping convicted criminals out to them to fend for themselves.

In 1828 the Buenos Airean government granted a concession to a German-born immigrant, Lewis Vernet, and his partners for the use of the fishery, cattle and tracts of land in Eastern Malvina in order to manage and provide for the subsistence of the settlement. In 1831 Vernet had found it necessary to detain by force the crews of some

Port Louis in the Falkland Islands, by Conrad Martens

North American sealers who had disobeyed his instructions, drawing upon him and his unfortunate colony the wrath of the United States. The US corvette *Lexington* sailed down to Port Louis, where Vernet's agent and partner Mr Brisbane and other innocent people were taken prisoner for transportation to Buenos Aires, and a fair amount of both property and buildings was destroyed by the Americans. While prolonged discussions on the rights and wrongs of the situation were in progress with an American chargé-d'affaires in Buenos Aires, the British government, following up warnings to the Argentinians on the importance of keeping the shipping in the Falklands under proper control, had instructed the naval Commander-in-Chief on the South American station to set the British flag flying once again in the Falklands.

On 2 January 1833 HMS *Clio* had hoisted and saluted the Union Jack at Port Louis, while HMS *Tyne* did the same at Port Egmont. A loquacious Irish storekeeper called Dickson had been left in charge of the flag at Port Louis, and had been instructed to hoist it on Sundays and when vessels came into port. He reported to FitzRoy on the *Beagle* on 2 March that plenty of beef, rabbits and geese were to be had. On 3 March an English merchantman arrived from Buenos Aires bringing back Mr Brisbane. In his account of the past history of events in the Falklands, FitzRoy regretted never having met Mr Vernet, but said that he had sympathy with his position. Brisbane was very pleased to meet the officers of the *Beagle*, one of whom had once taken part in rescuing him from a shipwreck, and he was quickly on good terms with them. On the following day, FitzRoy was taken by Brisbane to visit the settlement, and was shown how it had been 'ruined' by the Captain and crew of the USS *Lexington* in 1831. On returning to the *Beagle*, FitzRoy found to his great sorrow that his excellent clerk, Mr Hellyer, had gone out shooting and had been drowned, entangled in the kelp close to the shore while trying to recover a duck that he had shot. His grave may still be seen on Duclos Point, near the *Beagle*'s anchorage at Johnson's Harbour.

During the next few days, Charles explored the area north of the sound, which he found very dreary, with nothing but a brown wiry grass growing on the peat, and except for snipe and rabbits scarcely any animals. He collected the grass and a specimen of what had to be regarded as the largest tree, sometimes growing two or three feet high, one or two of the snipe and other common land birds, and the only hawk that he saw, a species of harrier. On the beaches he found quite a number of what he identified as a species named by Linnaeus *Corallina inarticulata* encrusting the rocks, and from a careful examination under his microscope of their anatomy and mode of reproduction concluded that they were undoubtedly plants rather than animals.* He also found a species of sea slug, *Doris*, which impressed him because of the immense size of its egg ribbon, enabling a single individual to lay more than 600,000 eggs. He made no comment at the time on the biological purpose that might be served by such a

* This specimen was indeed a type of plant growing in water known as a coralline alga, whose modern name is *Amphiroa exilis*, and of which the type specimen collected by Charles is now preserved in the Herbarium of Trinity College Dublin.

lavish production of eggs, but it was one of the features sometimes ensuring the survival of a species that was later recognised as important in his theory of evolution.

He then made a useful start on a systematic study of the geology of East Falkland Island by using the transect method that he had been taught twenty months previously by Adam Sedgwick in North Wales, and collecting the minerals and examining the strata for an irregular line crossing the head of Berkeley Sound from north to south.* After the night of 10 March, when there was yet another gale that threatened even the ships safely at anchor, all the boats were away for a week on surveys, and Charles was left to himself to enjoy 'one of the quietest places we have ever been to'. In his pocket book he wrote some brief memoranda to himself on a variety of topics:[85]

> March 2.
> To what animals did the dung beetles in S. America belong – Is not the closer connection of insects & plants as well as this fact point out closer connection than Migration.
>
> Scarcity of Aphidians?
>
> Vide Annales des Sciences for Rio Plata.
>
> The peat not forming at present & but little of the Bog Plants of Tierra del F; no moss; perhaps decaying vegetables may slowly increase it. Beds ranging from 10 to one foot thick.
>
> Great scarcity in Tierra del of Corallines, supplanted by Fuci; Clytra prevailing genus.
>
> Procure Trachaea of Upland Goose.

* In a lecture to the Geological Society delivered in 1846 he said: 'In crossing the eastern island in a N.N.W. and S.S.E. direction, in a line intersecting the head of Berkeley Sound, we find north of it several low, parallel, interrupted, east and west ranges, with strata all dipping a little west of south, at angles varying between 20° and 40°. South of Berkeley Sound the first range we come to is a short one, rising like the others through the clay-slate formation.'

March 1833

Tuesday 12th –

Examine Balanus in fresh water beneath high water mark.

Horses fond of catching cattle – aberration of instinct.

Examine pits for peat. Specimen of do – Have there been any bones ever found &c or Timber.

Are there any reptiles? or Limestone?

21st

Saw a cormorant catch a fish & let it go 8 times successively like a cat does a mouse or otter a fish; & extreme wildness of shags.

Read Bougainville.

In 1784, from returns of Gov. Figueroa, buildings amounted to 34, population including 28 convicts, 80 persons, & cattle of all kinds 7,774.

22nd

East of basin, peat above 12 feet thick resting on clay, & now eaten by the sea. Lower parts very compact, but not so good to burn as higher up; small bones are found in it like Rats – argument for original inhabitants; from big bones must be forming at present, but very slowly. *Fossils in Slate;* opposite points of dip & mistake of stratification: What has become of Lime?

It will be interesting to observe differences of species & proportionate Numbers; what also appear characters of different habitations.

Migration of Geese in Falkland Islands as connected with Rio Negro.

March 28th –

Emberiza in flocks.

Send watch to be mended.

Enquire period of flooding of R. Negro & Plata.

Is the cleavage of M. Video (an untroubled country) very generally vertical, or what is the dip? –

At this time, Charles had not yet begun to think seriously about the evolution of new species, and still believed conventionally in

A fossiliferous rock (from Armstrong[107])

Paley's doctrine of a benevolent Creator who had set up specific Centres of Creation. Yet these notes reveal that his thoughts were already turning in significant new directions to consider problems in behaviour and ecology, and ways in which geographical separation might be related to the distribution of animals.

Charles's slight boredom with the Falklands was relieved by an exciting discovery that he made one day during a walk to the old buildings of the Port Louis settlement. He found near the settlement, as he wrote to his sister Caroline and to Henslow, 'a number of fossil shells in the very oldest rocks which ever have organic remains'. This rock was in beds of slaty yellow-red micaceous sandstone, and contained thickly packed Palaeozoic fossils of lamp shells (Brachiopoda), today assigned to the Devonian period some 410 million years ago. The stark contrast between the bleak and impoverished state of the land now around him, and the seas teeming with life that had existed there so long ago, was not lost on Charles.

On 26 March the schooner *Unicorn* commanded by William Low joined the vessels anchored in Berkeley Sound. On board her were survivors from yet another vessel, an American sealer wrecked off Tierra del Fuego by the storm of 13 January. Low was a Scottish trader and sea captain who had been working for a number of years in the stormy waters around Patagonia, the Falkland Islands, Tierra del Fuego and the south coast of Chile. He had the reputation of being the toughest buccaneer in those parts, and 'the terror to all small vessels', but was found by the *Beagle* to be a most helpful pilot and adviser. FitzRoy had convinced himself that the *Beagle*'s surveys could be completed more quickly and efficiently if she had a consort to keep company with her, which could easily be rigged for working with few hands, and was adapted for carrying cargoes. On seeing *Unicorn*, he promptly fell for her, and wrote:

> A fitter vessel I could hardly have met with, one hundred and seventy tons burthen, oak built and copper fastened throughout, very roomy, a good sailer, extremely handy, and a first-rate sea-boat. Her only deficiencies were such as I could supply, namely a few sheets of copper, and an outfit of canvas and rope. A few days elapsed, in which she was surveyed very carefully by Mr May [the *Beagle*'s carpenter], and my mind fully made up, before I decided to buy her, and I then agreed to give six thousand dollars (nearly £1,300) for immediate possession. Being part owner, and authorized by the other owners to do as he thought best with the vessel in case of failure, Mr Low sold her to me, payment to be made into his partners' hands at Monte Video.

To keep up old associations, this addition to FitzRoy's little detachment was named *Adventure* after the larger vessel commanded by Captain P.P. King in 1826–30. The transaction had to be carried out quickly, like the chartering of the cock-boats *Paz* and *Liebre*, and without the delay that seeking prior approval from London would inevitably have entailed. But although it contributed materially to the success of the coastal surveys that FitzRoy had been appointed to make, the unfriendly Lords of the Admiralty did not at all approve, the politics and policies of this august body being firmly Whig at that time, while FitzRoy was indelibly a Tory.

In a private letter to Beaufort describing his latest acquisition, FitzRoy opened with a nice comment on the new Beaufort Scale for strength of wind (see p.31), of which the *Beagle* was the first user:

March 1833

> I like your addition to, or subtraction from the log book very much, and have adopted it. You would have smiled at hearing some of my Shipmates saying, during the last cruize, 'if Captain Beaufort were here now he'd call this *fifteen*' (alluding to the *wind numbers*) – we certainly had a large share of *eleven*, at the *least*. The numbers & letters are as familiarly used now as could be desired, & no one would willingly return to the old plan. The abbreviations for Soundings &c were much wanted, and are very welcome.

After explaining why he had bought the schooner, FitzRoy closed with a passionate plea:

> Now *pray fight my battle* & get me *twenty super-numerary Seamen* for the Beagle – *fifteen* AB & *five first class* ratings – it will save my pocket *so very much*, & have them you know I must, either for his *Majesty* or *myself*. I feel as if we could now get on fast again, & much more securely, by having so fine a craft to carry our *luggage, provisions, boats,* &c &c. I mean to make her a regular '*Lighter*'.[86]

Beaufort always came loyally to FitzRoy's defence, but evidently had to fight some hard battles to obtain the additional crew members that were needed to man the *Adventure*. FitzRoy himself wrote to the Admiralty explaining his action by saying: 'I believe that their Lordships will approve of what I have done, but if I am wrong, no inconvenience will result to the public service, since I alone am responsible for the agreement with the owner of the vessels, and am able and willing to pay the stipulated sum.' The subsequent Minute across the corner of his letter nevertheless ran: 'Do not approve of hiring vessels for this service, and therefore desire that they may be discharged as soon as possible.'[87] In Valparaiso eighteen months later, FitzRoy went through a period of acute depression when he was obliged to sell off the *Adventure*, thanks to the injustice and wholly deplorable lack of gratitude with which he had been treated in London. The fact that the Second Secretary of the Admiralty, John Barrow,[37] had well-established interests in other directions may also not have helped FitzRoy's case.

April 1833

It had originally been agreed that the crew of about thirty men from the French whaler which had been wrecked in the storm of 12–13 January would be taken by the *Beagle* to Monte Video, and that FitzRoy would buy the stores that had been saved. To this had to be added transport of the crew of the American sealer rescued by Low. But there were then too many men to be carried by the *Beagle* alone, so they were shared out with the *Adventure*, which on 4 April sailed north for the Rio Negro, temporarily under the command of Mr Chaffers, followed two days later by the *Beagle*.

CHAPTER 11

Collecting Around Maldonado

FitzRoy's aim was to intercept Wickham before he had set off with Stokes on a surveying cruise further to the south, but exact timing of such a rendezvous was hard to achieve. Off the mouth of the Rio Negro on 14 April the *Adventure* was duly found, and was immediately dispatched to Maldonado on the north bank of the Rio Plata to await the *Beagle*'s arrival. The next day the *Beagle* met a local trading schooner which passed on the welcome news that all had been well with the two cock-boats a few days earlier, when they had departed southwards down the Gulf of St Matthias to St Joseph's Bay, a nicely sheltered expanse of water on the north side of the Valdés Peninsula. But to their disappointment, and even more to that of their French passengers, there was still no sign of Wickham and his boats at St Joseph's Bay. Charles was vouchsafed a tantalising glimpse of cliffs abounding with fossil shells, but there was no time to examine them properly. A consolation was that a large troop of porpoises were sporting round the *Beagle*, and a female was successfully harpooned. After Charles had carefully measured it up, FitzRoy painted a watercolour. Its total length was just over five feet, so that only its head could be preserved, but from Charles's measurements and FitzRoy's portrait, George Waterhouse in due course identified it at the Zoological Society of London as belonging to a new species which he named *Delfinus fitzRoyii*.

Gales and dangerous soundings within St Joseph's Bay now forced FitzRoy to abandon his search for Wickham, and make sail for Maldonado. Sometimes with foul winds, and sometimes with half a gale in which 'the *Beagle*'s decks fully deserved their nickname of a

FitzRoy's dolphin

"half tide rock", so constantly did the water flow over them', they arrived at Monte Video on 25 April. Charles commented in his notebook, presumably with reference to continued strife between the Blancos and the Colorados, that 'During our absence, things have been going on pretty quietly, with the exception of a few revolutions.'* He then went off to visit Augustus Earle, who had remained at Monte Video during the whole of their cruise in the hope of recovering his health, though sadly without any success. Having rid themselves of their French passengers, and picked up all their parcels and mail, they weighed anchor and with a fresh breeze made sail for Maldonado, where they found *Adventure* at the anchorage, 'all safe and snug'.

Charles's first task was to procure lodgings in the little town. Even in the houses at which he later stayed, which sometimes belonged to wealthy men owning many hundreds of cattle, the bedrooms were devoid of any furniture except for a bed, and his saddle-cloths had to

* Charles' wry comment turned out to be an apt one, for the conflict in which the *Beagle* had participated eight months earlier (see pp.95–7) continued to dominate politics in Uruguay for many years. Civil war broke out again in 1836, and the Blancos, aided by Argentine forces, besieged Monte Video from 1843 until 1852, when the Colorados, aided by Brazil, won control of the country. Between 1865 and 1870 alliances shifted, and Uruguay was allied with Brazil and Argentina in a war against Paraguay. With brief interruptions, the Colorados then remained in power in Uruguay for more than a century.

serve as bed-clothes. Hampered at first by his 'vile' Spanish, he enlisted two travelling companions who had the necessary qualifications of being well armed with pistols and sabres, and of having plenty of friends and relatives in the country. Driving before them a troop of fresh horses, a very luxurious way of travelling that eliminated the danger of having a tired or lame one, they set out to explore the country around the town of Las Minas to the north of Maldonado.

The countryfolk and gauchos turned out to be wholly ignorant of geography, often thinking that England, London and North America were all the same place. Charles's small pocket compass was greatly admired everywhere he went, as was his ability to light a 'promethean' – a recently invented type of match used before the introduction of phosphorus matches – by biting it between his teeth. His habits of washing his face every morning, and of wearing a beard on board ship, were other practices that gave rise to suspicion about him. However, this was allayed by his ability to distinguish between venomous and harmless snakes, and since he paid for the hospitality extended to him, his delight in breaking stones and collecting insects could be forgiven. 'Being able to talk very little Spanish,' he wrote, 'I was looked at with much pity, wonder, & a great deal of kindness.'

In this first of his excursions into the interior, extending to a *pulperia* (inn) about seventy miles from Maldonado near the Rio Polanco, Charles was happy to have had a preliminary look at the geology of the hilly band of country some miles in width running east and west between the coast and Las Minas. It was formed of siliceous clay-slate with some quartz, rock and limestone, and near the *Beagle*'s camp white marble was to be found, from which lime was manufactured. Further to the north there was an alternation of masses of granite and of some quite extensive areas of marble.

Everywhere there was the endless grass plain, with its beautiful flowers and birds, the same hedges of cactus, and the same entire absence of trees. Although it was delightful to ride over so much turf, Charles sometimes found himself thinking with pleasure of iron-shod horses and dusty roads. Even the beggars here, it was said, took a horse rather than walking the shortest distance. And it was wonderful to see the boys riding on barebacked colts and chasing each other over hill and dale, with the horses twisting about in a manner of which no one till he had seen it would believe them to be capable.

When he returned to Maldonado on 20 May, Charles had a proposal to put to his father that he explained in a letter to his sister Catherine:

> The following business piece is to my Father: having a servant of my own would be a really great addition to my comfort. For these two reasons, as at present, the Captain has appointed one of the men always to be with me, but I do not think it just thus to take a seaman out of the ship: & 2^d when at sea, I am rather badly off for anyone to wait on me. The man is willing to be my servant & all the expences would be under sixty £ per annum. I have taught him to shoot & skin birds, so that in my main object he is very useful. I have now left England nearly 1 & ½ years, & I find my expences are not above 200£ per annum, so that it being hopeless from time to time to write for permission I have come to the conclusion you would allow me this expense.

His faith in his father's generosity was not misplaced, so Syms Covington, shown on the list of the ship's supernumeraries at the start of the voyage as Fiddler and Boy to the poop cabin, became Charles's servant, and remained with him as assistant, secretary and servant until 1839.[88] He then migrated to Australia, where he died in 1861. Charles had been accepting responsibility for clothing him since the beginning of the year. Although he wrote later to Catherine that 'My servant is an odd sort of person; I do not very much like him, but he is, perhaps from his very oddity, very well adapted to all my purposes.' As the voyage proceeded, much of the collecting was delegated to Covington, and during its closing months he had the important task of drawing up long lists of the specimens in various categories. His services over the years were of great value to Charles, and were warmly appreciated.

May 1833

Charles devoted the next month principally to the collection of several quadrupeds, no fewer than eighty kinds of birds, and a number of reptiles including nine species of snake. He wrote many pages of lively notes on the characteristic appearance and behaviour of the birds of the district, complicated only for the English reader by the fact that he was obliged to identify them in terms of the nearest European family with which he was familiar. In some cases this was fine – an owl is an owl both in England and in Patagonia – but for many of the passerines (the smaller songbirds) there was no directly equivalent South American family, although there were many species that, as he was careful to point out, occupied an ecological niche very similar to those of their European cousins. A good example was provided by the birds that had adopted parasitic habits like the European cuckoo.

Here he again encountered that remarkable bird the scissor-beak *Rhyncops nigra* Linn., of which he wrote:

> Rhyncops: base of bill & legs "vermilion red". This curious bird was shot at a lake from which the water had lately been drained & abounded with small fish. They were in flocks: I here saw what I have heard is seen at sea: these birds fly close to the water with their bills wide open, the lower mandible is half buried in the water. They thus skim the water & plough it as they proceed: the water was quite calm & it was a most curious spectacle to see a flock thus each leave on the water its track: they often twist about & dexterously manage that the projecting lower mandible should plough up a small fish, which is secured by the upper. This I saw as they flew close to me backwards & forwards as swallows: they occassionally left the water, then the flight was wild, rapid & irregular: they then also uttered a harsh loud cry: The length of the 1st remige must be very necessary to keep the wing dry: the tail is most used in steering their flight: It appears to me their whole structure, bill weak, short legs, long wings, appear to be more adapted for this method of catching its prey than for what M. Lisson states,[89] viz. that they open & eat Mactræ buried in the sand.

June 1833

He added a few months later:

> I have stated that at M: Video, when these birds are in large flocks on the sand banks, that they seem to go out to sea every night. Now if I were to conjecture, I should imagine that they fished at night, when their only method of catching prey would be by thus furrowing the water: it is probable that they eat other animals besides fishs; & many, for instance Crustaceæ, come to the surface chiefly at night. It would be curious to note whether the lower mandible is well furnished with nerves as an organ of touch. I imagine these birds fishing by day in a fresh water lake an *extra*ordinary circumstance, & depended solely upon the myriads of minute fish which were jumping about.

Later still he mentioned that the London anatomist Richard Owen had dissected the head of a specimen brought home by him in spirits, but had not found any specially rich nervous supply to the lower part of the bill.

The largest mammals that Charles found and shot were deer that were common on the 'mammillated' plain round the prominent hill called the Pan de Azucar. Their most notable characteristic was having an intolerably offensive smell, strong enough almost to create nausea. This impregnated the whole air for a distance of half a mile or more from a buck. The odour was still very perceptible a year and a half later on a pocket handkerchief in which the skin had been carried, although Charles had since washed it many times.

Other mammals of interest were the capybaras, the largest living rodents, weighing up to fifty kilograms, which were abundant on the

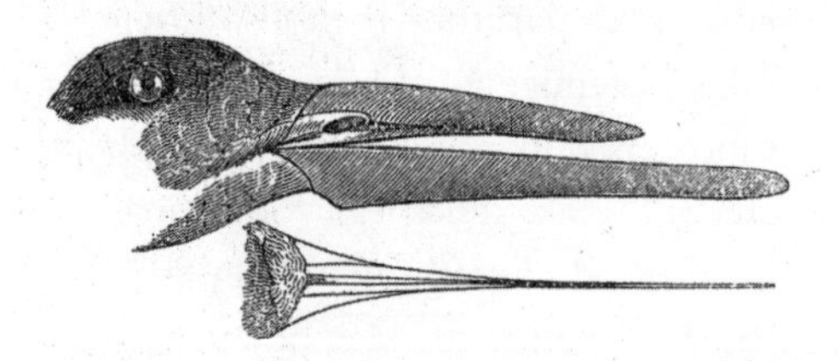

The head of the scissor-beak *Rhyncops*

borders of the lakes in the neighbourhood of Maldonado. There were also the guinea-pigs or cavies locally called Aperea, which lived in huge numbers in the sand-dunes and cactus hedges, and especially in marshy places covered with aquatic plants. Yet another burrowing rodent of the family Ctenomyidae was the Tucotuco, named after its peculiar nasal noise repeated four times in succession, whose mazes of burrows sometimes undermined the ground so much that the hoofs of the horses sank deeply into it. To complete the catalogue, eight kinds of mice were captured, usually in traps baited with cheese or a piece of meat, although one came from a place so wet that it must have had aquatic habits, and another was recorded as having been caught by a bird in the *Beagle*'s camp.

Less attention than usual was paid by Charles to the insects of Maldonado, but he could not resist making an interesting and forward-looking note on what was always one of his favourite topics, the role of dung beetles such as those of the genus that he called *Aphodius*:

> After being accustomed to the great numbers of Coprophagous insects in England, it was at first with surprise that I here found the ample repast afforded by the immense herds of horses & cattle almost untouched. Aphodius (Specimen No. 1181) is the only good exception: this insect amongst the sand dumes [his habitual spelling] burrows holes beneath Horse dung: Aphodius (1225) I have only observed once under very old dung. Any other Aphodii which I have taken have been wandering. It is curious to enquire what animal (No 1181) belonged to before the introduction of horses. All the larger animals here, such as Guanaco, Deer, Capincho [Capybara], have dung in the form of pellets, which must be of a very different nature with respect to insects. M. Video was founded in 1725, it is said the country abounded with Vicunnas [small relatives of the guanaco]. Cattle & horses have perhaps only abounded for about 80 years. This absence of Coprophagous beetles appears to me to be a very beautiful fact; as showing a connection in the creating between animals as widely apart as Mammalia & Insects. Coleoptera [beetles], which when one of

July 1833

them is removed out of its original Zone, can scarcely be produced by a length of time & the most favourable circumstances.

The same subject of investigation will recur in Australia: If proofs were wanting to show the Horse & Ox to be aboriginals of great Britain I think the very presence of so *many* species of insects feeding on their dung, would be a very strong one.

On 29 June, Charles went back to his cabin on the *Beagle* with his 'menagerie', to find that he had become such a complete landsman that he knocked his head against the deck, and felt the motion of the ship even in harbour. The next month was spent by all the crew on refitting the new *Adventure* for the next cruise to the south. By a curious chance, Captain King's old *Adventure* had a few years earlier buried thirty tons of excess ballast close to the well on the island of Goritti near Maldonado. The Brazilians who were in possession of this part of the country at that time had done their best to find this valuable ballast, but had not thought to look for it in such a public place. So there it still lay safely, to be recovered for the new *Adventure*.

On 20 July, the packet boat carrying the mail fired its guns as it passed the *Beagle* and *Adventure* at anchor, to indicate that it was on its way to Rio. The *Beagle* was hastily unmoored so that FitzRoy's urgent letters and Charles's collections could be transferred to the packet. After the usual problems with the fickleness of the weather, the *Beagle* got under way four days later, bound once more for the mouth of the Rio Negro in Patagonia.

CHAPTER 12

A Meeting with General Rosas on the Ride from Patagones to Buenos Aires and Santa Fé

On 3 August the *Beagle* arrived off the mouth of the Rio Negro, a large river flowing south-eastwards into the Atlantic from the cordilleras of the Andes. About twenty miles upstream lay Patagones, now known as El Carmen de Patagones. The settlement had been founded by the Spaniards about fifty years earlier, when it became the most southerly outpost on the American continent inhabited by 'civilised' people. Buenos Aires was some five hundred miles to the north, and the endless grass plains of the Pampas spreading between the Rio Negro and the Rio Plata had formerly been quite thickly occupied by Araucanian and related tribes of Indians, once living in villages holding several thousands, but now scattered about in smaller groups. In order to oust them altogether from their tribal lands to provide more grazing for herds of cattle, the Argentinians were fighting a fierce war of extermination against them. But they were superb horsemen and formidable adversaries, so the struggle was a bitter one, conducted in an even more ruthless fashion than the comparable conflict being fought in the Midwest of North America.

The Argentinian forces were currently commanded by General Juan Manuel de Rosas (1793–1877). Born in Buenos Aires, he had amassed great wealth as a cattleman and exporter of beef during the period when the conservative Argentinian aristocracy was establishing its independence from Spain, and from 1829 to 1832 he was

General Juan Manuel de Rosas

Governor of Buenos Aires Province. In 1833 he was appointed to lead a campaign against the Indians in the south of Argentina, and had established his camp close to another large river, the Rio Colorado, about thirty miles from its mouth. Two years later, having achieved his objectives, he was reinstated as Governor of Buenos Aires with dictatorial powers, and in alliance with strongmen in the other provinces soon made himself dictator of the whole country. Charles said after meeting him, as will be described: 'I was altogether pleased with my interview with the terrible General. He is worth seeing, as being decidedly the most prominent character in S. America.' But like others of his kidney, Rosas eventually exhausted his dominance over his countrymen, and in 1852 was removed from power, curiously to spend the rest of his life in exile at Swaythling in Hampshire.

FitzRoy met Stokes and the officers with him, and received a very satisfactory report on their surveys with the little cock-boats *Paz* and *Liebre*. Charles was impressed to see how with only one-inch planks, *Paz* and *Liebre* could nevertheless survive the gale that nearly sank the *Beagle*, for their small size enabled them to run before the sea instead of being struck by it. He was then deposited at Patagones to start out on the first of his long rides on horseback, while FitzRoy took the *Beagle* to survey some outer banks that the cock-boats could not cover because of especially difficult conditions of wind and sea.

Charles took a long walk ashore to have a preliminary look at the geology of the level and sterile plains characteristic of the district, which were terminated by a 120-foot cliff falling to the sea. The sandstone was so full of salt that it readily formed natural salinas or salt-pans. In the winter the grand salina near the town of Patagones was nothing more than a large shallow lake of brine, which in the summer dried out to become a large field of snow-white salt. This could be harvested and used for flaying hides, though paradoxically it was too pure for preserving meat.* Patagones was built on the cliffs facing the river, with many of the houses actually excavated in the sandstone. Beside the road leading to it, Charles saw the ruins of several fine *estancias* (houses of landowners), and was given a foretaste of the fierceness of the fighting that was continuing to take place in the country through which he proposed to ride. A survivor of one past encounter told him how several hundred Indians of a tribe headed by a *cacique* (chieftain) called Pinchera had appeared in two bodies on a neighbouring hill, and after dismounting and taking off their fur mantles had advanced naked and with great steadiness to attack the *estancia*, armed with *chusas* (bamboo lances) ornamented with ostrich feathers and pointed by sharp spearheads. The quivering of the *chusas* as the Indians drew close was remembered with horror. They halted, and the *cacique* told the besieged Spaniards to give up their arms or he would cut all their throats. As this would

* Pure sodium salts of chloride or sulphate are preferred for human consumption, but for the preservation of meat or cheese it was best to add sea salt containing some gypsum, hydrated calcium sulphate.

in any case have been the result of their entrance, the answer was given by a volley of musketry, and the attack was resumed. The day was saved for the Spaniards by the fact that the corral surrounding the *estancia* was held together by iron nails rather than leather thongs that could have been cut by the knives of the Indians. Many of the wounded Indians were carried away by their companions, and it was only when an under-*cacique* eventually fell that a bugle sounded the retreat. This was an awful pause for the Spaniards, for apart from a few cartridges all their ammunition had been expended. But in an instant the Indians jumped on to their horses, and galloped out of sight. On another occasion the fight was shorter, for a gun loaded with grapeshot was available to defend the corral, and after thirty-nine Indians had been laid on the ground, the rest of the party quickly withdrew.

Patagones was now occupied mainly by pure-blooded Indians and Spaniards, with a few of mixed blood. A supposedly friendly Indian tribe of the *cacique* Leucanee had their *toldos* (huts) outside the town to protect it. The government supported them by giving them all the old horses to eat, but what their character might have gained by a lessening of their ferocity, was lost by their total immorality. All Charles could find to say in their favour was that to his surprise their taste in dress was admirable.

Wholly undeterred by the alarming stories, Charles's plan was to obtain a troop of horses, and make his way to General Rosas's encampment on the Rio Colorado, guided by Mr Harris, owner of the *Paz* and *Liebre*. Five Spanish gauchos added themselves to the party, preferring to ride in company across the waterless and desolate country. After they had passed a famous old tree growing at a high point in the land and worshipped by the Indians as some kind of divinity, an unfortunate cow was spotted by the gauchos, and was quickly lassoed and slaughtered for supper.

Charles wrote:

> This was the first night which I passed under the open sky with the gear of the Recado [saddle] for a bed. There is high enjoyment in the

independence of the Gaucho life, to be able at any moment to pull up your horse and say here we will pass the night. The death-like stillness of the plain, the dogs keeping watch, the gipsy-group of Gauchos making their beds around the fire, has left in my mind a strongly marked picture of this first night, which will not soon be forgotten.

After two such nights they reached the Rio Colorado, and after waiting for a huge troop of mares to swim back across the river, for the flesh of mares was the only food eaten by the soldiers when venturing into the interior, they themselves crossed in a canoe. The General's camp was a square of three or four hundred yards close to the river, formed by wagons, artillery and a miscellany of huts. His soldiers were nearly all cavalry of mixed races between Negro, Indian and Spaniard – as villainous and *banditti*-like an army as could be imagined. Charles called on Rosas's secretary to present the letter of recommendation for himself as naturalist of the *Beagle* that he had brought from the government of Buenos Aires. This was very well received by the General, and the secretary returned 'all smiles and graciousness'.

Charles spent the next three nights at the *rancho* or thatched hut of an old Spaniard who had served with Napoleon against Russia. During the first day his main amusement was to watch the families of the General's Indian allies, said to number about six hundred, when they came to the *rancho* to buy small articles, sugar and herbs. The men were tall, and recognisably from a race related to the Fuegians. Some of the young women, known as *chinas*, were not bad-looking, with their two long plaits of black hair, bright eyes and well-formed limbs. When travelling, two or three *chinas* would mount one laden horse, using a broad band round its neck as a stirrup. Their duty was to pack and unpack the horses, and to make the tents for the night. Their chief occupation indoors was to knock together, and so make more round, two of the *bolas*, the balls used for lassoing game to eat, catching horses and attacking enemies to spear them.

The following day General Rosas sent a message inviting Charles to come and see him. His reputation for ferocity was well earned, and had helped to make him the wealthiest cattle-rancher and grower of corn in the country, owning some 300,000 cattle and seven hundred square miles, which he governed strictly according to his own laws, with several hundred workmen or *peons* well trained to resist the attacks of the Indians. He could rival any of his gauchos in horsemanship, and could not be bettered at the art of jumping from a doorway on to the back of an unbroken colt, and resisting its most strenuous efforts to unseat him. He habitually wore the dress of a gaucho, and was said to have done so when calling upon Lord Ponsonby, the British Minister in Buenos Aires, five years earlier, on the grounds that it was the costume of the country and therefore the proper and most respectful dress to choose. He was immensely popular with his men, and maintained despotic discipline over them. When one of the two mad buffoons that he kept about him like a mediaeval monarch went too far with his importunities, he ordered the man to be staked out on the ground and stretched for several hours like a hide being dried. No amount of pleading would relax the punishment.

In conversation General Rosas nevertheless proved himself to be enthusiastic, sensible and very grave. The interview passed off without a smile, but Charles obtained what he wanted, a passport and order for the government post-horses, which were given to him 'in the most obliging and ready manner'. When the General, some months earlier, left Buenos Aires with his army, he had struck in a direct line across the unknown country, and in his march left at wide intervals a '*posta*' of five men with a small troop of horses, so as to be able to send expresses to the capital. With their aid, Charles travelled first to Bahia Blanca, and then on to Buenos Aires.

Early on the morning of 16 August, Charles set off on the first leg of his journey, hoping to meet the *Beagle* at Bahia Blanca, but not certain exactly when she would be there. Unfortunately Mr Harris was unwell and could not come with him. Charles first rode beside the river past the *toldos* – little round ovens covered with hides, with a tapering *chusa* stuck in the ground beside the entrance – of General Rosas's Indian allies, divided into groups corresponding to the

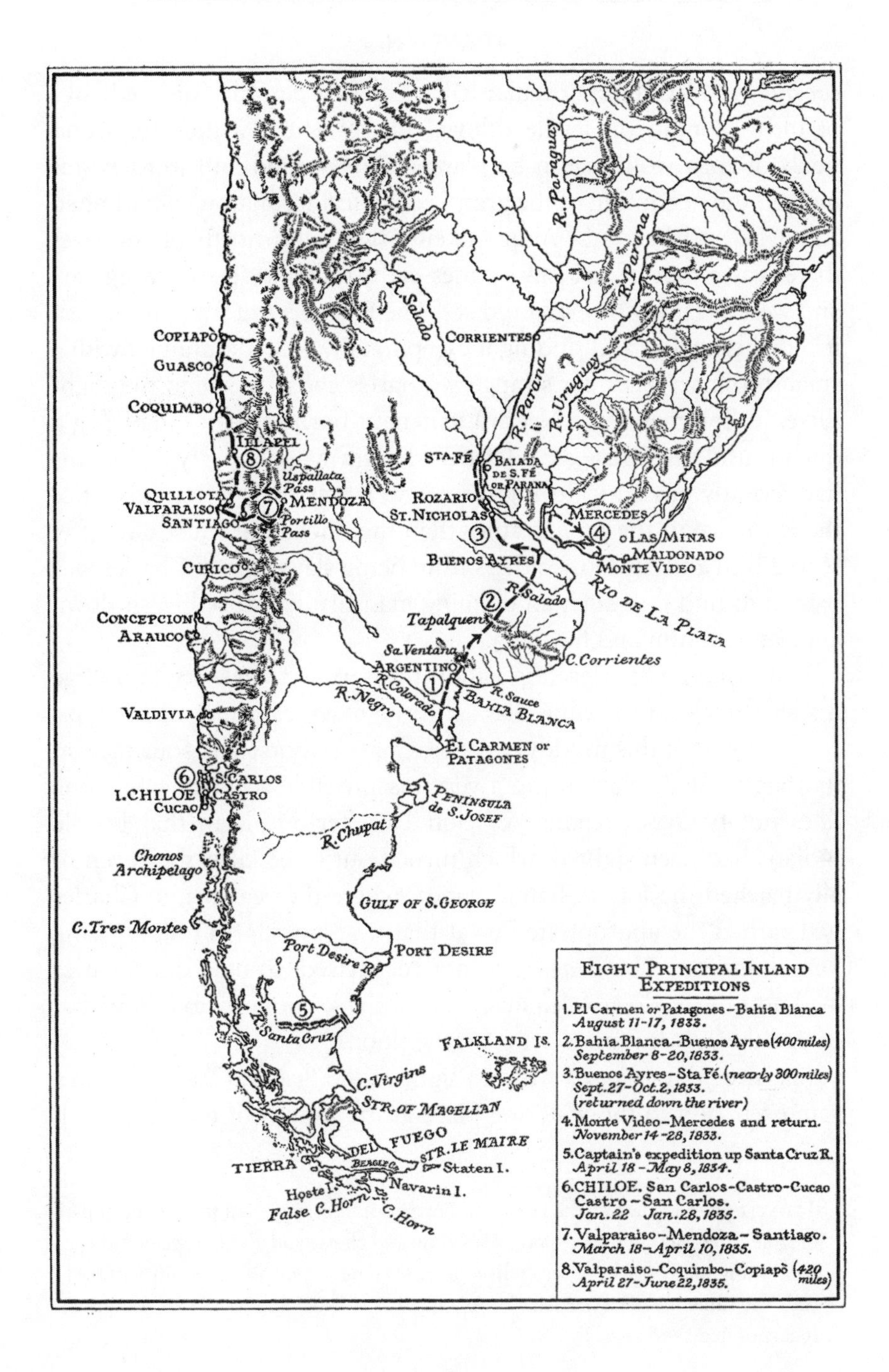
R. Paraguay
R. Parana
R. Salado
Corrientes
R. Parana
R. Uruguay
Copiapò
Guasco
Coquimbo
Illapel
Sta Fé
Baiada de S. Fé or Parana
Uspallata Pass
Quillota
Valparaiso
Santiago
Mendoza
Portillo Pass
Rozario
St. Nicholas
Mercedes
Las Minas
Maldonado
Monte Video
Buenos Ayres
Curico
R. Salado
Rio de La Plata
Concepcion
Arauco
Tapalquen
Sa. Ventana
Argentino
C. Corrientes
R. Colorado
R. Negro
R. Sauce
Bahia Blanca
Valdivia
El Carmen or Patagones
S. Carlos
I. Chiloe
Castro
Cucao
Peninsula de S. Josef
R. Chupat
Chonos Archipelago
Gulf of S. George
C. Tres Montes
Port Desire R.
Port Desire
R. Santa Cruz
Falkland Is.
C. Virgins
Str. of Magellan
Tierra del Fuego
Str. Le Maire
Beagle Ch.
Staten I.
Hoste I.
Navarin I.
False C. Horn
C. Horn
Eight Principal Inland Expeditions
1. El Carmen or Patagones – Bahia Blanca
August 11–17, 1833.
2. Bahia Blanca – Buenos Ayres (400 miles)
September 8–20, 1833.
3. Buenos Ayres – Sta Fé. (nearly 300 miles)
Sept. 27 – Oct. 2, 1833.
(returned down the river)
4. Monte Video – Mercedes and return.
November 14–28, 1833.
5. Captain's expedition up Santa Cruz R.
April 18 – May 8, 1834.
6. CHILOE. San Carlos – Castro – Cucao
Castro – San Carlos.
Jan. 22 Jan. 28, 1835.
7. Valparaiso – Mendoza – Santiago.
March 18 – April 10, 1835.
8. Valparaiso – Coquimbo – Copiapò (420 miles)
April 27 – June 22, 1835.

cacique who led that particular tribe. The first *posta* lay on the banks of the Colorado on fertile diluvial plains where willow trees and fields of corn would soon be planted. The second and third *postas* marked the beginning of the grand geological formation extending to Santa Fé and beyond.* About twenty-five miles north of the river there was a belt of red sand dunes eight miles wide stretching east and west as far as the eye could see. The fourth *posta* was situated on its northern edge, and finding it exceptionally well maintained with a small room set aside for strangers, Charles chose to spend the night there. Its commander, a Negro Lieutenant born in Africa, had dug a ditch round the house as a defence, though when a party of Indians had recently passed it during the night, it was their failure to notice the existence of the *posta*, rather than the ditch, that must have preserved him and his four soldiers from being slaughtered. Charles was pained to find that such an obliging man firmly refused to sit down and eat with him and his party.

Following an exhilarating gallop they arrived next day at a large marsh stretching to Bahia Blanca through several miles of swamps. Near the end of this muddy expanse, Charles's horse fell, sousing him thoroughly with black mire, 'a very disagreeable accident when one does not possess a change of clothes'. After an alarm that hostile Indians had been sighted, which turned out to be false, they eventually reached the fort at Bahia Blanca, where the year before Charles had earned the appropriate but at that time unhelpful title of being '*un naturalista*'. The *Beagle* had not yet arrived, so next day Charles 'had nothing to do, no clean clothes, no books, nobody to talk with. I envied the very kittens playing on the floor.'

Six more days passed with no sight of the *Beagle*. Charles made a vain excursion to the watering place at the mouth of the bay, which

* The vast plain known as the Pampean Formation that occupied the northern part of Argentina was covered by a reddish, limy and clayey soil, the Pampean mud, containing many traces of seashells and corals. There were also quantities of basically the same material hardened into what was called Tosca rock, that sometimes incorporated sand as well.

was enlivened by his guide's account of how, two months before, when he had been hunting with two other soldiers a few miles from the fort, a party of Indians had lassoed, speared and killed his companions, and wounded him. He had managed to keep ahead of them until they gave up the pursuit within sight of the fort. After that the *Comandante* had issued an order forbidding individuals to leave the fort, and Charles understood why his guide had kept such a close eye on a deer which appeared to have been frightened from some other quarter. The horses loaned from the fort were such miserable creatures that Charles and the guide only got back the next day, having had no more than an armadillo and part of a kid to eat, and nothing to drink.

Having bought a fine young horse for £4.10*s*, Charles spent a day riding about the neighbouring plains. Tired of doing nothing, he then hired the same guide for a visit to Punta Alta, better provided this time with food and water. When they were near their destination, the gaucho spotted three people on horseback, and dismounted to have a closer look, saying that they did not ride like Christians, and that nobody could leave the fort. The three joined company, and two dismounted while the third rode over the hill out of sight. The gaucho said, 'We must now get on our horses, load your pistols,' and looked to his sword. Charles asked, 'Are they Indians?' '*Quien sabe?* (Who knows?)' he replied. 'If they are no more than three, it does not signify.' Charles thought his uncommon coolness rather too good a joke, and asked whether they should not return home. His guide answered, 'We *are* returning, only near to a swamp into which we can gallop our horses as far as they can go, and then trust to our own legs. So that there is no danger.' The riders were kept under close observation for some while, until at last the gaucho, bursting into laughter, exclaimed '*Mugeres*! (Women)', whom he had recognised as the wife and sister-in-law of the Major of the fort's son, hunting for ostrich eggs. Charles was puzzled as to why he had taken so long to realise his mistake, but forgave him, and they rode on to the place at Punta Alta where so many fossil bones had been unearthed the year before.

August 1833

Charles spent some time marking the places where more bones were lying, and they then proceeded to eat their dinner in peace. There was a beautiful sunset, and everything was deliciously quiet and still. But appearances were false, and an hour after retiring to bed on their *recados*, very heavy rain began. In the morning it had not stopped, so they started on their return. Soon they saw the fresh track of a puma, an animal that was relatively rare and was sometimes called a lion in South America. They started to follow it, but the dogs seemed to be aware of their intentions, and declined to cooperate. There were also some skunks or '*zorillas*' around, but the dogs well knew that the foetid oil ejected by these animals would make them sick with copiously running noses, and they too were therefore left well alone.

On their arrival at the fort, they found that their fellow traveller Harris had come in with the news that a few days earlier Indians had murdered every soul in one of the *postas*. The perpetrators were suspected to be members of a tribe led by the *cacique* Bernantio, who were on their way to join General Rosas. Harris had met both the tribe and an officer from Rosas carrying a summary message saying that 'if Bernantio failed to bring the heads of the murderers, it should be his bitterest day, for not one of his tribe should be left in the Pampas'.

The next day *Comandante* Miranda set out with three hundred men to accompany Bernantio's tribe and follow the track, or '*rastro*', of the murderers. They were highly skilled trackers, and could follow a trail that was a week or more old. If the suspected murderers were guilty, they were all to be massacred; if not, the *rastro* was to be followed if need be to its end in Chile. Miranda's troops were mostly Indians, and Charles described the scene of their bivouacking and drinking the warm and steaming blood of the cattle that were eaten for supper as savage in the extreme. It transpired later that the *rastro* proved Bernantio to be innocent, and that the murderers had succeeded in escaping into the open pampas.

On 24 August the *Beagle* was at last sighted, 'its figure curiously altered by the refraction over the widely extended mud banks'. There was too much wind to send a boat ashore that day, so FitzRoy sat down and wrote a short letter to Charles that throws a vivid light on the closeness of their relationship:

Beagle. off M. Megatherii.

Saturday. 24th.

My dear Philos*

Trusting that you are not entirely expended, though half starved, occasionally frozen, and at times half drowned, I wish you joy of your campaign with Gen[l]. Rosas, and I do assure you that whenever the ship pitches (which is *very* often as you *well* know) I am extremely vexed to think how much *sea practice* you are losing; and how unhappy you must feel upon the firm ground.

Your home (upon the waters) will remain at anchor near the Montem Megatherii until you return to assist in the parturition of a Megalonyx measuring seventy two feet from the end of his snout to the tip of his tail – and an Ichthyosaurus† somewhat larger than the Beagle.

Our wise ones say that you are not enough of an Archimedes to accomplish the removal of this latter animalcule.

I have sent, by Chaffers [the Master on the *Beagle*], to the Commandant. On *your* account, and on behalf of *our* intestines, which have a strange inclination to be interested by beef.

If you have already departed for the Sierra Ventana – tanto mejor – I shall stay here, at the old trade – "quarter-er-less four".

Sancho goes with Chaffers in case you should require his right trusty service.

Send word when *you* want a boat – *we* shall send, *once* in *four* days.

Take *your own time* – there is abundant occupation here for *all* the *Sounders*, so we shall not growl at you when you return.

Yours very truly

Rob[t]. FitzRoy

---

* Charles was now known by all on the ship as 'the Philosopher'.

† An extinct shark-like aquatic reptile whose bones, unlike those of Megatherium and Megalonyx, Charles did not excavate in South America.

September 1833

P.S. I do not rejoice at your extraordinary and outrageous peregrinations because I am envious – jealous, – and extremely full of all uncharitableness. What will they think of at home of "Master Charles" "I do think he be gone mad" – Prithee be *careful* while there's no *fear* – says the saw.

PS. 2d. (*Irish* fash* Have you yet heard from Henslow – or about your collections sent to England?[90]

On 26 August a boat with Mr Chaffers arrived from the ship, and Charles recorded that 'we waited till the evening for a cow to be killed, to take fresh meat on board. We did not start till late, but the night was beautiful & calm. The ship had moved her berth, & we had a long hunt after her, but at last arrived on board at ½ after one o'clock.'

The whole of the twenty-seventh was consumed 'in telling my travellers tales'. Charles then went in the *Beagle*'s yawl to Punta Alta and left his new servant Syms Covington and another man to spend a few days excavating the bones that he had marked. The site was a quiet one, whose 'very quietness is almost sublime, even in the midst of mud banks & gulls, sand hillocks & solitary vultures'.

There followed a week of frustration while Charles struggled to find a *vacciano* or guide to replace a man who had let him down. This was of vital importance, for as he wrote to his sister Caroline, 'Travelling is very cheap in this country; the only expence [*sic*] is procuring a trusty companion, but in that depends your safety, for a more throat-cutting gentry do not exist than these Gauchos on the face of the world.' He heard many 'curious' anecdotes about the Indians: 'alarming' might have been a better word, but he never betrayed the slightest trace of nervousness about what he was doing. In one recent battle, two hundred soldiers had been sent into the distant foothills of the Andes in pursuit of a group of 112 women and children and men who had almost all been taken or killed, at a cost

* The second postscript precedes the first in the manuscript.

of only one wounded Christian. Any women who appeared above twenty years old were invariably massacred in cold blood, but when Charles hinted that this might be considered rather inhuman, the reply was, 'Why what can be done, they breed so.' There were stories of the noble manner in which the young men acting as messengers or ambassadors for their tribes would face their death without breathing a syllable of information that might injure the cause of their people. Yet some of their leaders, the *caciques*, were willing to save their own lives by betraying the plans of their allies. General Rosas's plan was to kill all the stragglers, and drive the rest to a common point, where with the help of the Chileans they could be attacked in a body. This would be done in the summer when the plains were waterless, and there were very few directions in which the Indians could escape. South of the Rio Negro, the General had a treaty with a tribe called the Tehuelches that they should be paid a certain sum for the slaughter of every Indian crossing the river, but that if any were let past, they themselves would be exterminated. Remembering that his friends might in the future become his enemies, Rosas saw to it that his Indian allies were thinned by being placed in the front lines during his battles. Success in this bloody warfare and butchering of all the Indians east of the Andes would release a huge area for the production of cattle, and the valleys of the Rios Negro, Colorado and Sauce for growing corn. Charles concluded that the country would then be in the hands of white gaucho savages instead of copper-coloured Indians, 'the former being a little superior in civilization, as they are inferior in every moral virtue'. He later noted in the *Journal of Researches* that 'Since leaving South America we have heard that this war of extermination completely failed.' The Indians were nevertheless drastically reduced in numbers, and the process that had begun in 1535 when the first colonists of the Rio Plata had introduced horses to the Pampas was complete.

One day Charles saw a soldier striking fire with a piece of opaque cream-coloured flint that had evidently once been an arrowhead, and which was said to be common on the island of Churichoel. It was between two and three inches long, and therefore twice as large as those still used by the Fuegians. It was known that none of the Pampas Indians apart from a small tribe in forested land in the Banda Oriental, to the north of Monte Video, now used bows and arrows, having had no further use for such ineffective weapons once they had learnt to ride horses and catch animals with the aid of their lassos. So the large arrowheads were antiquarian relics used for hunting by the Indians before 1535, and had gone out of use only when horses had appeared on the scene. This helped to convince Charles that the horse was not an original inhabitant of the Pampas.

On 8 September the guide and passport for government horses had at last arrived from General Rosas, and Charles set off on his four-hundred-mile ride to Buenos Aires. The second *posta* was on the banks of the Rio Sauce, a deep and rapid little river at that time of year, that arose in the distant Andes and dried out at times when there was no snow water to feed it. Charles's first objective was the Sierra de la Ventana, a mountain visible from the anchorage at Bahia Blanca whose height was estimated by FitzRoy by triangulation as 3500 feet. It had been a favourite refuge of the Indians in the recent fighting, but few of the soldiers knew anything about it, and Charles was probably the first European ever to ascend it. With fresh horses and a soldier for a guide, he reached the second of its four peaks the following afternoon after a difficult struggle and several attacks of cramp. He decided that honour was satisfied, and nothing would be gained by trying to reach the tops of the two slightly higher peaks. The mountain was a rather bleak structure composed of white quartz rock associated in places with a little glossy clay-slate, and the view from the top was uninspiring. By sunset they were back at their bivouac by an easier road, and Charles wrote that after 'drinking much mattee [herb tea] & smoking several little cigaretos, I made up my bed for the night. It blew furiously, but I never passed a more comfortable night.'

In the morning they 'fairly scudded before the gale' to the Sauce *posta*, where after reminiscences that cannot have been very encouraging to Charles about the fierceness of the recent fighting on the Sierra de la Ventana the night was spent.

October 1833

The ride northwards continued without very notable incidents. At one point they met a party of friendly Indians of Bernantio's tribe, on their way to a *salina* for salt. They saw the *posta* where the Lieutenant had recently been found with eighteen *chusa* wounds on his body, and all his men dead. Two days were devoted by Charles, while they waited for a party of General Rosas's soldiers on their way to Buenos Aires to catch them up, to geologising, bird-watching, and demonstrations of the skill of the gauchos at hunting with their whirling *bolas*. They passed a spot where there had recently been a storm with hailstones the size of small apples that had killed numbers of deer and ostriches. At supper at the eighth *posta* on the Rio Tapalguen, Charles was 'suddenly struck with horror that I was eating one of the very favourite dishes of the country, viz a half formed calf long before its time of birth. It turned out to be the Lion or Puma; the flesh is very white & remarkably like Veal in its taste.' The twelfth *posta* was the first at which they saw cattle and a white woman, for it was one of General Rosas's great *estancias* covering some six hundred square miles of land. Between the twelfth and fourteenth *postas* they had to cross a stretch of water coming above the horses' knees, riding like Arabs with the stirrups crossed and their legs cocked up. From the fifteenth to the twentieth *postas* they crossed a uniformly rich green plain, with an abundance of cattle, horses and sheep, and here and there a solitary *estancia* with its Ombu [umbra] tree.

On 20 September Charles arrived in Buenos Aires, and went to stay with the merchant Edward Lumb, who was most helpful to him in obtaining supplies, and arranging for the shipment to England of his fossil specimens. Here Charles 'soon enjoyed all the comforts of an English house', had a very pleasant rest, and set about obtaining letters of introduction and a passport for his ride northwards to Santa Fé, on the banks of the Rio Parana. Syms Covington arrived from Monte Video, and was dispatched to Mr Lumb's *estancia* to shoot and skin some birds. On 26 September Charles wrote a letter that has not survived to FitzRoy in order to enquire about the *Beagle*'s plans. A week later FitzRoy posted a long reply from Monte Video, written like the previous one in a breathless but very friendly style:

October 1833

Beagle, Monte Video
4th October 1833
My dear Darwin,

Two hours since, I received your epistle, dated 26th and most punctually and immediately am I about to answer your queries. (mirabile!!)

But firstly of the first – my good Philos why have you told me nothing of your hairbreadth scapes & moving accidents. How many times did you flee from the Indians? How many precipices did you fall over? How many bogs did you fall into? – How often were you carried away by the floods? and how many times were you kilt? that you were not kilt *dead* I have visible evidence in your handwriting, as well as in a columnar paragraph in Mr Love's unamiable paper. You did not tell me whether you received the blank papers safely – you informal homo – how am I to feel certain that I have not signed what may blast my *immaculate* reputation? Harris carried the Packet which contained them and promised to deliver them faithfully. How Sancho by Mr Hood's [the British Consul-General in Montevideo] assistance, contrived so to mismanage as to reach Bs Ayres some days after Harris – Quien sabe? [Who knows?] In it were 5 "Skimpy" lines, as Capt Beaufort would call them & a promise of better behaviour. Since the date of that note the Beagle has been two days at Maldonado – one day here and about a week between this & Cape Corrientes. – Not having any Stone pounders on board – nor any qualified person (the *Mate* being absent) – I could not think of landing, – so *you* have yet a *chance*, – "de verus" (it *blew* strong & prevented landing.) I believe you have heard from Mr Parry [a British merchant at Montevideo] and are aware of his loss. – If you have not heard from him – your *ally* (!! of bone stealing fame) will have informed you. Shocking as it was to him, and his family, but to him, most particularly, I am in hopes that better times will be found by our good friend Parry, – in consequence of his being a single man. Warm hearted and friendly as she was – and friendly to the utmost extent of her means – she had her share of woman's weakness and woman's failings. Robert Parry is gone to

October 1833

Self-portrait of Conrad Martens (Mitchell Library)

school in England in the "Mary Worrall", Merchant man, – to be placed at a school, – the young daughters are going to B. Ayres – also to school – Mr P. intends to give up his house and turn "bachelor, in lodgings", – a wise resolve, though painful indeed to the Father of a family, – think what a change in a domestic circle.

If Mr P. has written as he intended you have heard of Mr Martens – Earle's Successor, – a *stone pounding artist* – who exclaims *in his sleep* "*think* of *me* standing upon a pinnacle of the Andes – or sketching a Fuegian glacier!!!" By my faith in Bumpology, I am sure you will like him, and like him *much* – he is – or I am wofully mistaken – a "rara avis in navibus, – Carlo que Simillima Darwin". – Don't be jealous now for I only put in the last bit to make the line scan – you know very well your degree is "rarissima" and that *your* line runs thus – Est avis in navibus Carlos rarissima Darwin – but you will think I am cracked so seriatim he is a gentlemanlike, well informed man. – His landscapes are *really* good (compared with London men) though perhaps in *figures* he cannot equal Earle. – He is very industrious – and gentlemanlike in his habits, – (not a *small* recommendation).

Wickham gets on famously – really the "Lighter" will not merit *trifling* considerations – Mr Kent of the Pylades is at Gorriti – belonging to our squad. We have plenty of men, – and *good* ones; and all is prospering –

"*Well, but the conjunctions – the conjunctions*" I hear you saying –"*you have got to the end of a sheet of paper without telling me one thing that I wanted to know*".

– This is the 4th of October, – "*so the date of your letter tells me*" — well – hum — if – hum – but – we must consider – then – hum – tomorrow will be the sixth – "*Prodigious*"!! Do you know what I mean – "*to be sure*" so – and so & so – & hum hum hum & off goes the head!! –

I never will write another letter after tea – that green beverage makes one tipsy – besides it is such a luxury feeling that your epistle is not to go across the wide atlantick – and has only to cross the muddy Plata. It is so awful writing to a person thousands of miles off – when your conscience reproaches you with having been extremely negligent and tells you that six or eight or (oh – how awful) twelve months' "*History*" is due to your expectant and irate correspondent.

Still *you* get no answer – "*what is the Beagle going to do – will you tell me or not?*" –

Philos – be not irate – have patience and I will tell thee all.

Tomorrow we shall sail, for Maldonado – there we shall remain until the middle of this month, – thence we shall return to Monte Video – to remain quietly, *if possible*, until the end of the month, – I will try all I can to get away from the River Plate the first week in November but there is much to do – and I shall not be surprised if we are detained even until the middle of November. – However – weather is of such consequence, that every long day gained will tell heavily – and I hope & will try hard to be off *Early* in November – therefore do

October 1833

not delay your arrival *here* later than the *first few* days of November, at the *farthest*.

You say nothing about the "Journal of the expedition up the Rio Negro" – nor have you sent me the map of the province of Buenos Ayres – I pray you to *do the latter* – right speedily – and enquire about the *former* – from Mr Gore [British chargé-d'affaires in Buenos Aires] as well as the other man whose name I forget (Señor – Don – or Colonel Something, or somebody.) – but in writing to Mr Gore I mentioned it – so he will know it – I wish to compare the map with our charts – previous to sending them away – in order to "connive" a little, as your *friend* Mr Bathurst says.

Roberts ([the pilot] of the *Liebre*) passed our bows *this morning* on board of the "Paz" bound to Rio Negro with a cargo of *tobacco*. He did not honour us with a visit – nor did he ask for Chico – respecting the former, he was somewhat rude, and as to the latter rather wise I think. –

Adios Philos – Ever very faithfully yours. Robt FitzRoy[91]

In the meantime, Charles had set out on 27 September to ride to Santa Fé across flat plains that from a distance looked as though they might provide good pasture, but were in fact covered so thickly with huge thistles six feet tall that few animals or birds were to be seen. On the evening of the twenty-ninth he arrived at the town of St Nicholas, for his first glimpse of the Rio Paraná, and next morning crossed a small river called the Arrozo del Medio to enter the province of Santa Fe, bounded to the east by the Paraná and to the west by the province of Cordoba. He had been warned that nearly all the inhabitants of the province were 'most dexterous thieves', which they immediately proved by stealing his pistol.

He entered Rosario, where he had a letter of introduction to a most hospitable Spaniard who was kind enough to re-arm him with 'this most indispensable article'. He galloped on into the town, famed for the large size of its church and the virtue and hospitality of its friars, which was built on a level plain bounded by sixty-foot cliffs, sometimes vertical and red in colour, or else broken masses covered with cacti and mimosa trees, descending to the Paraná. This was a river that drained a huge area of country, though its impressive size was disguised at first sight by the many spindle-shaped islands scattered across it.

Starting by moonlight that evening, Charles arrived at dawn at the Rio Carcavána, a river also called the Saladillo because its water was so brackish. He spent the day searching for fossil bones in the cliff, and quickly found a very large and perfect cutting tooth of *Toxodon*. There were also two immense skeletons protruding from the cliff close to one another, but which disintegrated into fragments when he tried to dig them out. He managed, however, to extract sufficient pieces of one of the huge molar teeth to establish that the remains belonged to a mastodon, another of the extinct ungulates that had once been present in large numbers on the plains of South America.* The men who took Charles to the cliff said that they had long been puzzled to explain how the skeletons could have got to where they were lying, and had concluded that they must have been burrowing animals!

On 2 October Charles rode on past a very pretty village called Corunda into country that was not infrequently ravaged by bands of Indians. Charles's guide spotted with glee the dried body of an Indian suspended from the branch of a tree. Having changed the horses at a well-manned *posta*, they arrived at the town of Santa Fé, where Charles, feeling somewhat feverish from over-exertion, was glad to find an unfurnished room in which he could retreat to bed for a couple of days. On 5 October he spent four hours crossing the Paraná to Santa Fé Bajada, now known as Paraná, capital of the Province of Entre Rios. Still unwell, he then decided to return as quickly as possible to Buenos Aires in order to be in good time for

* *Toxodon* was the commonest extinct ungulate or grazing mammal in South America in the Tertiary Pliocene and Quaternary Pleistocene periods, around three million years ago. It was built like a short-legged rhinoceros about nine feet in length, probably with a short snout so that it looked like a gigantic guinea-pig. Mastodon was another ungulate of a similar size, distantly related to the modern elephant.

Reconstruction of *Toxodon*, length about nine feet (Romer)

the *Beagle*'s planned departure for the south. He was unable to hire a boat of his own, and therefore took a place on a *balandra*, a single-masted vessel of a hundred tons displacement. But the *balandra* was in no hurry to depart, and Charles spent the next five days on the cliffs of the Paraná picking up fossil shells, and admiring the humming birds hovering round the beautiful flowers.

The *balandra* made its way in a leisurely fashion down the Paraná. Charles spent most of the time in bed, for the ceiling of his cabin was too low for him to sit up. He faced new perils on this journey, for when he landed on one of the islands he found 'the most indubitable & recent sign of the tiger', and had to beat a quick retreat. There was good reason to fear that the numerous jaguars that inhabited the thickets, and sometimes added woodcutters and young oxen to their normal diet of capybaras, would not have hesitated also to include '*un naturalista*' on the menu. Mosquitoes were another hazard, and when Charles exposed the back of his hand for five minutes, he reckoned that fifty were soon sucking his blood. Fortunately they cannot have been malarious. Less dangerous were the scissor-beaks flying rapidly up and down the stream, ploughing the surface with their lower mandibles, and occasionally seizing a small fish. Other birds of interest were some small kingfishers not dissimilar to the European variety, green parrots in large flocks that sometimes ravaged the corn fields, and insectivorous birds that in flight resembled swallows until they made abrupt turns by opening and shutting their two long tail-feathers in a scissor-like fashion.

After a week the *balandra* reached the mouth of the Parana, and Charles managed to get a canoe and land at St Fernando, about twenty miles to the north of Buenos Aires. His intention was then to ride into the city, but to his dismay he found himself once again in the middle of a revolution in which a party of men basically in sympathy with General Rosas had lost patience with the Governor of Buenos Aires, and were tightly blockading the city and all its ports. The next morning he was able to persuade General Rolor, commanding the northern division of the rebels, to give him a passport to their commander-in-chief at Quilmes on the opposite side of the city. Having with difficulty procured some horses, he made a great sweep round the city, and arrived at the main encampment of the rebels near Quilmes. At first they were unreceptive, but General Rosas's brother was among them, and Charles's account of his visit to Rosas's camp at the Rio Colorado soon thawed things out. He was given permission to enter the city on foot, and having satisfied the Governor's soldiers with an old passport, was exceedingly glad to find himself 'safe on the stones of B. Ayres'. Some days later, a message from Rosas announced that although he disapproved of the peace having been broken, he thought that the rebels had justice on their side. The Governor and his ministers at once resigned and made themselves scarce, a new government was elected, and the rebel soldiers were handsomely rewarded for their services. Charles concluded correctly that Rosas would ultimately be absolute dictator of the country.

Charles's troubles were, however, not quite over, for his servant Syms Covington had been collecting at the helpful Mr Lumb's *estancia* in the country, where one day 'he nearly lost his life in a quicksand & my gun completely'. A man had to be bribed to smuggle Covington through the belligerents, leaving his and some of Charles's collections and possessions behind, though luckily these could later be shipped to Monte Video. Conditions were uncomfortable in Buenos Aires, with all the shops closed, and the lawless soldiery robbing people at random. So on 2 November Charles was thankful to embark on the packet to Monte Video, and escape from so miserable a town.

CHAPTER 13

The Last of Monte Video

On arriving at Monte Video, Charles went on board the *Beagle*, and was surprised to learn that their sailing had been put off until the beginning of December, because all the information gathered by the schooners *Paz* and *Liebre* had still to be entered on the charts. Finding the poop cabin full of carpenters, he took a room ashore in order to make the most of this additional month in Uruguay.

His first expedition was a pleasant gallop westwards along the coast of the Rio Plata to the gently sloping Barrancas de St Gregorio. The geology was disappointing, but to cross the flooded Rio St Lucia, the horses were obliged to swim at least six hundred yards, and it was surprising to see what light work they made of it. After five days lost by 'true Spanish delay' in getting a passport and other papers, Charles set out with his *vaqueano* (cowhand) on a longer ride to see the Rio Uruguay flowing down from the north and forming the western border of the country, and its tributary, a second Rio Negro. Their progress was slow at first, for they had to cross the rivers Canelones, St Lucia and San José in boats, unlike the *peons* who slipped off backwards when their horses were out of their depth, and were towed across grasping the horses' tails. The naked men galloping about on naked horses were a fine spectacle that reminded Charles of the Elgin Marbles. After crossing the deep and rapid Rio Rozario he arrived at Colonia del Sacramento, an ancient town on the coast that had been the headquarters of the Uruguayan army during the recent war against Brazil. His view was that this had been:

November 1833

Monte Video from the *Beagle*'s anchorage, the first drawing made by Conrad Martens on the *Beagle*

A war most injurious to this country, not so much in its immediate effects, as in being the origin of a multitude of Generals, & all other grades of officers. More generals are numbered but not paid in the united provinces of La Plata than in Great Britain. These gentlemen have learned to like power & do not object to a little skirmishing. Hence arises a constant temptation to fresh revolutions, which in proportion as they are easily effected, so are they easily overturned. But I noticed here & in other places a very general interest in the ensuing election for the President; & this appears a good sign for the stability of this little country. The inhabitants do not require much education in their representatives; I heard some men discussing the merits of those for Colonia "that although they were not men of business, they could all sign their names". With this every reasonable man was satisfied.[92]

The following day Charles was invited by a local gentleman to visit his *estancia* near the town. It was in what was known as a '*rincon*', having an area of twenty square miles with one side fronted by the Rio Plata on which there was a port for small vessels, and two others guarded by impassable brooks. It contained three thousand cattle, eight hundred mares and 150 broken horses, plenty of water and limestone, many trees, a rough house, excellent corrals, and a peach orchard. Charles was interested to learn that the value of this property was little more than £2000. It was managed on the principle that the cattle always divided themselves into groups of less than a hundred animals that could be recognised by a few peculiarly marked individuals. These groups could be gathered together and counted weekly.

The low thick woods bordering the river harboured many jaguars, and the trees were grooved by fresh scratches a yard long. However Charles was unable to attract one into the open. Further north, on the road to Mercedes on the Rio Negro, the party stayed at a very

The outer wall of Monte Video, by Conrad Martens

large *estancia* occupying ninety square miles. It was run by a nephew of the owner in Buenos Aires, and a Captain who had escaped from the Argentinian army. Their knowledge of the outside world was as usual very limited, and finding that in England animals were not caught with a lasso, they concluded, 'Ah then, you use nothing but the bolas.'

> The Captain at last said he had one question to ask me, & he should be very much obliged if I would answer him with all truth. I trembled to think how deeply scientific it would be. It was "whether the ladies of Buenos Ayres were not the handsomest in the world". I replied, "Charmingly so". He added "I have one other question – Do ladies in any other part of the world wear such large combs". I solemnly assured him they did not. They were absolutely delighted. The Captain exclaimed "Look there, a man who has seen half the world, says it is the case; we always thought so, but now we know it". My excellent judgment in beauty procured me a most hospitable reception; the Captain forced me to take his bed, & he would sleep on his Recado.[93]

November 1833

Starting at sunrise they rode slowly on towards Mercedes. Here the country became more like the Pampas than the rest of the province. Immense beds of thistles growing higher than the head of a rider, and of the related cardoons (artichoke thistle) that were somewhat less tall, discouraged the raising of cattle. Arriving on 22 November at the *estancia* of the Berquelo near Mercedes, Charles spent the day geologising until in the evening its owner returned. They rode twenty miles to the Sierra del Pedro Flaco, from which he had a fine view of the Rio Negro with its rapidly running stream of blue water, nearly as large as its namesake in Patagonia. Back at Mercedes, he heard of some giants' bones that turned out to be *Megatherium*, though he could only dig out some fragments. Starting to ride back to Monte Video by way of San José on 26 November, he found at an *estancia* a part, very perfect, of the head of a *Megatherium*, which he bought for a few shillings. He reached the capital two days later, the distance, as charged by the post-chaise, being about seventy leagues (two hundred miles) across the rather uninteresting plains of the Rio de la Plata.

Near the English Gate in Monte Video, by Conrad Martens

November 1833

Monte Video and the harbour, by Conrad Martens

The *Beagle*'s charts were still not quite finished, and Charles stayed ashore for four more days making his preparations for the voyage to the south, and musing on the character of the inhabitants of the northern provinces. He considered the gauchos very superior to those who lived in the towns, being invariably most obliging and hospitable, modest about themselves while at the same time spirited and bold. He had however to admit that far too much blood was shed, and lives were lost, from knife fights in trivial quarrels. His views on the town-dwellers of the higher and supposedly better-educated classes were uncompromising, for he classed them as profligate sensualists who laughed at all religion, were open to the grossest corruption, and were entirely wanting in principle. Every public officer was receptive to bribery, the head of the post office was selling forged government francs, the Governor and Prime Minister openly plundered the state, and justice – where gold was in the case – was hardly expected. Charles's opinion was that before many years the country would be trembling under the iron hand of a dictator, and he wished it well enough to hope that this period was not too distant.

November 1833

The other side of the picture was the excellent taste in dress of all the women, the general good manners in all grades of life, and above all the remarkable equality of all ranks. At the Colorado, men who kept the lowest little shops used to dine with General Rosas. Many in the army could neither read nor write, yet all met on terms of perfect equality. All this might be expected in a new country, but the absence of gentlemen *par excellence* did strike Charles as a novelty.

CHAPTER 14

Christmas Day at Port Desire, and on to Port St Julian and Port Famine

At four o'clock in the morning of 6 December 1833, the *Beagle* got under weigh, and accompanied by the *Adventure* ran up the river to take in fresh water. With a fair wind they stood out of the river next day, and by the evening were in clear water, 'never I trust again to enter the muddy water of the Plata', noted Charles.

There had been some changes among the officers, for Lieutenant John Wickham now commanded the *Adventure*, and had with him Messrs Johnstone and Forsyth, and Mr Usborne as Under-Surveyor. Mr Kent from the *Pylades* had come to join Benjamin Bynoe as Assistant Surgeon. Augustus Earle had been obliged to return home because of ill health, and Conrad Martens had taken his place as the ship's official artist.

After a slow passage thanks to light winds, the *Beagle* and *Adventure* arrived on 23 December at Port Desire – Puerto Deseado on a modern map – on the coast of Patagonia. This was a good anchorage discovered by the Elizabethan adventurer and second English circumnavigator of the globe, Sir Thomas Cavendish, when he too spent Christmas there in 1586. That same day, Conrad Martens painted the *Adventure* at anchor with the ruins of a Spanish fort on the north side of the harbour in the background. The Spaniards had some years earlier attempted to establish a settlement there, but it had quickly succumbed to a total lack of water in the summer, and attacks by the Indians in the winter. The following day Charles went ashore and took a long walk to the north, where he found a veritable desert composed of gravel, with rather

Adventure and Spanish ruins at Port Desire, by Conrad Martens

little vegetation and not a drop of water. The dryness of the country was nevertheless redeemed by the fact that it supported a good many guanacos, the animals of which llamas are the somewhat smaller domesticated variety; one of them was shot by Charles to be eaten on Christmas Day.

On the surface of the plain, 247 feet above sea level according to Charles's barometric measurements, there were beds of oysters, mussels and other shells of the species currently found in the sea; and he was struck to see that the mussels still retained their blue colour. 'It is therefore certain,' he wrote, 'that within no great number of centuries all this country has been beneath the sea.' But in this he was mistaken, for the current view of geologists is that the 247-feet terrace at Port Desire belongs to the late Pleistocene epoch, with a minimum age of 120,000 years.[94] Charles was quite correct to conclude that a substantial elevation of the coast had taken place, but the process was much slower than he at first supposed, for the blueness of the mussel shells was not in fact a reliable index of their

A guanaco, by Martens

age.* The Christmas festivities opened with lunch in the gunroom, after which the officers and almost all the men from the two ships went ashore for athletic contests and a naval game called 'slinging the monkey' (see Plate 3). Normally the 'monkey' was a keg of rum suspended on a tripod to extract the last drop, but on this occasion it was a member of the crew slung up by the heels. The Captain distributed prizes to the best runners, leapers and wrestlers. Charles commented that 'These Olympic games were very amusing, it was quite delightful to see with what school-boy eagerness the seamen enjoyed them: old men with long beards & young men without any were playing like so many children, certainly a much better way of passing Christmas day than the usual one, of every seaman getting as drunk as he possibly can.' Conrad Martens painted a vivid watercolour of the scene, with the *Beagle* and *Adventure* in the background. It was initialled 'RF' in one corner, but did not earn FitzRoy's total

* As Charles had already appreciated when in 1845 he was writing his book on *Geological Observations* on South America, mussel shells are not merely coloured blue at their outer surface, but throughout their thickness, so that the colouring is often not lost until the whole shell has been worn away after hundreds of thousands of years.

approval, for the artist wrote below, 'Note Mainmast of Beagle a little farther aft, Miz. Mast to rake more.'

The yawl was sent under the command of Mr Chaffers, with half a dozen officers and men, Charles and Conrad Martens, and provisions for three days, to explore the head of the creek in search of the watering places mentioned in an old Spanish chart. On the first day, little was found but a small rill of brackish water. While waiting for the tide to allow them to proceed further, Charles took a walk inland, of which he wrote:

> There is not a tree, &, excepting the Guanaco, who stands on some hill top a watchful sentinel over his herd, scarcely an animal or a bird. All is stillness & desolation.[95]

In the evening they sailed a few miles further, and then pitched the tents for the night. By the middle of the next day, they could get no further in the yawl, but Mr Chaffers in the dinghy did find a small fresh-water river several miles higher up. Charles thought that this stream might arise in the Cordilleras, but FitzRoy considered it more likely to be filled locally in particularly rainy seasons. At the head of the inlet the views were very fine, with the red porphyry* rock rising from the water in steep cliffs, and forming spires in its very course. Here they bivouacked, and Conrad Martens recorded the scene (Plate 5).

Back at Port Desire, Charles busied himself with more geological studies, in the course of which he discovered an Indian grave at the top of a distant hill. It consisted of a heap of large stones, placed with some care at the foot of a ledge of rock about six feet high, three yards in front of which there were two huge rocks resting on one another, each weighing at least two tons. On the hard rock at the bottom of the grave was a layer of earth a foot deep, covered by a pavement of flat stones and then a heap of stones to fill the gap

* A rock formed by the solidification on cooling of molten volcanic lava to yield an igneous rock containing a multitude of large crystals in a finely crystalline matrix. A red porphyry was a major component of the base of the Andes.

December 1833

An Indian grave at Port Desire, by Conrad Martens

between the ledge and the two large rocks. An immense rock had been prised off the ledge to rest on top of the others. Remains of small fires and the bones of some horses near the grave suggested that it had been visited recently, but excavations beneath the rocks did not reveal any human bones.

The animals and birds that Charles collected around Port Desire were mainly the larger ones, though he made one important observation at the opposite end of the size range when large quantities of what he called the Corallina *Halimeda* were thrown up on the beach one day. From a careful examination of their anatomy under his microscope, he concluded correctly that this particular group of organisms, that had previously been included under the general label of Zoophytes whose classification was doubtful, were definitely plants rather than animals.* One of the very few animals adapted to survive the dryness of the desolate plains was the guanaco, which was sometimes seen to drink the brackish water in the rills and salt-pans. It is possible that guanacos may be able to save water in the same fashion as the camels in Africa and the Middle East to which

* *Halimeda* is a marine alga, i.e. a type of lower plant growing in salt water, being a green variety belonging to the phylum Chlorophyta. Charles's decision was based on a close examination of the articulations in the reproductive gemmules of the algae. On the same occasion he collected smaller quantities of another Corallina, a red-coloured species of alga called *Amphiroa*, belonging to the phylum Rhodophyta. As became clear later, these belonged to the group nowadays termed 'coralline' algae, because they grow symbiotically, i.e. in close partnership, with the reef-building hydrozoans that are indeed classified as primitive animals. But Charles's conclusions about their articulations were equally valid for both classes.

they are distantly related,[96] by allowing their body temperature to fall well below 37°C during the freezing cold nights, and then to rise well above 37°C during the hot day before perspiration and consequent loss of water starts. But although the problem is of considerable physiological interest, no detailed study of the manner in which the guanacos in South American deserts conserve water has yet been reported. Of some of their habits, Charles wrote:

> Frequently the sportsman receives the first intimation of their presence by hearing from a long distance the peculiar shrill neighing note of alarm; if he looks attentively, he will perhaps see the herd standing in a line on the side of a distant hill. On approaching, a few more squeals are given, & the herd set off, at an apparently slow but really quick canter, along some narrow beaten track to a neighbouring hill. If however by chance, he should abruptly meet a single Guanaco, or a herd; they will generally stand motionless & intently look at him – then perhaps move on a few yards, turn round & gaze again. What is the cause of this difference in their shyness? Do they mistake a man in the distance for their chief enemy the puma? Or does curiousity overcome their timidity? That they are curious is certain, for if a person lies on the ground & plays strange antics, such as throwing up his feet in the air, they will almost always approach by degrees to reconnoitre him. It is an artifice which has been repeatedly practised by our sportsmen; it has moreover the advantage of allowing several shots to be fired, which are all taken as parts of the performance.[97]

On another occasion Charles shot a large guanaco which must have weighed more than two hundred pounds. 'Two males were fighting furiously & galloping like race horses with their ears down & necks low; they did not see me & passed within 30 yards; & then I settled the contest by shooting the Persecutor.'

When Charles was at the Rio Negro in the north of Patagonia, he had often heard the gauchos talking of a very rare bird that they called the *Avestruz petise*, '*petiso*' being a slang word used in Argentina for a very short person. This bird was similar to the common ostrich, but smaller, darker and with its legs feathered lower down. Their eggs were of much the same size, but slightly different in shape, and coloured with a tinge of pale blue. However Charles had yet to see the bird for himself.

In a letter to his brother Henry, Conrad Martens mentioned that while at Port Desire, 'as there was but little to be done in the way of sketching, I used generally to take out my gun, and was fortunate enough one day to bring home an ostrich, the only one indeed that as yet we had been able to kill, altho great numbers had been seen. It was a young one, and excellent eating.'[98] But its interest was more than gastronomic, for let Charles take up the story:

> When at Port Desire in Patagonia (Lat. 48°), Mr Martens shot an ostrich; I looked at it, and from most unfortunately forgetting at the moment the whole subject of the Petises, thought it was a two-third grown one of the common sort. The bird was skinned and cooked before my memory returned. But the head, neck, legs, wings, many of the larger feathers, and a large part of the skin had been preserved. From these a very nearly perfect specimen has been put together, and is now exhibited in the museum of the Zoological Society.[99]

The bird destined to become the type-specimen of *Rhea darwinii* (Plate 4), when in early 1837 Charles's collections were classified by the ornithologist John Gould in London, became the first example of the important principle later known by zoologists as 'representative species', according to which the geographical spread of newly evolved species to adjacent areas of a continent might result, as in this case in the presence of the larger species *Rhea americana* in northern Patagonia with its range overlapping that of the smaller *Rhea darwinii* in southern Patagonia.

While the *Adventure* remained at Port Desire for some adjustments to be made to her sails, the *Beagle* worked out of the harbour to sail southwards down the coast. In the course of this manoeuvre she struck rather heavily on an awkwardly located rock in the middle of the channel, exactly where FitzRoy remembered that she had run into trouble during a previous visit in 1829. Lieutenant Sulivan dived down under the keel to inspect the damage, which seemed not to be severe although he came up bleeding from several scratches received from the jagged copper. To make doubly sure, FitzRoy repeated the dive himself.

The *Beagle* sailed on for two days surveying the coast of a continuation of the tableland seen at Port Desire, interrupted in places by great valleys where the land had been removed. On 9 January 1834 FitzRoy anchored the ship off the entrance to the harbour of Port St Julian, and landed Charles to explore the shore while he made soundings of the bar. Conrad Martens remained on board and painted a watercolour of the view seen from the harbour bar (Plate 6).

Puerto San Julian had been the scene of two historic incidents, for it was here on 31 March 1520 that the great Portuguese navigator Ferdinand Magellan had in desperation taken refuge with the Spanish fleet under his command in a serious state of disarray after a severe tempest. On 2 April two of his subordinate commanders, Luis de Mendoza and Gaspar Quesada, led an unsuccessful revolt against their Captain General, in the course of which Mendoza was killed in a fight and Quesada was outwitted by Magellan and placed in irons. Mendoza's body was then drawn, quartered, and impaled upon a gibbet set on the beach at a place later known as Gallows Point. At a court martial held in the great cabin of the flagship *Trinidada*, Quesada was found guilty of treason, was beheaded and quartered, and his parts were impaled alongside those of Mendoza. Magellan remained at Puerto San Julian for a further six months, repairing his ships and waiting for the southern summer to continue his voyage. Before sailing, he erected a cross on the highest hill to the north-west of the anchorage, Monte Cristo or Mount Wood, noted by FitzRoy as an excellent landmark for Port St Julian, and seen at the right-hand end of Conrad Martens's picture. On 28 November 1520, after passing through the straits named after him, Magellan's three ships emerged into the ocean that he called the 'Pacific' because of its calmness, at least on that day. In March 1521 he discovered the Philippines, but was killed soon afterwards in a fight with the natives on the island of

January 1834

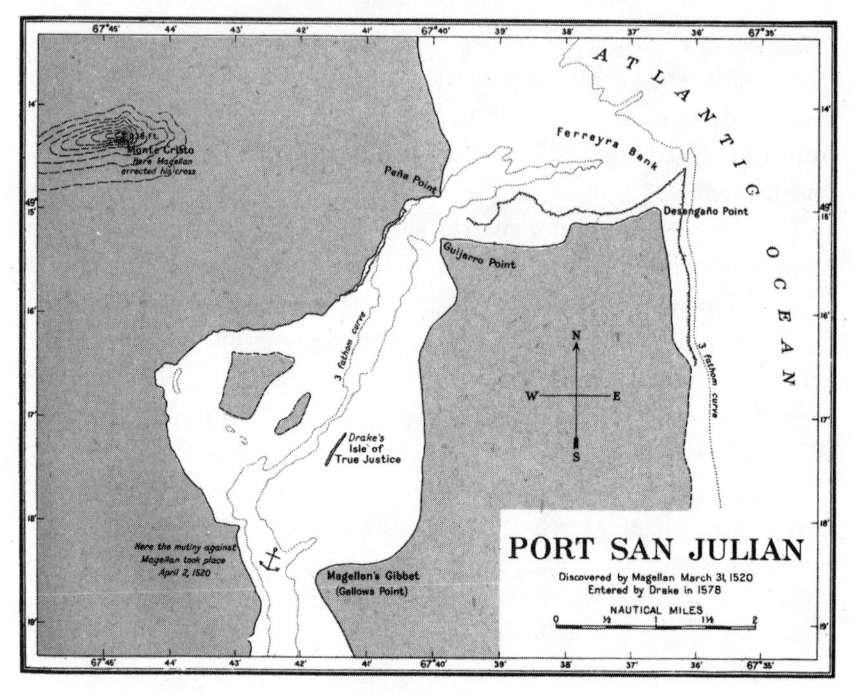

Mactan. Ironically it was the Spanish navigator Sebastián del Cano, who had participated in Mendoza's revolt but had been pardoned, who completed the circumnavigation on 6 September 1522 by sailing the *Vittoria* round the Cape of Good Hope and so back to Seville.

The first English circumnavigator, Sir Francis Drake, having been commissioned by Queen Elizabeth to harry the Spanish colonies on the west coast of South America, landed at Port St Julian on his way to round Cape Horn in 1578. One of Drake's chosen companions, Thomas Doughty, had been accused of acts of insubordination and attempts to foment mutiny, and was tried before a jury of forty drawn from the ships of Drake's fleet. Having been found guilty, Doughty was offered a choice of three sentences – a return to England for retrial, to be marooned at Port St Julian, or to be beheaded. Surprisingly he chose execution, and was duly beheaded on a block at the foot of Magellan's gibbet. Drake carved his name in Latin on a rock at his Isle of True Justice, in order 'that it might be better understood by all that should come after us'. But whether or not justice had in fact been well served became a matter of widespread controversy. Drake sailed on in the *Pelican*, later renamed the *Golden Hind*, and returned to England in September 1580 via the Cape of Good Hope, heavily laden with spices and captured Spanish treasure.[101]

Of his first landing at Port St Julian on 9 January, Charles said that he found some most interesting geological facts. As he explained in his next letter to Henslow, written in March from the Falkland Islands:

> Since leaving the R. Plata, I have had some opportunities of examining the great Southern Patagonian formation. I have a good many shells; from the little I know of the subject it must be a Tertiary formation for some of the shells & (Corallines?) now exist in the sea – others I believe do not. This bed, which is chiefly characterised by a great Oyster is covered by a very curious bed of Porphyry pebbles, which I have traced for more than 700 miles. But the most curious fact is that the whole of the East coast of South part of S. America has been elevated from the ocean since a period during which Muscles [an archaic spelling of 'mussels' habitually used by Charles] have not lost their blue colour. At Port St Julian I found some very perfect bones of some large animal, I fancy a Mastodon* – the bones of one hind extremity are very perfect & solid. This is very interesting as the Latitude is between 49° & 50° & the site is so far removed from the great Pampas, where bones of the narrow toothed Mastodon are so frequently found. By the way, this Mastodon & the Megatherium I have no doubt were fellow brethren in the ancient plains. Relics of the Megatherium I have found at a distance of nearly 600 miles apart in a N & S line.

* His identification from the rather few bones that he had found was incorrect. It was not until the *Beagle's* return to England in 1836, when they were examined by Professor Richard Owen, that they were found to belong not to a mastodon but to a hitherto unknown extinct animal named *Macrauchenia*, at first thought to be related to llamas and camels, but later identified as a litoptern, a long-necked three-toed grazing animal distantly related to the South American tapir.

January 1834

In Tierra del Fuego I have been interested in finding some sort of Ammonite* (also I believe found by Capt. King) in the Slate near Port Famine; on the Eastern coast there are some curious alluvial plains, by which the existence of certain quadrupeds in the islands can clearly be accounted for. There is a sandstone with the impression of the leaves of the common Beech tree, also modern shells, &c &c. On the surface of which table land there are, as usual, muscles with their blue colour &c.

This is the *report* of my *geological section!* to you my President & Master. I am quite charmed with Geology but like the wise animal between two bundles of hay, I do not know which to like the best, the old crystalline group of rocks or the softer & fossiliferous beds. When puzzling about stratification &c, I feel inclined to cry a fig for your big oysters & your bigger Megatheriums. But then when digging out some fine bones, I wonder how any man can tire his arms with hammering granite. By the way I have not one clear idea about cleavage, stratification, lines of upheaval. I have no books which tell me much, & what they do I cannot apply to what I see. In consequence I draw my own conclusions, & most gloriously ridiculous ones they are, I sometimes fancy I shall persuade myself there are no such things as mountains, which would be a very original discovery to make in Tierra del Fuego. Can you throw any light into my mind, by telling me what relation cleavage & planes of deposition bear to each other?[100]

In his next letter home, also written from the Falkland Islands, to his sister Catherine, Charles was still under the spell of his geological studies, which he praised in terms that for him could not have been warmer:

There is nothing like geology; the pleasure of the first day's partridge shooting or first day's hunting cannot be compared to finding a fine group of fossil bones, which tell their story of former times with almost a living tongue.

* An extinct cephalopod mollusc with a typical spiral shell that was common in the Devonian period around four hundred million years ago.

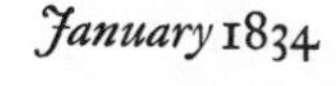

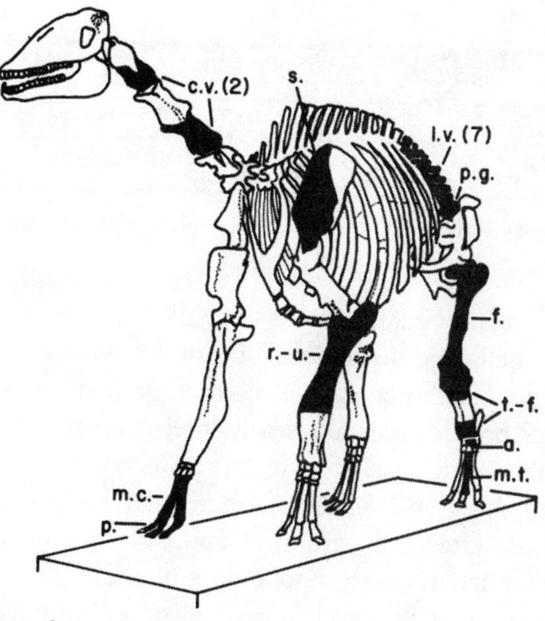

Reconstruction of *Macrauchenia* with the bones collected by Charles at Port St Julian shown in black

The problem that he found so fascinating was how to account for his observations on the origin of the arid plain of gravel in which there were huge oyster shells nearly a foot wide, and which stretched seven hundred miles down the southern coast of Patagonia, from the Rio Colorado to the Straits of Magellan. As he had seen at Port Desire, among the oysters were shells mostly of extinct species but some strongly resembling existing ones, and as before including mussels which misleadingly still retained some of their blue colour.

Charles then set out to examine the composition of the strata in the ninety-foot cliffs on the south side of the harbour at Port St Julian. At the top was a layer of earth (A A A) which contained the bones of the supposed mastodon in a pocket cutting through the gravel layer (B) of porphyritic pebbles with recent shells on its surface, into a twenty-foot layer of a soft white clayey material (C D) in which were never any fossils. Beneath this was a yellowish earthy sandstone (E) rich in Scutellae shells, and finally a greenish sandy

January 1834

Strata in the cliffs of the ninety-feet plain at Port St Julian

clay (F) incorporating large blocks of chalky sandstone, many of the large oyster shells, and the remains of countless barnacles.

Further inland, the land rose in three or four steps sometimes several miles broad, to form a nearly level plain stretching back all the way to the Cordilleras. This structure was typical of what Charles came to call the grand Tertiary formation of Patagonia, extending some seven hundred miles from the Rio Colorado to the Straits of Magellan, and one to four hundred miles in width. The terraces were uniformly covered with gravel consisting of well rounded pebbles of various sizes evidently derived from the porphyritic rocks at the base of the Andes. Beneath the gravel there was a layer of soft white clayey material most closely resembling decomposed feldspar,* but containing much gypsum and copious remains of plankton. Near the coast there were sometimes shells on the gravel. The whole rested on older and fossiliferous strata.

Charles immediately had some difficulties that he tried to argue out with himself in his Geology Notes. One was to explain the uniformity of the gravel bed all the way from the source of the rock in the Cordilleras to the sea, which could only have arisen from the action of tides over a long period of time to form one of the largest shingle beds in the world. Beneath the shingle was the white bed of mudstone only twenty feet thick in some places, and several hundred feet in others. As it was lighter than the shingle, perhaps the tide had washed it out faster than the pebbles and in more variable quantities.

* Feldspars are common minerals containing the silicate of aluminium mixed with various proportions of potassium, sodium or calcium silicate.

January 1834

An important point that Charles emphasised from the start was that the whole formation was horizontally stratified, with no sign of any faults or violent episodes during its upheaval. The whole huge mass had been raised to a height of three to four hundred feet within the period of the existing seashells.* As he learned later, the extinct shells from the late Tertiary period that he found in the bottom layer could only have lived in a depth of water not greater than about two hundred feet, yet now they were covered by sea-deposited strata eight hundred to a thousand feet thick. There had therefore also been a downward movement by several hundred feet of the seabed on which the shells must have lived. The uprising movement was interrupted by at least eight long periods of rest, during which the sea ate deeply back into the land, forming at successive levels long lines of cliffs that separated the different plains rising in steps behind one another. Charles looked further into this question when he was travelling up the Rio Santa Cruz.

An expedition made one day to the head of the harbour provided graphic evidence of the aridity of the gravel plains, as was explained by FitzRoy:

> One day Mr Darwin and I undertook an excursion in search of fresh-water to the head of the inlet, and towards a place marked in an old Spanish plan "pozos de agua dulce", but after a very fatiguing walk not a drop of water could be found. I lay down on the top of a hill, too tired and thirsty to move farther, seeing two lakes of water, as we thought, about two miles off, but unable to reach them. Mr Darwin, more accustomed than the men or myself to long excursions on shore, thought he could get to the lakes, and went to try. We watched him anxiously from the top of the hill, named in the plan "Thirsty Hill", saw him stoop down at the lake, but immediately leave it and go on to

* The period of time over which these movements of the land had indeed taken place is now known to cover more than a million years, and the four-hundred-foot marine terrace that can be traced stretching down the coast of Patagonia from the Rio Colorado to Tierra del Fuego dates from the early Pleistocene epoch.

> another that he also quitted without delay, and we knew by his slow returning pace that the apparent lakes were "salinas". We then had no alternative but to return if we could, so descending to meet him at one side of the height, we all turned eastward and trudged along heavily enough. The day had been so hot that our little stock of water was soon exhausted, and we were all more or less laden with instruments, ammunition or weapons. About dusk I could move no further, having foolishly carried a heavy double-barrelled gun all day, besides instruments, so choosing a place which could be found again, I sent a party on and lay down to sleep; one man, the most tired next to me, staying with me. A glass of water would have made me quite fresh, but it was not to be had. After some hours, two of my boat's crew returned with water, and we very soon revived. Towards morning we all got on board, and no one suffered afterwards from the over-fatigue except Mr Darwin, who had had no rest during the whole of that thirsty day – now a matter of amusement, but at the time a very serious affair.[102]

Charles spent the next two days 'very feverish' in bed, but then ventured out again, finding some fine fossil shells, a large Spanish oven built of bricks, and on the top of a hill a small wooden cross. He did not record whether the hill in question was Mount Wood, but if it was, the cross might have been Magellan's, for in the exceptionally dry environment the wood could possibly have survived for more than three hundred years.

On 19 January the surveys had been completed, and the *Beagle* sailed back to join the *Adventure*, now equipped with a new square topsail, at Port Desire. Three days later the *Adventure* departed to complete a survey of West Falkland Island, and the *Beagle* sailed south to the Straits of Magellan, coming to an anchor on the twenty-ninth at St Gregory Bay on the north shore. Here they found a large tribe of Patagonian Indians with the primitive *toldos* (tents) painted by Conrad Martens (Plate 7). Through their frequent contacts with the sealers, they spoke a lot of Spanish and some English. The men were up to six feet in height, and broad to match. All the people wore large mantles of guanaco skins, had long black hair, and with their

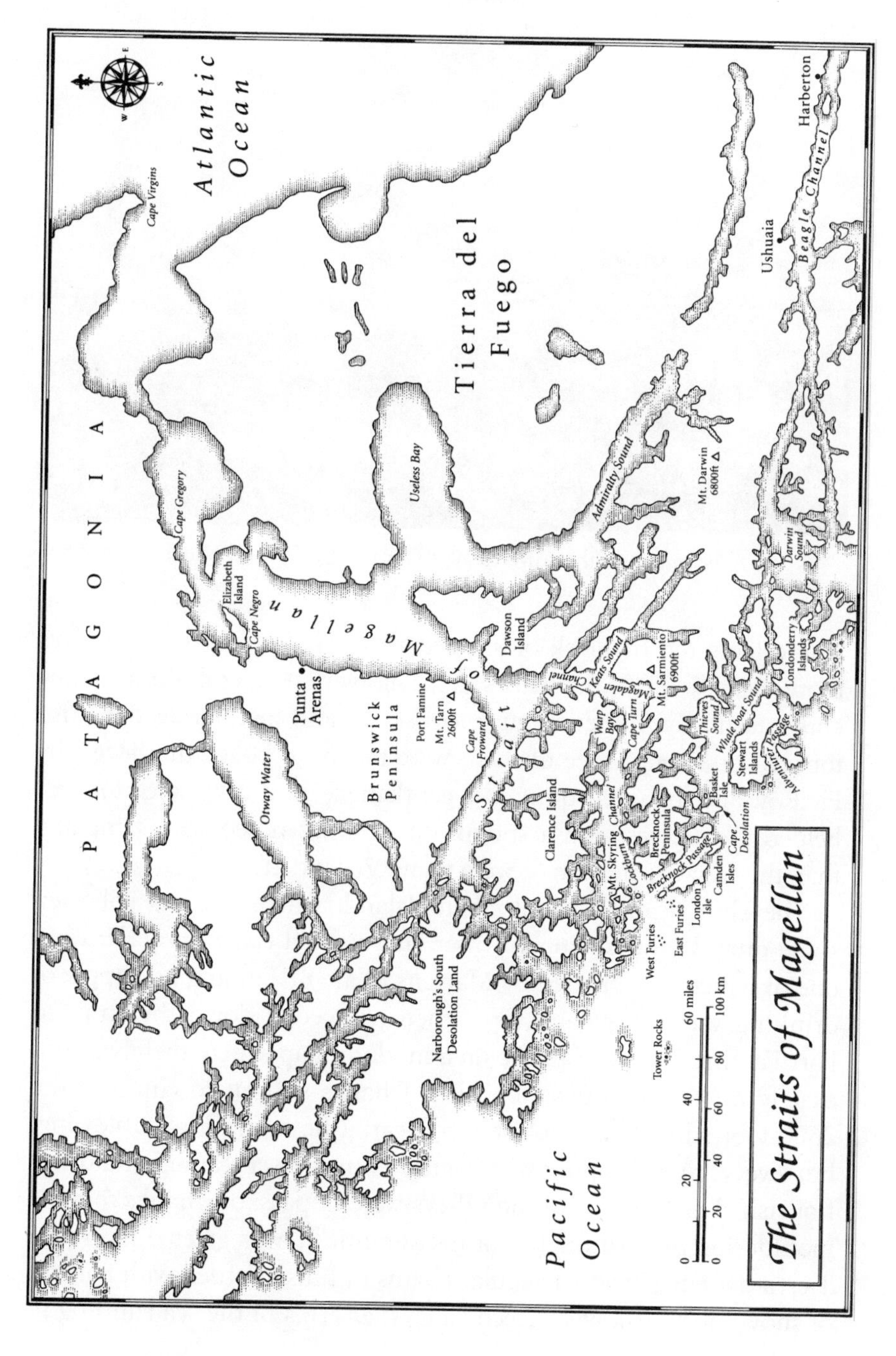
The Straits of Magellan
PATAGONIA
Atlantic Ocean
Pacific Ocean
Tierra del Fuego
Cape Virgins
Cape Gregory
Elizabeth Island
Cape Negro
Magellan
Strait of
Useless Bay
Punta Arenas
Brunswick Peninsula
Port Famine
Mt. Tarn 2600ft
Cape Froward
Otway Water
Dawson Island
Admiralty Sound
Mt. Darwin 6800ft
Keats Sound
Magdalen Channel
Mt. Sarmiento 6900ft
Cape Turn
Warp Bay
Clarence Island
Mt. Skyring
Cockburn Channel
Brecknock Peninsula
Brecknock Passage
Thieves Sound
Whale boat Sound
Stewart Islands
Adventure Passage
Basket Isle
Cape Desolation
Camden Isles
London Isle
East Furies
West Furies
Tower Rocks
Narborough's South Desolation Land
Londonderry Islands
Darwin Sound
Ushuaia
Beagle Channel
Harberton
0 20 40 60 miles
0 20 40 60 80 100 km

Adventure and the ruins of Port Famine, by Conrad Martens

whole faces red or black or spotted with white, painted themselves more heavily than the tribes fighting with General Rosas. They enthusiastically bartered guanaco skins and ostrich feathers for tobacco or sugar. Three of them were taken on board the *Beagle* by FitzRoy, and 'behaved quite like gentlemen, used a knife & fork, & helped themselves with a spoon'. But Charles noted with sympathy that 'they felt the motion & were therefore landed'.

The ship beat up to Elizabeth Island, where Charles collected some orchids and a slipper-flower later named *Calceolaria darwinii*, one of which was painted by Martens (Plate 8). Finding no supply of drinking water anywhere, they then proceeded down the strait to Port Famine, Puerto Hambre on a modern map, where there was not exactly a water shortage, for when Charles set out to climb to the 2600-foot summit of Mount Tarn nearby, he found a gale blowing, but 'everything dripping with water; even the very Fungi could not flourish'. The bottoms of the valleys were impassable thanks to huge mouldering tree trunks, but at the summit he was greeted by 'a true Tierra del Fuego view; irregular chains of hills, mottled with patches of snow; deep yellowish-green valleys; & arms of the sea running in all directions'. After collecting some shells in the rocks near the summit his descent was easier, 'for the weight of the body will force a passage through the underwood, & all the slips & falls are in the right direction'.

There was little to see of the remains of the settlement of Port Famine, abandoned by the Spaniards 240 years before. But its tragic history deserves to be recalled.

When Francis Drake, 'the scourge of God', launched his attacks on the rich and unprotected Spanish colonies on the west coast of South America in 1579, Philip II of Spain decided that measures would have to be taken to seal off the Straits of Magellan. In due course an armada of twenty-three ships carrying 3500 men under the command of the more politically than navigationally inclined Admiral Don Diego Flores de Valdez, with Don Pedro de Sarmiento y Gamboa, a famous navigator and historian of the Incas in Peru as his second-in-command, sailed from San Lúcar in September 1581. From the start the weather was against them, and only sixteen ships reached Rio de Janeiro, where they rested in winter quarters for six months. There followed more mishaps, although the weather was no worse than usual at the latitude of the River Plate; the basic problem was that the Spanish ships could not cope with wind and weather as well as those built 250 years later. In 1583 only five ships remained when at last the armada entered the Straits of Magellan. A storm then drove them back into the Atlantic, and after beating about for a month they again ran for Rio, though Don Diego kept on running all the way back to Spain.

Happy to be rid at last of his incompetent Admiral, Sarmiento found some relief ships in Rio, and entered the straits for the second time in December 1583, still with five ships. Yet again the weather intervened, and after one of the ships had been wrecked, and four hundred men and thirty women had been put ashore in an unsuccessful salvage attempt, Sarmiento's new second-in-command proceeded to decamp for Brazil with three of the largest and best-provisioned ships, leaving his Admiral with one small pinnace, the *Maria*, all the men and women on shore, and provisions for no

January 1834

Mount Sarmiento, by Conrad Martens

more than eight months. Still undeterred, and under fierce attack by the natives, the indomitable Sarmiento left Lieutenant Viedma, his last and only faithful second-in-command, to establish a settlement named La Ciudad del Nombre de Jesus in Possession Bay near Cape Virgins at the entrance to the straits, and marched inland with a hundred men to found the Ciudad del Rey Felipe. A strong fort was planned, to be protected by cannon taken from the foundered ship, but its construction was well beyond the capability of the remaining men, and after quelling a mutiny and executing the ringleaders, Sarmiento sailed back in the *Maria* in order to rejoin Viedma in Possession Bay. Once more storms intervened and blew Sarmiento out to sea, where all he could do after surviving the wrecking of the *Maria* in its turn was to try to seek further help in Brazil. This was refused him, and on his way back to Spain to make a plea to Philip II he was captured by three English ships off the Azores, taken to London, and brought before Elizabeth. For once his luck seemed to have changed, for, moved by his story, the Queen ordered his release. But no sooner had he been set free than he was captured by the French, and only ransomed and returned to Spain five years later.

The final twist in this tragic story was that Viedma, at the end of his

7. Patagonians at Gregory Bay, by Conrad Martens.

8. Caryophyllia on Elizabeth Island, by Conrad Martens.

9. Fuegians and the *Beagle* at Portrait Cove, by Conrad Martens.

10. Shooting guanacos on the banks of the Rio Santa Cruz, by Conrad Martens.

11. Mount Sarmiento from Warp Bay, by Conrad Martens.

12. Condors preying on a dead guanaco, by Conrad Martens.

13. Church in San Carlos, Chiloe, by Conrad Martens.

endurance at the outer settlement, Nombre de Jesus, marched inland with his men and women to rejoin the party at Rey Felipe. The supplies there were quite inadequate, and two hundred men had to be turned out to shift for themselves. When after the winter ended Viedma called the roll, only fifteen men and three women were still alive. In January 1587, Sir Thomas Cavendish anchored at Rey Felipe after taking a Spanish ship on which he found the last of Viedma's survivors. The settlement was deserted, and he named the place Port Famine because so many Spaniards had died there. It was the last good anchorage for vessels standing through the Straits of Magellan before reaching Cape Froward and turning north-west. The truly heroic role of Sarmiento in his efforts to obey his King by founding the Ciudad del Rey Felipe was commemorated by giving his name to the highest mountain in Tierra del Fuego, a snow-covered peak 7330 feet in height, situated fifty miles to the south of Port Famine.

It was once believed by some that a lost and wealthy city had been founded by Viedma's soldiers in the north of Tierra del Fuego. This never existed, but a legacy of the Spaniards far more valuable than gold came about when in 1535 Captain Pedro Mendoza released seven stallions and five mares at a spot near where Buenos Aires now stands. Three hundred years later their countless descendants had spread all over the southern part of the continent, providing not only General Rosas's Indian adversaries with their mounts, but also Charles with some of his.

CHAPTER 15

Goodbye to Jemmy Button and Tierra del Fuego

On 13 February the *Beagle* sailed out of the Straits of Magellan through the Narrows, and after spending ten days surveying the bleak eastern coast of Tierra del Fuego, rounded Cape St Diego, passed southwards through the Strait of Le Maire and on the evening of 24 February came to an anchor close to Woollaston Island, and at a comfortable distance to the north-west of Cape Horn.

The following morning, Charles went ashore to get to the tops of some of the hills, and on the way met a group of canoe-using Yahgan Indians, about whom he reflected:

> We pulled alongside a canoe with 6 Fuegians. I never saw more miserable creatures; stunted in their growth, their hideous faces bedaubed with white paint & quite naked. One full aged woman absolutely so, the rain & spray were dripping from her body; their red skins filthy & greasy, their hair entangled, their voices discordant, their gesticulation violent & without any dignity. Viewing such men, one can hardly make oneself believe that they are fellow creatures placed in the same world. I can scarcely imagine that there is any spectacle more interesting & worthy of reflection than one of these unbroken savages. It is a common subject of conjecture, what pleasure in life some of the less gifted animals can enjoy? How much more reasonably it may be asked with respect to these men.
>
> To look at the Wigwam; any little depression in the soil is chosen, over this a few rotten trunks of trees are placed & to windward some tufts of grass. Here 5 or 6 human beings, naked & uncovered from the

wind, rain & snow in this tempestuous climate, sleep on the wet ground, coiled up like animals. In the morning they rise to pick shell fish at low water; & the women winter & summer dive to collect sea eggs; such miserable food is eked out by tasteless berrys & Fungi. They are surrounded by hostile tribes speaking different dialects; & the cause of their warfare would appear to be the means of subsistence.

Their country is a broken mass of wild rocks, lofty hills & useless forests, & these are viewed through mists & endless storms. In search of food they move from spot to spot, & so steep is the coast, this must be done in wretched canoes. They cannot know the feeling of having a home – & still less that of domestic affection, without indeed that of a master to an abject laborious slave can be called so. How little can the higher powers of the mind come into play: what is there for imagination to paint, for reason to compare, for judgement to decide upon. To knock a limpet from the rock does not even require cunning, that lowest power of the mind. Their skill, like the instinct of animals is not improved by experience; the canoe, their most ingenious work, poor as it may be, we know has remained the same for the last 300 years. Although essentially the same creature, how little must the mind of one of these beings resemble that of an educated man. What a

Two Fuegians fishing from a canoe, by Martens

scale of improvement is comprehended between the faculties of a Fuegian savage & a Sir Isaac Newton.

Whence have these people come? Have they remained in the same state since the creation of the world? What could have tempted a tribe of men leaving the fine regions of the North to travel down the Cordilleras, the backbone of America, to invent & build canoes, & then to enter upon one of the most inhospitable countries in the world. Such & many other reflections must occupy the mind of every one who views one of these poor savages. At the same time, however, he may be aware that some of them are erroneous. There can be no reason for supposing the race of Fuegians are decreasing, we may therefore be sure he enjoys a sufficient share of happiness (whatever its kind may be) to render life worth having. Nature, by making habit omnipotent, has fitted the Fuegian to the climate & productions of his country.

It was not until 28 September 1838 that Charles recorded[103] having read the passage in Malthus on *Population* concerned with the effects of competition for food in man that triggered his arrival at the Theory of Natural Selection. But his sentence here about the cause of warfare between neighbouring tribes of Fuegians had surely anticipated the point.

Fuegians off Wollaston Island near Cape Horn, by Conrad Martens

March 1834

1 March was spent replenishing wood and water at a delightful cove where Charles was more complimentary about the natives than about the first canoe-full, and said that they 'were very quiet & civil & more amusing than any monkeys'. As in the previous year, their main employment was to beg for everything that they saw, saying repeatedly '*Yammer-scooner*.' Conrad Martens drew portraits of one of them with a canoe, and family groups in a wigwam and a canoe (see Plate 9). When FitzRoy used a combination of two of these drawings as the frontispiece for *Narrative* 2 he told the engraver, as may be seen overleaf, to add a dog in the foreground. The Fuegians valued their dogs very highly, and trained them to assist in the hunting of otters and guanaco, and even in fishing. They were certainly European in origin, and had presumably been introduced by the Spaniards when Buenos Aires was settled. But although Charles usually took a close interest in introduced animals, he surprisingly never mentioned the dogs of the Fuegians.

FitzRoy now made what Charles described as the bold attempt of beating against the westerly winds and proceeding up the Beagle Channel to Jemmy Button's country in Ponsonby Sound. On 5 March they anchored at Woollya, where the huts that had been built for York, Jemmy and Fuegia stood empty with their gardens trampled over, although some potatoes and turnips were there to be dug up. After a little while canoes approached which turned out to contain Jemmy, his brother and his wife. When the *Beagle* had left Jemmy he had been very fat, and very particular about his clothes and shoes. Now he was barely recognisable, thin and pale as he was, with long hair hanging over his shoulders and a scrap of blanket round his waist. But when FitzRoy had taken him on board and dressed him, he sat down to dinner below, using his knife and fork properly, speaking as much English as ever and behaving impeccably. He was very glad to see all his old friends, especially Mr Bynoe the surgeon and FitzRoy's Coxswain James Bennett, who had looked after him in England. FitzRoy thought Jemmy must have been ill, but he said he was 'hearty, sir, never better' and had not been ill, even for a day, was happy and contented, and had no wish whatever to change his way of

A Fuegian with his dog at Portrait Cove, by T. Landseer after Martens

Fuegians alongside the ship, by Conrad Martens

life. He had 'plenty fruits', 'plenty birdies', 'ten guanaco in snow time', and 'too much fish'. Surprisingly his companions, his brothers and their wives, and a good-looking young woman who turned out to be his wife, used broken English words in talking to him. He brought two finely dressed otter skins as gifts for FitzRoy and Bennett, two spearheads for Charles, and a bow and a quiverful of arrows for the schoolmaster at Walthamstow with whom he had once lodged. His wife was presented with shawls, handkerchiefs and a gold-laced cap, and laden with gifts the canoes in due course went ashore.

It was on this day that Conrad Martens made a drawing of the view in Ponsonby Sound, looking through the Murray Narrows towards the north bank of the Beagle Channel, with Hoste Island on the left. Charles bought the watercolour development of the scene from Martens for three guineas in Sydney, and it hung at Down House for many years.

The next morning, Jemmy had breakfast with FitzRoy, and explained what had happened soon after the *Beagle*'s departure in February 1833. A large party of the Ona tribe, Jemmy's hated 'Oens-men' from the east of Tierra del Fuego, had come overland to Woollya, and after stealing everything that was portable, had forced

Jemmy's people to take refuge on the small islands. York had valiantly defended his and Fuegia's hut, but in doing so he had become a menacing figure in Jemmy's eyes – 'York very much jaw', 'pick up big stones', 'all men afraid'. Later, having built a suspiciously large canoe, York had invited Jemmy and his family to 'look at his land', and had taken them westwards to Devil Island, at the junction of the north-west and south-west arms of the Beagle Channel. There they met York's brothers and other members of the Alacaloof tribe (spelled 'Alikhoolip' in FitzRoy's account) who inhabited the western end of Tierra del Fuego. While Jemmy was asleep, York and the Alacaloofs decamped with Fuegia in the middle of the night, taking all of his possessions in their big canoe, and leaving him defenceless and naked. He had returned to 'his own island' rather than Woollya, feeling that it offered him a safer refuge from the predatory Oens-men.

FitzRoy was left with a feeling that a beginning had been made towards civilising the Fuegians by gaining their confidence, but that his own efforts were on too small a scale. Nevertheless, as he wrote:

Fuegians at Jemmy Button's island, by Martens

March 1834

> I cannot help still hoping that some benefit, however slight, may result from the intercourse of these people, Jemmy, York and Fuegia, with other natives of Tierra del Fuego. Perhaps a shipwrecked seaman may hereafter receive help and kind treatment from Jemmy Button's children; prompted, as they can hardly fail to be, by the traditions they will have heard of men of other lands; and by an idea, however faint, of their duty to God as well as their neighbour.

Charles wrote that

> Every soul on board was as sorry to shake hands with poor Jemmy for the last time, as we were glad to have seen him. I hope & have little doubt he will be as happy as if he had never left his country; which is much more than I formerly thought. He lighted a farewell signal fire as the ship stood out of Ponsonby Sound, on her course to East Falkland Island.[104]

The troubled later history of Jemmy Button has been well related elsewhere,[105] and there is only space here to recall its main features. It was tied up closely with the Patagonian Missionary Society, founded in London in 1841 by a former naval officer, Allen Gardiner, an adventurous character with a penchant for carrying out missionary work in remote parts of the world. His first move to set up a mission in the Straits of Magellan was frustrated, as FitzRoy might have warned him, by the descent on him of hordes of Fuegians stealing everything on which they could lay their hands. He made an ill-prepared attempt ten years later to operate offshore in Tierra del Fuego from two twenty-six-foot metal launches, but this time he and his whole group were lost when their boats were wrecked in a storm. As a memorial to Gardiner, a ship named after him was built by subscription in England, and sailed out to the Falkland Islands in 1854 with the object of running the Patagonian Missionary Society from there. A suitable site for its headquarters was found at Keppel Island off the north coast of West Falkland Island. But the project ran quickly into trouble because of legal objections to its very existence raised by the

Governor of the islands in Port Stanley, and acute personality clashes between the Captain of the *Allen Gardiner*, William Parker Snow, and the other missionaries.

Early in November 1855, the *Allen Gardiner* sailed across to Tierra del Fuego, passed through the Murray Narrows and anchored in Ponsonby Sound. Nothing had been heard of Jemmy Button for twenty-two years, and Snow was astonished when a bedraggled little native, portly and naked, stood up in a canoe and shouted, 'Jam-mes Button, me; where's the ladder?' Learning that an 'Ingliss' lady was on board, Jemmy next demanded a pair of trousers, and had dinner with Captain and Mrs Snow. His English had been surprisingly well remembered, and it transpired that he had two wives and several children. In accordance with the policy of the Missionary Society for conversion of the Fuegians, Snow pressed Jemmy to return to Keppel Island with him, but received a firm refusal. In the morning, fighting off more Fuegians anxious as usual for presents, the *Allen Gardiner* sailed back to the Falklands.

During the next couple of years, the Missionary Society struggled on, facing dissension at home, official opposition in the Falklands about its aims, and the refusal of Snow to accept the policy of bringing Fuegians back to Keppel Island. Eventually the Revd George Packenham Despard was appointed as the first Mission Superintendent, Snow was replaced by a tough new Captain, Robert Fell, and the facilities at Keppel Island for housing some Fuegians were improved. In April 1858, Despard sailed over to Woollya, and with some difficulty persuaded Jemmy Button to bring his wife and three of his children back to Keppel Island for 'five moons'. The visit was not, however, a great success, and Jemmy quickly became a disenchantment because he was a good deal less than helpful in teaching the missionaries to speak his language, Yamana, and could never be persuaded to do any domestic work. At the end of November the family were taken back on the *Allen Gardiner* to Woollya, and after some trouble with bad weather leading to wild tantrums on Jemmy's part, and quarrels over the number of presents to be given to the Fuegians, the party disembarked.

Despard next had to find a fresh group of Fuegians to return to Keppel Island with him. Some volunteers were assembled, but there was immediately a quarrel with Jemmy, who knew that there was a large store of clothes in the hold, and demanded that they should be given to him and his relatives. Offering biscuits to the Fuegians in payment for their help in building a large hut on the shore to become a place of worship was not a success, because one biscuit for shouldering a load of timber up the ladder quickly became inadequate, and clothes had instead to be given. On Christmas Day the hut was finished, and fifty-one Fuegians attended a party at which four huge plum puddings and a can of treacle were served. Despard then read out a list of the Fuegians who had been selected to come to Keppel Island, all of whom shouted '*Ow-a*' ('yes'), and were told to board the ship two days later, when they would be 'purified and dressed'. However, there followed a complicated set of quarrels arising from a continued lack of mutual understanding between the parties, and it was not until New Year's Day that the *Allen Gardiner* finally set sail for the Falklands with nine Fuegians on board. Jemmy Button, rewarded for his help with the negotiations by a barrel of biscuits, some pieces of pork and twelve pounds of sugar, waved them goodbye.

The new group of Fuegians contained no one with any experience of life outside Tierra del Fuego, and no interpreter, so they took a little while to settle down on Keppel Island. Kept hard at work around the settlement, an activity for which in fact they had never signed up, under tight discipline as far as searching for their accustomed diet of shellfish from the beach and birds from the fields was concerned, and their behaviour strictly supervised, they nevertheless seemed reasonably contented. The men, except for the teenaged boys Ookoko and Lucca, remained somewhat resistant to education, and in April, after only three months at the settlement, were already beginning to look forward anxiously to their promised return home in October for the start of the 'wild bird egg' season.

On 28 September 1859 the *Allen Gardiner* was ready to depart for Woollya, and the Fuegians lined up on the jetty heavily laden with a variety of large bundles. Despard was suspicious at the extent of their baggage, and ordered his men to carry out a search, which quickly uncovered a large selection of valuable tools, Captain Fell's leggings, lines belonging to the missionaries' fishing net, the necks of geese and animal entrails, and pots and canisters of biscuits. This pilfered material was duly confiscated, leaving the natives deeply resentful at having been found out. Bad weather then kept the ship at anchor for a week, during which tempers were increasingly frayed, and there was a fight between the Fuegians and the ship's cook Alfred Coles. On 7 October the *Allen Gardiner* put in to Port Stanley to spend six days loading up with coal destined later for the settlement, despite the growing impatience of its unhappy passengers to be taken straight home. While the women and children were shown off in Port Stanley, and given presents of clothing, biscuits and sweets that doubtless were well received, there were disturbing reports of increasing friction between the Fuegian men and the missionaries, which was not dispelled by yet more stormy weather that in the end delayed arrival of the *Allen Gardiner* at Woollya until Wednesday, 2 November. Robert Fell recorded in his diary that Jemmy Button came on board that day naked and as wild-looking as ever, and was soon complaining at the shortage of presents given to him. The Fuegians then began to disembark, but Fell had been advised that members of the ship's crew had reported the loss of some of their possessions, and ordered another search. A fight at once broke out in which Fell himself was attacked, and the angry natives departed in their canoes, leaving behind them bundles in which the missing articles were indeed found.

During the next four days, the crew of the *Allen Gardiner* went ashore to complete the building of a house and to start digging a garden for it. There was no positive interference from the Fuegians, but their numbers increased steadily, and by Saturday more than seventy canoes had arrived. On Sunday, in accordance with Despard's instructions the ship's longboat was lowered into the water and all the crew except for the cook Alfred Coles went ashore in their smartest dress with 'Mission Yacht' emblazoned across their chests.

The singing of a hymn began, but then from the ship Coles saw a large group of natives armed with wooden clubs and stones rush in violent assault on the house, and soon eight men lay dead upon the shore. Picking up a gun and three loaves of bread, Coles rowed safely away from the massacre in a small boat, and managed to survive among the Yahgans in the neighbourhood of Woollya for the next four months, robbed of everything as usual, but treated with kindness by the women, by Ookoko and Lucca, and by Jemmy Button and his brothers who had been on Keppel Island.

By February 1860 it had become clear that some mishap must have overtaken the *Allen Gardiner*, and William Smyley, a tough American seafarer, was commissioned by Despard to search in the *Nancy* for the lost schooner. Arriving at Woollya on 1 March, Smyley found the *Allen Gardiner* at anchor, stripped of everything removable. A canoe containing a white man came alongside the *Nancy*, and Alfred Coles was taken on board with Jemmy Button hard on his heels. Jemmy at once headed for the galley to find some bread, and after giving Coles some food and wrapping him in a blanket, Smyley listened to his tragic story. At its conclusion, he hastily cut the painter of Jemmy's canoe, and sailed with his two witnesses back to Port Stanley.

The news of the massacre of the crew of the *Allen Gardiner* at Woollya, and the report of the Governor of the Falklands on his cross-questioning of Jemmy Button, aroused consternation and strong controversy both in Port Stanley and in London. Jemmy's attempts to shift all the blame on to the Oens-men were not accepted, neither could he be decisively cleared of any responsibility for what had happened. There was justified criticism of the manner in which the Missionary Society had mishandled the bringing of the groups of Fuegians to Keppel Island without any real appreciation of their reactions to the process. Following a formal inquiry into the affair held by the Society in 1861, Despard left the Falklands for England, and eventually settled in Australia. His post as Superintendent of the Patagonian Mission was successfully filled by the former Secretary of the Society, Waite Stirling, who was ably assisted by Despard's stepson, Thomas Bridges. When in February 1864 Stirling and Bridges paid a visit to Woollya in the *Allen Gardiner*, they were greeted with the news that a malignant disease – probably measles or some other virus – had recently stricken the people, and that Jemmy Button was dead.

Three years later the first steps were taken towards the establishment of a mission on the shores of the Beagle Channel facing the Murray Narrows at what in due course became the modern town of Ushuaia. On his return from England to take holy orders, Thomas Bridges landed with his wife and small daughter at Ushuaia on 1 October 1871. Here he worked with notable success for sixteen years, succeeding at long last in communicating properly with the Fuegians, not least through his production of an English–Yamana dictionary containing no fewer than thirty-two thousand words. Their language was a sophisticated one in which a single letter attached to a word could mean an additional sentence. One early misunderstanding cleared up by Bridges was that '*yammer-scooner*' did not convey a command to 'give me', but rather a request to 'be kind to me'. Another was that the Fuegians' reputation for cannibalism, which Charles was falsely accused of having spread, had resulted from FitzRoy's Indians having replied to questions in the manner that seemed to be expected of them, rather than with the truth, which was that they would never, under any circumstances, eat raw flesh. Lastly, Bridges found that Despard's claim that he had at least taught the Indians to praise God in their own tongue was not well based. Seeking for a translation for the word 'praise', Despard had complimented a Fuegian on his good linguistic progress and at the same time had tapped him on the shoulder. The consequence was that the first words in the service sung initially by the Fuegians were 'slap God'.

In recognition of his pioneer work, Thomas Bridges was in 1887 awarded by the government of Argentina a fifty-thousand-acre plot of land on the shores of the Beagle Channel some forty miles east of Ushuaia, where he could establish a farm and provide employment for a community of native Fuegians. This was named Estancia Harberton after his wife's family home in Devonshire, and here he lived until his death in 1898, after which the sons who had been born and brought up among the Fuegians continued in the tradition that he had established. One of them, E. Lucas Bridges, moved north to Viamonte on the east coast of Tierra del Fuego in order to live with the Ona Indians, and wrote a classical account[106] of the manner in which after the initially hamfisted attempts of the Patagonian Missionary Society to convert the Fuegians had ended so tragically, his father's sympathetic and understanding approach had proved to be so much more effective.

Of the four Fuegians taken back to England by FitzRoy in 1830, Fuegia Basket was the youngest and survived the longest. After she had helped York Minster to betray Jemmy Button in 1833, she had returned with York to the Alacaloofs in western Tierra del Fuego, and had two children by him. Some time after that, York had been killed in retaliation for murdering a man. In 1841, an English ship cruising in the Straits of Magellan had come across a native woman who asked, 'How do? I have been to Plymouth and London.' In 1842, Charles heard from Bartholomew Sulivan, by then a naval Captain engaged on a survey of the Falkland Islands, that a sealer in those same waters had met a native woman able to talk some English, and who must have been Fuegia Basket. She had lived, a term thought by Charles probably to bear a double interpretation, on board for a few days.

In 1873 a party of Yahgans from the western coast of Tierra del Fuego visited Thomas Bridges in Ushuaia, and among them was Fuegia Basket. He found her strong and well, as thickset as ever, and with many teeth missing from a mouth that was large even for a Fuegian. On her memory being tested, she recollected London and the wife of the schoolmaster Mr Jenkins, whose special charge she had been, and also Captain FitzRoy and the good ship *Beagle*. She recalled few more English words than 'knife', 'fork' and 'beads', and when Mrs Bridges produced her eldest children Mary and Despard, she seemed very pleased and said 'Little boy, little gal.' She had, however, forgotten the art of sitting on a chair, and when offered one she squatted beside it on the floor. She reckoned herself to be over fifty years of age, and was accompanied by a new husband aged eighteen,

February 1883

Fuegians spearing fish at the water's edge, by Conrad Martens

a discrepancy in ages considered by the Yahgans to be a sensible practical arrangement.

On 19 February 1883, Thomas Bridges met Fuegia Basket for the last time when making a visit to London Island near the Brecknock Peninsula. She was then sixty-two, in a weak condition, and nearing her end. She was back with her own people, two brothers with children of their own, and Bridges satisfied himself that she was unlikely to fall victim to the Fuegian practice of *tabacana*, hastening the end of aged relatives by strangulation. This was a form of euthanasia kindly meant, carried out openly with the approval of all except the unconscious victim. Might it conceivably have been the custom mentioned misleadingly by Jemmy Button to FitzRoy, which gave rise to the mistaken notion that cannibalism was sometimes practised in Tierra del Fuego?

In 1833 the total number of Fuegian Indians was probably around eight thousand. Seventy years later there were fewer than two hundred, mainly thanks to the ravages of measles and other virus diseases introduced by white people. Today it is doubtful whether any pure-blooded Indians remain in Tierra del Fuego.

CHAPTER 16

Second Visit to the Falkland Islands

On 10 March 1834, the *Beagle* arrived in the middle of the day at Berkeley Sound, having made a rapid passage from Ponsonby Sound scudding before a gale of wind. As FitzRoy had already been warned, they found a state of affairs rather different from that which they had left the previous year, although order had successfully been restored by the Royal Navy. Lieutenant Smith, who had been installed at Port Louis as acting Governor, came on board and related the tragic story.

On the morning of 26 August 1833, the Scottish sea-captain William Low, from whom FitzRoy had bought the *Adventure*, and who had been living at Port Louis in Lewis Vernet's house with their friend Mr Brisbane, set off with four men on a sealing excursion. The moment his boat was out of sight, the gauchos and three Indian prisoners who had been left by the Argentinians as a garrison for the settlement, and who had mutinied against their officer and murdered him, proceeded further with their revolt. They attacked and killed Mr Brisbane, Dickson, who had been left in charge of the store, and three other men. The rest of the settlers, thirteen men, three women and two children, stayed with the murderers for a couple of days, and then escaped to a small island in the sound where they kept themselves alive on birds' eggs and fish until the fortunate arrival on the scene about a month later of HMS *Challenger*, under the command of Captain Seymour, bringing Lieutenant Smith and four seamen who had volunteered from HMS *Tyne* to remain at the Falklands.

Finding that the *Challenger* was a well-armed ship of war, the murderous gauchos hastily left the ruins of the government house

Berkeley Sound and Port Louis in the Falklands, by Conrad Martens

where they had been camping out, and fled into the interior, taking all the fifty tame horses with them. Lieutenant Smith was at once dispatched with four midshipmen and a dozen marines, guided by a gaucho called Luna who was willing to turn King's evidence against his companions if rewarded with a pardon, to root out the murderers. But after four days of heavy rain, and having narrowly missed two chances of capturing some of them, Smith was obliged to return to Port Louis. Leaving the marines as additional protection for Smith, Captain Seymour had then sailed on in the *Challenger* to his next assignment. By the time the *Beagle* arrived, Lieutenant Smith had captured all but one of the murderers, and had restored the government house to some kind of order, with two cows that gave about two gallons of milk daily. Charles commented that, 'Surrounded as Mr Smith is with such a set of villains, he appears to be getting on with all his schemes admirably well.'

March 1834

FitzRoy wrote:

> When I visited the settlement it looked more melancholy than ever; and at two hundred yards' distance from the house in which he had lived, I found, to my horror, the feet of poor Brisbane protruding above the ground. So shallow was his grave that dogs had disturbed his mortal remains, and had fed upon the corpse. This was the fate of an honest, industrious and most faithful man: of a man who feared no danger, and despised hardships. He was murdered by villains, because he defended the property of his friend; he was mangled by them to satisfy their hellish spite; dragged by a lasso, at a horse's heels, away from the houses, and left to be eaten by dogs.

Early on the morning of 16 March, Charles set out to explore the country to the south-west of Berkeley Sound with six horses and the only two gauchos not directly concerned with the murder, though he suspected that they had had a very good idea of what was going to take place. He was able to make copious notes on the complicated geological structure of the country, and of the conspicuous stone-runs of angular quartzitic boulders that ran along many of the valleys and down their sides.[107] But the weather was atrocious, and there was little wildlife to be seen, except for a few snipe, some flocks of geese, herds of cattle, and far to the north a troop of wild horses. The gauchos exhibited their skill at capturing the cows and even a huge and aggressive bull with their lassos, and Charles was impressed by the manner in which they were then able quickly to make a hot fire with the bones of a newly killed bullock. The meat roasted with its skin ('*carne con cuero*') was delicious, and would have graced a London banquet. The place on the isthmus crossing the end of Choiseul Sound where they camped was near the settlements now named Darwin and Goose Green.

Back in Berkeley Sound, the *Adventure* sailed to continue her survey, but the *Beagle* was retained for ten days until a French whaler with a mutinous crew, a bad leak and a damaged rudder came in, and was with some difficulty bullied by FitzRoy into agreeing to transport to Rio de Janeiro the two murderers in irons, and the gaucho Luna who was due to turn King's evidence. However, the Frenchmen were spared in the end from carrying out this task by the well-timed arrival of a man-of-war to take charge of all the prisoners.

Charles passed his time 'very evenly. One day hammering the rocks; another pulling up the roots of the Kelp for the curious little Corallines which are attached to them.' He was one of the founding fathers of the important branch of biology named forty years later 'ecology', the science of the relation of living organisms with their environment. His claim to be so regarded rests on writings such as his splendid and far-seeing description of the flourishing community of different animals dependent on the huge growths of the kelp, a giant seaweed, in the southernmost oceans:

> The Zoology of the sea is I believe generally the same here as in Tierra del Fuego. Its main striking feature is the immense quantity & number of kinds of organic beings which are intimately connected

with the Kelp. This plant (the Fucus giganticus of Solander) is *universally* attached on rocks. From those which are awash at low water & those being in fathom water, it even *frequently* is attached to round stones lying in mud. From the degree to which these Southern lands are intersected by water, & the depth in which Kelp grows, the quantity may well be imagined, but not to a greater degree than it exists. I can only compare these great forests to terrestrial ones in the most teeming part of the Tropics; yet if the latter in any country were to be destroyed I do not believe that nearly the same number of animals would perish in them as would happen in the case of Kelp. All the fishing quadrupeds & birds (& man) haunt the beds, attracted by the infinite number of small fish which live amongst the leaves: (the *kinds* are not so very numerous, my specimens I believe show nearly all).

Amongst the invertebrates I will mention them in order of their importance. Crustaceæ [lobsters, crabs and shrimps] of evry order swarm, my collection gives no idea of them, especially the minute sorts. Encrusting Corallines & Clytia's [the moss animals, bryzoans] are excessively numerous. *Every leaf* (excepting those on the surface) is *white* with such Corallines or Corallinas [coralline algae] & Spirobeæ [fan worms] & compound Ascidiæ [tunicates or sea squirts]. Examining these with a strong microscope, infinite numbers of minute Crustaceæ will be seen. The number of compound & simple Ascidiæ is a very observable fact, as in a lesser degree are the Holuthuriæ [sea cucumbers] & Asterias [starfish]. On shaking the great entangled roots it is curious to see the heap of fish, shells, crabs, sea-eggs, cuttle fish, star fish, Planariæ [flatworms], Nereidæ [freely crawling polychaete worms], which fall out. This latter tribe I have much neglected. Among the Gasteropoda [snails], Pleurobranchus [a comb jelly] is common: but Trochus [top snails] & patelliform shells [limpets] abound on all the leaves. One single plant form is an immense & most interesting menagerie. If this fucus was to cease living, with it would go many: the Seals, the cormorants & certainly the small fish & then sooner or later the Fuegian man must follow. The greater number of the invertebrates would likewise perish, but how many it is hard to conjecture.[108]

March 1834

Here Charles had continued the studies that he had begun long ago with Robert Grant in the Firth of Forth, and had been pursuing further in Patagonia and Tierra del Fuego, by collecting many new species of bryozoan coating the kelp. Nor had he neglected the land animals, for he had gone carefully into the question of whether the black rabbits of the Falklands were an indigenous species, and had concluded correctly that they must have been introduced. He commented that on the other hand this was not the case for the Falkland foxes, whose excessive tameness was in danger of leading to their extermination by the settlers – as indeed did happen later – from what he still regarded in 1834 as a 'centre of creation'.

Another field of biology in which he was one of the pioneers was in behavioural studies, as he made clear by stressing the importance of behaviour in distinguishing between different birds of prey. He also recorded the amusing behaviour of a jackass penguin in crawling through the grass with the aid of its little wings so as to look like a four-footed animal, and then throwing its head back to bray exactly like a donkey. And a very curious bird was the logger-headed duck known as a steamer or racehorse from its extraordinary method of splashing and paddling along, and which croaked just like a tropical bullfrog.

CHAPTER 17

Ascent of the Rio Santa Cruz

The *Beagle* sailed back to the mainland, and on 13 April dropped her anchor at the mouth of the Rio Santa Cruz, not far from the Straits of Magellan. FitzRoy then prepared to lay the ship ashore for a tide, in order to check that the rock at Port Desire had not damaged the copper enough to permit the hungry worms of the Pacific to eat their way through an unprotected plank. It was found that several feet of the false keel under the fore-foot of the *Beagle* had indeed been knocked off, but this was quickly repaired. Charles collected some barnacles and small seaweeds attached to the hull, and Conrad Martens made his only drawing of the ship at close quarters. She was soon afloat again on her mooring.

FitzRoy now decided that the abortive attempt made in 1829 by Captain Pringle Stokes to go up the Rio Santa Cruz in too heavy a boat deserved to be repeated, for virtually nothing was known about the river. So on 18 April a party of twenty-five men commanded by FitzRoy set off in three whale-boats. They carried provisions for three weeks, and were well enough armed to defy a host of Indians, although in the event they saw no more of them than evidence of their having occasionally passed through the district. Except in the lowest and tidal region of the river, its width was between three and four hundred yards up to the highest point that was reached, and its fine milky blue water flowed at four to six knots per hour throughout its length. Against such a strong current, neither oars nor sails were effective. The boats were therefore tied astern of one another, with two men in each boat to steer, and were pulled upstream by a tow-rope to which a series of collars had been attached. The party,

April 1834

The *Beagle* laid ashore at Rio Santa Cruz; engraving after Martens

including *every one*, as Charles was careful to emphasise, was divided into two teams which pulled alternately for an hour and a half. The officers ate the same food, and slept in the same tent as their men. With a well organised routine, the average distance in a straight line that was covered each day was about ten miles, though the distance walked was often twice as much.

When he was not tugging on the tow rope, Charles was often on the heights above the river with another rifleman, on the lookout for guanacos, which apart from a good number of foxes, a few ostriches and an occasional puma, were the only large animals to be seen. He was presumably the frock-coated figure depicted by Conrad Martens on the afternoon of 2 May (Plate 10), with 'bright light catching on the distant mountains', near the point furthest west of their ascent up the river. The commonest of the birds that he collected were from flocks of sparrows (*Fringilla gayi*), finches (*Chlorospiza melanodera*) and meadowlarks (*Sturnella militaris*), with mockingbirds (*Mimus patagonicus*) singing amongst the spiny bushes. There were also some short-billed snipe (*Tinochorus rumicivorus*), three kinds of furnarii (*Eremobius phœnisurus, Opetiorhynchus vulgaris, O. patagonicus*), creepers (*Synallaxis*), a wren (*Troglodytes*), a warbler (*Muscisaxicola nigra*), condors and caracaras.

Charles's rifle would not, however, have been the only tool of which he made use on the Rio Santa Cruz, because one of his principal objectives was to continue his study of the elevation of the extensive gravel-capped plains of the eastern coast of Patagonia. He needed for this purpose his faithful geological hammer, and his two aneroid barometers for measurement of the height of the river above sea level, and of the height of the successive terraces and neighbouring hills above the level of the river. Unlimbering them at the river's mouth, he quickly found that the 430-feet plain at Port St Julian (see p.192) had continued without a break for the fifty miles to the northern side of the mouth of the Rio Santa Cruz, though its height there had fallen to 330 feet, rising to 355 feet at the southern side.

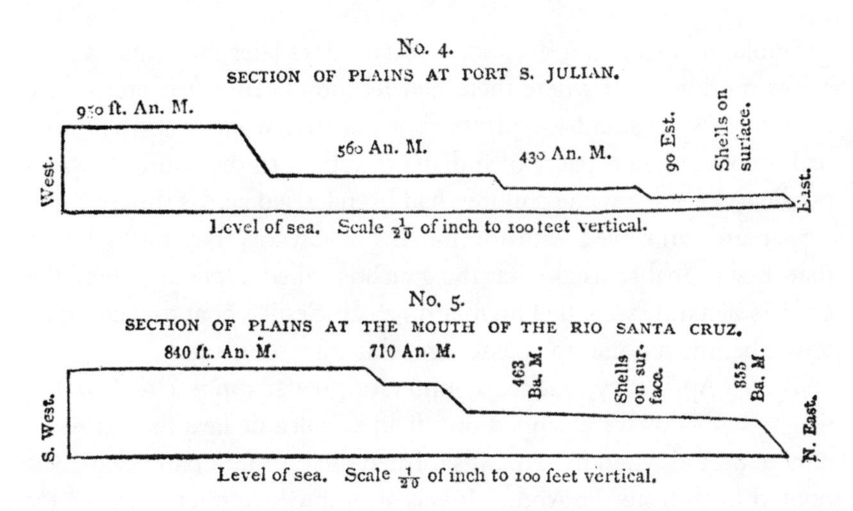

On the second day an old boathook bearing the King's mark was found at the edge of the river, and one of the boat's crew who had been on the *Beagle* in 1829 remembered its loss at that time. In the evening there was a sharp frost, and the dew froze so fast on the roof of the artificial horizon of FitzRoy's sextant when he attempted to observe the moon's meridian altitude, that its index error was shifted. Next morning Charles's fishing net, stowed in the bottom of the boat, was frozen too hard to be usable. The ascent continued through unexciting country, for 'the level plains of arid shingle support the same stunted & dwarf plants, in the valleys the same thorn-bearing bushes grow, & everywhere we see the same birds & insects'.

FitzRoy wrote in his diary:

> As we were going along the bank of the river, we saw some dark-coloured animals crossing the stream at a distance, but no one could guess what they were, until the foremost of them reached the shore, and rising upon his stilt-like legs showed himself an ostrich. Several of those birds were swimming across. I had no idea that so long-legged a bird, not web-footed, would, of its own accord, take to the water and cross a rapid stream. There were six or seven following one another.

Smoke was seen at a distance, and two days later they came across a soft muddy place where there had recently been a fire, and where there were very small footprints showing that women and children had accompanied a party of Indians travelling to the south. It was a puzzle to know how the children had been ferried across the river, but a Spaniard who lived with the Indians at Gregory Bay told Charles that they probably used what the gauchos called a '*pilota*', where the corners of a hide were tied up to form a coracle-like boat that could be towed behind a horse, the passengers clutching to its tail.

On 26 April they reached country of greater interest to Charles, where the plains were capped by a field of volcanic lava that 'at some remote period when these plains formed the bottom of an ocean, was poured forth from the Andes'. It was an impressive phenomenon, for further up the river the thickness of the lava field was over three hundred feet, although it was still far from its source. Moreover, the most southerly volcanic rocks previously known in the Andes were three hundred miles to the north. FitzRoy's description of the 'Basalt Glen' painted by Conrad Martens was:

> A wild-looking ravine bounded by black lava cliffs. A stream of excellent water winds through amongst the long grass, and a kind of jungle

at the bottom. Lions (pumas) shelter in it, as the recently torn remains of guanacoes showed us. Condors inhabit the cliffs. Imperfect columns of a basaltic nature give to a rocky height the semblance of an old castle. It is a scene of wild loneliness fit to be the breeding-place of lions.[109]

When Martens' 'Basalt Glen' was used to illustrate FitzRoy's published account of the voyage, the pumas were duly inserted by the engraver.

Charles recorded the next day that:

I shot a condor, it measured from tip to tip of wing 8 & ½ feet, from beak to tail 4 feet. They are magnificent birds; when seated on a pinnacle over some steep precipice, sultan-like they view the plains beneath them. I believe these birds are never found excepting where there are perpendicular cliffs. Further up the river where the lava is 8 & 900 feet above the bed of the river, I found a regular breeding place; it was a fine sight to see between ten & twenty of these Condors start heavily from their resting spot & then wheel away in majestic circles.[110]

Basalt Glen on Rio Santa Cruz; engraving by T. Landseer after Martens

The banks of the Rio Santa Cruz, by Martens

But FitzRoy had a less complimentary comment to make on the condors:

> Two guanacoes were shot by Mr Darwin and Mr Stokes. They covered them up with bushes, and hastened to the boats to ask for assistance. Some of our party went with them to bring in the animals, but the condors had eaten every morsel of the flesh of one animal. The other they found untouched, and brought to the boats. Four hours had sufficed to the condors for cleaning every bone. When our party reached the spot, several of those great birds were so heavily laden that they could hardly hop away from the place.[111]

Conrad Martens meanwhile confined himself to painting a picture of condors preying on a dead guanaco (Plate 12).

The ascent continued steadily, and soon they 'hailed with joy the snowy summits of the Cordilleras, as they were seen occasionally peeping through their dusky envelope of clouds'. Conrad Martens made several drawings of the column of men hauling the boat, a devel-

Pulling the boats up the Rio Santa Cruz with a distant view of the Andes, by Martens

opment of one of which he sold in Sydney to Charles for three guineas, to hang later at Down House. But the river was very tortuous, and in many places there were huge blocks of slate and granite that must have been washed down from the Andes. On 4 May the long north and south range of the mountains with steep and pointed cones clad with snow was fully in view, but FitzRoy decided that since rations were running short, the boats could be taken no further. Accompanied by a large party he set off to walk a few miles more to the west, but after crossing the desert plain at the head of the valley of the Santa Cruz, the bases of the mountains were still not visible. At this point they were about 140 miles from the Atlantic, and perhaps sixty from the nearest inlet of the Pacific. And they were probably close to Lake Argentino, which connects Lakes Viedma and San Martin.

Charles had therefore completed his section across the great Patagonian Tertiary formation, and could draw a plan of the terraces on either side of the Santa Cruz valley. From the facts that the plain at the head of the valley was tolerably level, but with many sand-dunes on it like those on the coast of a sea, and escarpments widening like two great bays facing the mountains, he concluded that not too long ago the course of the Rio Santa Cruz had formed a strait crossing the continent, with two similar straits to its south cutting through Tierra del Fuego. This was supported by his finding a very old and worn shell of a currently existing species in the bed of the river at an altitude of 410 feet. More such shells were found lower down the valley, still 105 miles from the Atlantic.

Before sunrise on 5 May, they began their descent of the Santa Cruz, and driven by the current, in three days had equalled fifteen days of strenuous tracking in the other direction, and were back at the head of the tidal stretch of the river. Most of the party felt that rather little had been achieved, though they were probably the first Europeans to have seen the southernmost part of the Cordilleras, and the sportsmen had shot ten guanacos, several condors, a very large puma, and a wild cat. Not least, Charles had his geological section safely in the bag.

CHAPTER 18

Through the Straits of Magellan to Valparaiso

On 12 May the *Beagle* put to sea, and sailed down to the entrance of the Straits of Magellan, where she was rejoined by the *Adventure*, whose surveys in the Falklands were complete. Charles was depressed to find how the constant dry weather in Patagonia was replaced 120 miles to the south by as constant clouds, rain, hail, snow and wind; but in ten fathoms of water off Cape Virgins he was cheered up by finding a new species of bryozoan, red in colour, that possessed a novel type of organ in continual movement like the nodding of the vulture's beak capsules that he had observed once before (see p.100).

He gave a graphic description of this important discovery in his notes:

> The extraordinary organ the bristle is drawn up at (H) [see drawing]. It is about 1/20th inch long; arched, serrated on outer margin, supported on basal concave side by ridge: connected to its cell by a hinge, & has a membranous appendage or vessel (K) leading into cell or polypier:* These bristles stand out at right angles, on the outer edge of

* Bryozoans are, like hydrozoans such as jellyfish, tiny animals that live in large colonies. They consist of a great many separate cells or zooids, among which the autozooids that take care of feeding and digestion have a retractable polyp crowned with tentacles or setæ for gathering food particles. A heterozooid carrying a bristle mounted on muscles that control its movements is known as a vibraculum, while a heterozooid equipped with a vulture's beak capsule is an avicularium.

the cells: I was perfectly astonished when I first saw every bristle in one branch suddenly with great rapidity collapse together on the branch, & one after the other (apparently by their elasticity) regain their places. Directly other branches commenced, till the whole Coralline, driven by these long oars, started from side to side on the object glass. The motion of the Coralline & the setæ was visible to the naked eye: a bit of Coralline being dried on blotting paper, yet for a short time in the air moved its bristles. Irritation would almost always cause the movement in a branch, & when one branch began, *generally* the others followed. They likewise moved (even after being kept a day) spontaneously. Any one bristle being forcibly moved, re-took its position & would move by itself. The Coralline placed on its face must entangle the bristles, they often made violent efforts to free themselves: Generally the bristles on each side of a branch moved together, but one side sometimes would remain collapsed for a longer time than another: this generally was only a second or two. The bristle was never depressed much below the rt angle: when collapsed on branch the concave & smooth side was on the branch, & in the extreme cells, the bristles were mingled with the spines. A bristle, when detached, never moved, the power must lie in the hinge. Polypus sometimes protrudes its arms during the motion of the bristles. The above facts are very important as showing a co-sensation & a co-will over whole Coralline. I think the bristle is not directly connected with the Polypus. What is its use? As the serrated edge is external it cannot be to collect food: as the motion is most vigorous & necessarily first towards the branch, it cannot be to drive away enemies or impurities. The motion must cause currents. Does it give warning to the Polypus that danger is at hand? When collapsed it does *not* protect mouth of cell.[112]

The specimen whose behaviour and structure Charles described so elegantly was identified some years later as a bryozoan belonging to the species *Caberea mimima*, and it may still be seen, its bristles immobilised by death, in its bottle of alcohol in the museum of the Department of Zoology at Cambridge University. The movable

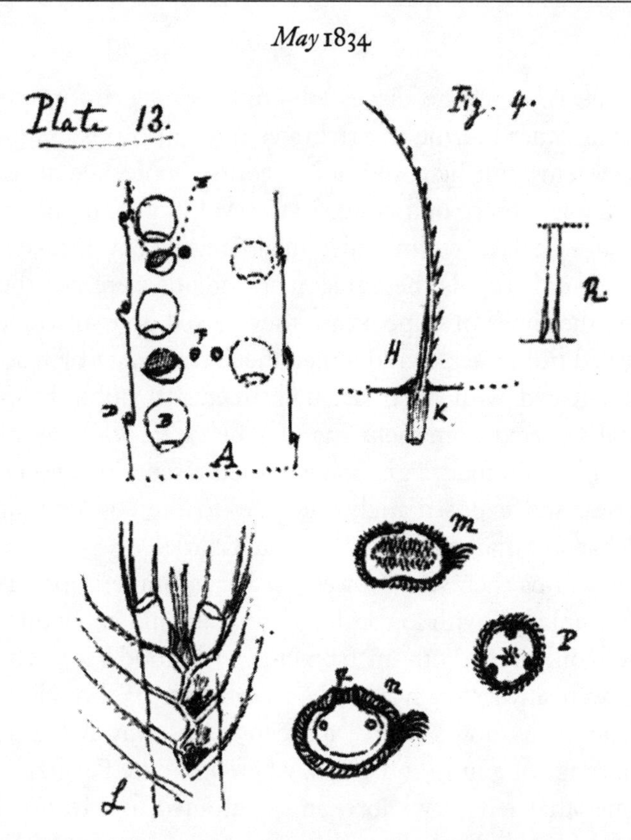

Drawings of the structure of the bryozoan *Caberea minima* with its moveable vibraculum

bristle is now known as a 'vibraculum', and Charles was correct in suggesting that the power that moves it is in what he called the hinge, for beneath it there are located tiny gyrator, abductor and adductor muscles. He was also right to reject the idea that the vibracula might serve to drive away enemies, though it is now accepted that in some species they do have a role in preventing rubbish from accumulating on the backs of the colony, while in others they may operate by generating external water currents and guiding food particles to the autozooids. His most perceptive observation was that the whole animal must have a 'co-sensation' and a 'co-will', or in other words a nervous system capable of receiving information about the state of its surroundings, and initiating an appropriate response of the vibracula.

Later in the voyage[113] he discussed similar evidence for supposing that various other marine invertebrates must also possess primitive nervous systems, but he could not of course appreciate at that time the crucial importance of this step in the evolution of higher animals.

The *Beagle* and *Adventure* called in at Gregory Bay to take in some water. Their old friends the Patagonian Indians were not there, but passing to the south of Cape Negro they picked up two seamen who had deserted from a sealer and joined the Patagonians, by whom they had been treated with 'their usual disinterested noble hospitality'. Accidentally parted from them, the two men were walking down the coast to find a ship that might come to their rescue, living on mussels and berries, and exposed night and day to the constant rain and snow. 'What will man not endure!' wrote Charles.

The two ships then spent a week at Port Famine in order to rate their chronometers, with fog added to the rain and snow. Since there were men on shore with instruments and goods for which the Fuegians were always ready to make a grab, FitzRoy thought it advisable to mount a show of force to keep them away. A big gun was fired at a range of a mile and a half, whereupon the Fuegians replied with stones that fell very short, and continued to advance. A boat was sent to fire musket balls wide of them: they replied with arrows which too fell short, and when laughed at, they 'shook their very mantles with fury'. Only when they saw the musket balls strike and cut the trees did they fall back. The next day they built a barricade of rotten trees, and with a good supply of stones for their slings, advanced still closer. Every time a musket was pointed at them, they pointed an arrow in reply. Since, however, the confrontation could only end in useless bloodshed, the sailors were finally ordered to withdraw.

The ships sailed out of the Straits of Magellan by a route newly discovered by William Low, taking them on 9 June down the Magdalen Channel past Mount Sarmiento. Conrad Martens made a superb drawing later developed as a watercolour (Plate 11), beneath which he wrote: 'The grand glacier, Mount Sarmiento. The mountain rises to about 3 times the height here seen, but all is here hidden by

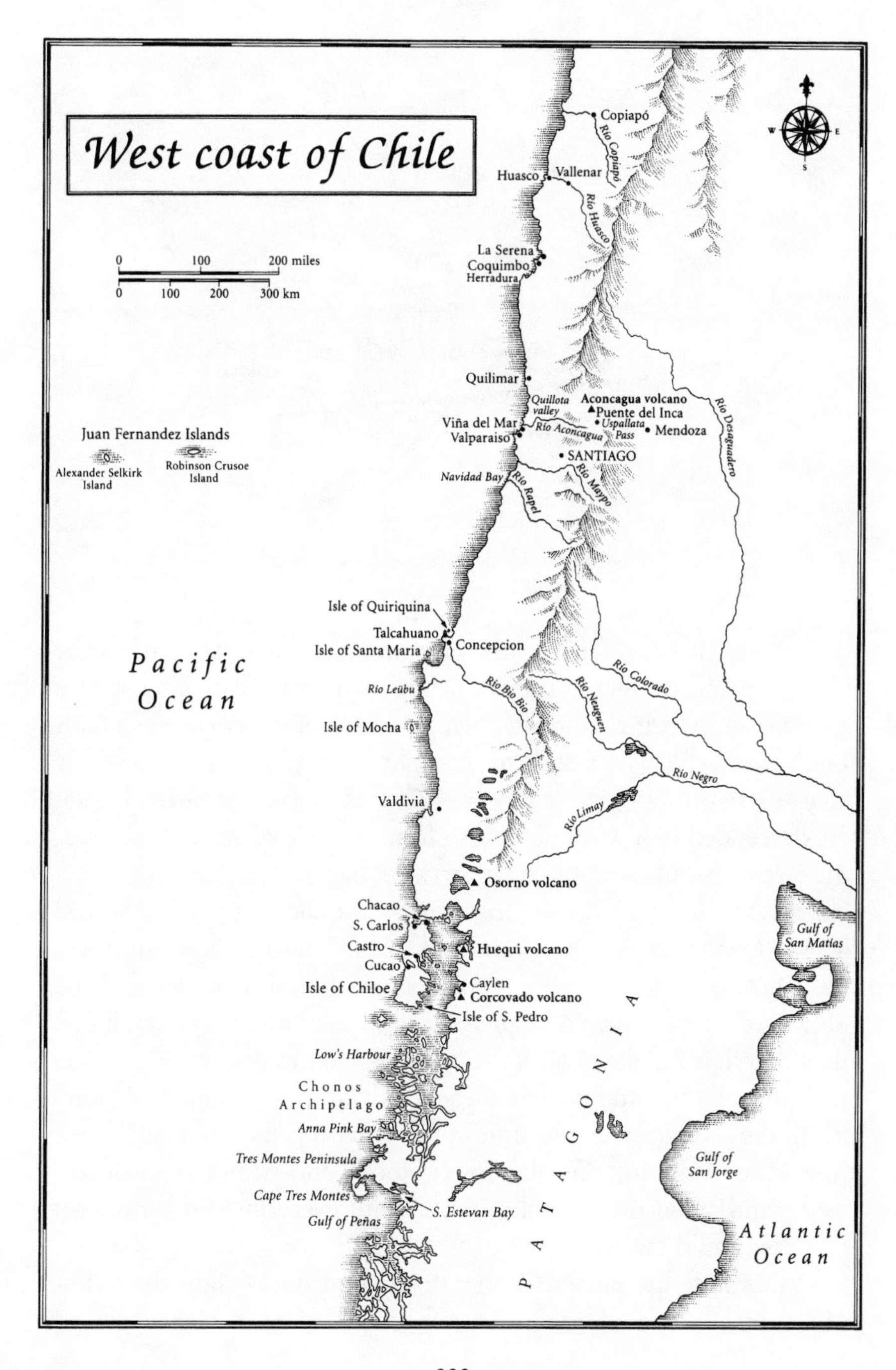
West coast of Chile
0 100 200 miles
0 100 200 300 km
Copiapó
Río Copiapó
Huasco
Vallenar
Río Huasco
La Serena
Coquimbo
Herradura
Quilimar
Quillota valley
Aconcagua volcano
Puente del Inca
Uspallata Pass
Mendoza
Río Desaguadero
Viña del Mar
Valparaiso
Río Aconcagua
SANTIAGO
Navidad Bay
Río Rapel
Río Maypo
Juan Fernandez Islands
Alexander Selkirk Island
Robinson Crusoe Island
Isle of Quiriquina
Talcahuano
Isle of Santa Maria
Concepcion
Pacific Ocean
Río Leübu
Río Bio Bio
Río Neuqen
Río Colorado
Isle of Mocha
Río Negro
Valdivia
Río Limay
Osorno volcano
Chacao
S. Carlos
Castro
Cucao
Huequi volcano
Isle of Chiloe
Caylen
Corcovado volcano
Isle of S. Pedro
Gulf of San Matías
Low's Harbour
Chonos Archipelago
Anna Pink Bay
PATAGONIA
Tres Montes Peninsula
Gulf of San Jorge
Cape Tres Montes
S. Estevan Bay
Gulf of Peñas
Atlantic Ocean

A woman of Chiloe weaving, by Martens

dark misty clouds – a faint sunny gleam lights the upper part of the glacier, giving its snowy surface a tinge which appears almost of a rose colour by being contrasted with the blue of its icy crags – a faint rainbow was likewise visible to the right of the glacier, but the whole was otherwise very grey & gloomy.' Charles noted that 'Several glaciers descended in a winding course from the pile of snow to the sea, they may be likened to great frozen Niagaras, & perhaps these cataracts of ice are fully as beautiful as the moving ones of water.' The following day the *Beagle* turned to windward to beat out down the Cockburn Channel, but finding at dusk that at its far end the only cove where it would be possible to anchor was a very small one that they left for the *Adventure*, were obliged to spend an anxious night beating to and fro in an area of only four square miles. At daybreak the *Adventure* emerged from her cove. Together they sailed out past Mount Skyring and the Tower Rocks, and with a fresh northwest wind stood out into the Pacific with every inch of canvas set that they could carry.

On 27 June the purser George Rowlett, at thirty-eight the oldest officer on board, died from a complication of diseases. He had served on the *Adventure* during the previous voyage, and was an old and respected friend of everyone. On the following day the funeral service was read by the Captain on the quarterdeck, and Rowlett's body was committed to the seaman's grave, that 'ever-changing and mysterious main'.

Later that day the *Beagle* anchored at San Carlos, nowadays known as Ancud, at the northern end of the island of Chiloe, off the south coast of Chile. It had originally been FitzRoy's intention to sail directly to refit at Coquimbo, a port 850 miles further to the north. But an endless succession of northerly gales obliged him first to think of proceeding to Concepción, and finally to come in at San Carlos. Here they remained for ten days, while the main-boom of the *Adventure* was repaired, fresh provisions and good rates for the chronometers were obtained, and the men were refreshed after having endured the rigours of the Pacific in such an unfriendly mood.

Charles took some walks around the island, and stayed for a few days in San Carlos, during one or two of which the weather, to the astonishment of the inhabitants, was 'very fine', and a glimpse was caught of the volcano of Osorno. Usually the rain was incessant, and

Breast Plowing at Chiloe, engraving after P.P. King and Conrad Martens

Maria Mercedes and Don Manuel de Chiloe, by Conrad Martens

nothing could be seen of the Cordilleras. Charles commented on the local population that:

> They all appear to have a great mixture of Indian blood, & widely differ from almost every other set of Spaniards in not being Gauchos. The country is so thickly wooded that neither horses or cattle seem to increase much. Potatoes & pigs & fish are the main articles of food; the obtaining these requires labour, & has consequently induced a different set of manners from what is found in other parts of S. America.[114]

Chiloe enjoyed a uniform temperature and an atmosphere saturated with moisture. Its volcanic soil was very fertile, and the densely growing evergreen trees of the temperate rainforest, clad with an arborescent sort of grass jointed like a bamboo up to a height of thirty or forty feet, and surrounded by large ferns, gave the woods a truly tropical look. The only road in the interior of the island, connecting San Carlos with the former capital Castro, had been made by the old Spaniards from trunks of trees roughly squared and placed

Cottages on Chiloe, black from the smoke issuing from their roofs, by Martens

side by side. Otherwise, the country was only inhabited around its coasts, with roads so bad that boats were needed for communication. San Carlos was a straggling, dirty little village, in which the houses were entirely built of planks split from the trunks of 'alerce' trees (a member of the cypress family later classified by Joseph Hooker as *Fitzroya patagonica*) growing in the region and on the slopes of the Andes. These planks were the staple export of the island, along with potatoes and hams, so that sadly the forests described by Charles have long since been cut down.

Conrad Martens made drawings of the landing place at Punta Arenas, the dense forests, the cottages in San Carlos described as 'black with smoke. The houses have no chimneys, and the smoke issues from all parts of the roof', and the women weaving.

CHAPTER 19

Valparaiso and Santiago

Having made a slow passage from San Carlos, the *Beagle* and *Adventure* anchored at Valparaiso late at night on 23 July. FitzRoy found himself so much in arrears in the computation and drawing of his charts that he decided to set immediately to work on them in Valparaiso. Wickham, who spoke Spanish and had been to Chile before, was accordingly sent to Santiago to show Beaufort's instructions to the Chilean authorities, and obtain their approval for FitzRoy's examination of the shores under their jurisdiction. This was cordially granted, and the *Beagle* received every official attention during the remainder of her visit to Chile. Messrs Stokes, King, Usborne and FitzRoy, whose occupation would be sedentary, and who would require more room as well as light and quiet than they could obtain on board the ship, took up their quarters on shore, while the other officers attended to the *Beagle*'s refit and provisioning.

Charles explored Valparaiso, where after Tierro del Fuego and Chiloe he felt the climate quite delicious, 'the sky so clear & blue, the air so dry & the sun so bright, that all nature seemed sparkling with life'. After Chile had gained independence in 1818, Valparaiso became an important port of call for ships rounding Cape Horn before the opening of the Panama Canal, and a number of merchants settled there, trading in British goods and the export of minerals, in particular copper ores. The town was built at the foot of a range of hills 1600 feet high, whose 'surface is worn into numberless little ravines, which exposes a singularly bright red soil between patches of light green grass & low shrubs'. The harbour was not large, and the shipping was crowded closely together. To the north-east there were some fine

A scene near Valparaiso, by Conrad Martens

glimpses of the Andes, the volcano of Aconcagua being especially beautiful. In his diary Charles wrote:

> The town of Valparaiso is from its local situation a long straggling place; wherever a little valley comes down to the beach the houses are piled up on each other, otherwise it consists of one street running parallel to the coast. We all, on board, have been much struck by the great superiority in the English residents over other towns in S. America. Already I have met with several people who have read works on geology & other branches of science, & actually take interest in subjects no way connected with bales of goods & pounds shillings & pence. It was as surprising as pleasant to be asked what I thought of Lyell's Geology. Moreover every one seems inclined to be very friendly to us, & all hands expect to spend the two ensuing months very pleasantly.[115]

Charles too moved to a residence in the town with Richard Corfield, one of the large number of English merchants then living in

The Quebrada Elias, Valparaiso, 'sketched from the balcony of the house in which resided Conrad Martens'

Valparaiso, who had overlapped with him at Shrewsbury School in 1818–19. Corfield had an attractive single-storey house with all the rooms opening into a quadrangle, and a garden that received a small stream of water from the slopes of the Andes for six hours every week. It was situated in the Almendral, a suburb built on a sandy plain that had once been a sea beach. Conrad Martens went to live in the Quebrada Elias with another highly competent topographical artist, a German called Johann Moritz Rugendas, and together they painted the view across the Bay of Valparaiso towards Viña del Mar (Plate 14), while Martens also sketched the shipping at the quayside and in the bay, and scenes in the town.

Charles took several long walks in the country, on which he had a revealing comment to make:

> The vegetation here has a peculiar aspect; this is owing to the number & variety of bushes which seem to supply the place of plants; many of them bear very pretty flowers & very commonly the whole shrub has

The bay of Valparaiso seen from a promontory, by Martens

a strong resinous or aromatic smell. In climbing amongst the hills one's hands & even clothes become strongly scented. With this sort of vegetation I am surprised to find that insects are far from common; indeed this scarcity holds good to some of the higher orders of animals; there are very few quadrupeds, & birds are not very plentiful. I have already found beds of recent shells yet retaining their colours at an elevation of 1300 feet; & beneath this level the country is strewed with them. It seems not a very improbable conjecture that the want of animals may be owing to none having been created since this country was raised from the sea.[116]

In both the 1839 and 1845 editions of the *Journal of Researches*, the evidence for the relatively recent elevation of the land was described in similar terms, but the final sentence was omitted. Charles was evidently determined to conceal for the time being the unorthodox directions in which his thoughts about the generation of new species were already developing.

After a brief expedition up the coast to visit the hacienda of

Quintero, the estate at one time owned by the Scottish Admiral Lord Cochrane, who had commanded the Chilean Navy in 1818–22 during the struggle of the country for independence, Charles rode off to visit the beautiful valley of Quillota, crowded with fruit orchards and olive groves. From there he ascended to the summit of the Campana or Bell Mountain, where he stayed a whole day, now commemorated by a plaque, and 'never enjoyed one more throughily, for who can avoid admiring the wonderful force which has upheaved these mountains, & even more so the countless ages which it must have required to have broken through, removed & levelled whole masses of them'. He spent the evening talking round the fire with the two Chilean *guassos* who had accompanied him, and recorded an illuminating comparison between them and the Argentinian gauchos with whom he had ridden so far across the Pampas:

> Chili is the more civilized of the two countries, & the inhabitants in consequence have lost much individual character. Gradations in rank are much more strongly marked; the Huasso* does not by any means consider every man his equal; I was quite surprised to find my companions did not like to eat at the same time as myself. This is a necessary consequence of the existence of an aristocracy of wealth; it is said that some few of the greater land owners possess from five to ten thousand pounds sterling per annum. This is an inequality of riches which I believe is not met with in any of the cattle-breeding countries to the eastward of the Andes. A traveller by no means here meets that unbounded hospitality which refuses all payment but yet is so kindly offered, that no scruples can be raised in accepting it. Almost every house in Chili will receive you for the night, but then a trifle is expected to be given in the morning; even a rich man will accept of two or three shillings. The Gaucho, although he may be a cut-throat, is a gentleman; the Huasso is in few respects better, but at the same time is a vulgar, ordinary fellow. The two men although employed much in the same manner are different in their habits & clothes; and

* The Spanish pronunciation of *guasso*.

> the peculiarities of each are universal in their respective countries. The Gaucho seems part of his horse & scorns to exert himself excepting when he is on its back. The Huasso can be hired to work as a labourer in the fields. The former lives entirely on animal food, the latter nearly as much on vegetable. We do not here see the white boots, the broard drawers & scarlet Chilipa, the picturesque costume of the Pampas; here common trowsers are protected by black & green worsted leggings; the poncho however is common to both. The chief pride of the Huasso lies in his spurs, these are absurdly large; I measured one that was six inches in the *diameter* of the rowel, & the rowel itself contained upwards of thirty points: the stirrups are on the same scale, each one consisting of a square carved block of wood, hollowed out, yet weighing three or four pounds. The huasso is perhaps more expert with the lazo than the gaucho, but from the nature of the country does not know the use of the bolas.[117]

Charles next visited the copper mines of Jajuel, high in the Andes. They were managed by a shrewd though uneducated Cornish miner, who had married a Spanish woman, and had no intention of returning to England, despite his intense admiration for the Cornish mines. He asked, now that George Rex was dead, how many of Rex's family were yet alive? Charles noted that this Rex was certainly a relation of Finis who wrote all the books. The copper that was mined was all shipped to Swansea for smelting, so the mine was a relatively peaceful place, with no great steam-engines to disturb the quiet of the surroundings. Charles enjoyed scrambling in the mountains, in which there was a chaos of huge angular or rounded fragments of a porphyritic claystone, traversed by once melted layers of greenstone,* and the strata were highly inclined and often vertical. The ground was dotted with numerous large cacti.

* Greenstone is a general term for any low-grade igneous rock rich in iron, magnesium and chlorite whose original structure has been reorganised by pressure or temperature.

August 1834

After leaving Jajuel, they crossed the basin of San Felipe and followed the main road into Santiago, the capital of Chile, built on the same model as Buenos Aires, but not as fine or large. Here Charles joined Corfield, 'who is going up to admire the beauties of Nature, in the form of Signoritas, whilst I hope to admire them amongst the Andes', and they stayed at the Fonda Inglese. On 28 August he sat down to write to FitzRoy in Valparaiso:

My dear Fitz Roy,

I arrived at this gay city late last night, and am now most comfortably established at an English Hotel. My little circuit by Quillota and Aconcagua was exceedingly pleasant. The difficulty in ascending the Campana is most absurdly exaggerated. We rode up 5/6ths of the height to a spring called the Aqua del Guanaco & there bivouacked for two nights in a beautiful little arbor of Bamboos. I spent one whole day on the very summit, the view is not so picturesque as interesting from giving so excellent a plan of the whole country from the Andes to the sea – I do not think I ever more thoroughly [his usual misspelling has been corrected by a copyist] enjoyed a days rambling. From Quillota I went to some copper mines beyond Aconcagua situated in a ravine in the Cordilleras. The major domo is a good simple hearted Cornish Miner – It would do Sulivan good to hear his constant exclamation "As for London – what is London? They can do anything in my country". I enjoyed climbing about the mountains to my hearts content; the snow however at present quite prevents the reaching any elevation – On Monday my Cornish friend and myself narrowly escaped being snowed in. We were involved in a multitude of snow banks, and a few hours afterwards there was a heavy snowstorm which would have completely puzzled us – The next morning I started for this place. I never saw anything so gloriously beautiful as the view of mountains with their fresh and brilliant clothing of snow – Altogether I am delighted with the Country of Chile – The country Chilenos themselves appear to me a very uninteresting race of people – They have lost much individual character in an *essay towards an approximation* to civilization.

August 1834

My ride has enabled me to understand a little of the Geology – there is nothing of particular interest – all the rocks have been frizzled melted and bedevilled in every possible fashion. But here also the "confounded Frenchmen" have been at work. A M. Gay has given me today a copy of a paper with some interesting details about the Geology of this province, published by himself in the Annales des Sciences – I have been very busy all day, and have seen a host of people. I called on Col. Walpole, but he was in bed – or said so. Corfield took me to dine with a Mr Kennedy, who talks much about the Adventure & Beagle; he says he saw you at Chiloe – I have seen a strange genius a Major Sutcliffe. He tells me as soon as he heard there were two English Surveying Vessels at Valparaiso, he sent a Book of old Voyages in the Straits of Magellan to Mr Caldcleugh [see p.276] to be forwarded to the Commanding Officer as they might prove of service – He has not heard whether Mr Caldcleugh has sent them to you – I told him I would mention the circumstance when I wrote. The Major is inclined to be very civil – I do not know what to make of him. He is full of marvellous stories; and to the surprise of every one every now & then some of them are proved to be true.

My head is full of schemes; I shall not remain long here, although from the little I have yet seen I feel much inclined to like it. How very striking & beautiful the situation of the city is – I sat for an hour gazing all round me, from the little hill of St Lucia. I wish you could come here to readmire the glorious prospect. I can by no means procure any sort of map. You could most exceedingly oblige me if you would get King to trace from Miers [1826] a little piece of the Country from Valparaiso to a degree south of R. Rapel – without any mountains. I do not think it will be more than ½ an hours work. I have some intentions of returning to Valparaiso by the Rapel. If you would send me this *soon* and half a dozen lines, mentioning, if you should know anything about the Samarangs movements; it would assist in my schemes very much.

Adios, dear FitzRoy / yr faithful Philos. /C.D.[118]

September 1834

Having arrived in Santiago by a circuit to the north, Charles decided to return to Valparaiso on 5 September by a longer route to the south. This first day took him across one of the famous suspension bridges where the road consisted of bundles of sticks supported by hides, and followed the curvature of the suspending ropes. It was not kept in good repair, and oscillated in an alarming fashion with the weight of a man leading a horse. In the evening he lodged at a very nice hacienda, where some charming signoritas quizzed him for entering a church to look about him, and asked why he did not become a Christian. He assured them that he was a sort of Christian, but they would not have it, saying, 'Do not your padres, your very bishops, marry?' They were particularly struck by the absurdity of a bishop having a wife, and did not know whether to be more amused or more horrified by such an atrocity.

Charles rode on through Rancagua, and stayed five days at the hot baths of Cauquenes, celebrated for their medicinal properties, trapped by heavy rain that made the river crossings temporarily impassable. At San Fernando he turned seawards to visit the gold-mines of Yaquil belonging to an American called Nixon, and though shocked at the severity of the labour imposed on the miners for very small wages, concluded that the agricultural workers were probably even worse off. The ore was sent down to the mills on mules, and on level roads each animal carried 416 pounds. Charles found it surprising that 'a Hybrid should possess far more reason, memory, obstinacy, powers of digestion & muscular endurance than either of its parents. One fancys art has here out-mastered Nature.'

In Mr Nixon's house a German called Renous was staying. Charles was amused by a conversation between Renous and an old Spanish lawyer, when Renous asked the lawyer what he thought of the King of England 'sending out me to their country to collect lizards & beetles & to break rocks. The old Gentleman thought for some time & said "it is not well, hay un gato encerrado aqui ('there is a cat shut up here'); no man is so rich as to send persons to pick up such rubbish; I do not like it; if one of us was to go & do such things in England, the King would very soon send us out of the country." ' And yet the old lawyer belonged to one of the most intelligent classes. A few years earlier, Renous himself had left some caterpillars at a house in San Fernando for a while, to hatch into butterflies. This had been talked about in the town, whereupon the priests and the Governor concluded that some heresy was afoot, and when Renous returned he was at once arrested.

Charles now began to feel seriously unwell, and pressed on with his ride back to Valparaiso across the coastal grass plains. Despite his sickness, however, he managed to collect, from beds of Tertiary rocks eight hundred feet thick near the mouth of the Rio Rapel, some fossil shells that could only have lived in shallow water, thus confirming that there must in the past have been a slow subsidence of the sea-bottom. On 26 September he reached Casa Blanca, and was able to summon a carriage to take him back to Mr Corfield's house in Valparaiso, where he remained in bed until the end of October.

One of his anxieties was quickly resolved, when he learnt that FitzRoy had just sent to Portsmouth with HMS *Samarang* a consignment of very valuable specimens consisting of two casks containing bones and stones, a box with six small bottles, and a large jar.

Writing to his sister Caroline on 13 October, Charles said:

> I have been unwell & in bed for the last fortnight, & am now only able to sit up for a short time. As I want occupation I will try & fill this letter. Returning from my excursion into the country I staid a few days at some Gold-mines & whilst there I drank some Chichi a very weak, sour new made wine, this half poisoned me. I staid till I thought I was well; but my first days ride, which was a long one again disordered my stomach, & afterwards I could not get well; I quite lost my appetite & became very weak. I had a long distance to travel & I suffered very much; at last I arrived here quite exhausted. But Bynoe with a good deal of Calomel & rest has nearly put me right again & I am now only a little feeble. I consider myself very lucky in having reached this place, without having tried it, I should have thought it not possible; a man has a great deal more strength in him, when he is unwell, than he is aware of . . .

October 1834

You will be sorry to hear the Schooner, the *Adventure*, is sold; the Captain received no sort of encouragement from the Admiralty & he found the expense of so large a vessel so immense he determined at once to give her up [see pp.250–1]. We are now in the same state as when we left England with Wickham for 1st Lieut, which part of the business anyhow is a good job. We shall all be very badly off for room; & I shall have trouble enough with stowing my collections. It is in every point of view a grievous affair in our little world; a sad tumbling down for some of the officers, from 1st Lieut of the Schooner to the miserable midshipmans birth, & many similar degradations. It is necessary also to leave our little painter, Martens, to wander about ye world. Thank Heavens, however, the Captain positively asserts that this change shall not prolong the voyage – that in less than 2 years we shall be at New S. Wales . . .

I have picked up one very odd correspondent, it is Mr Fox the Minister at Rio. (It is the Mr Fox, who in one of Lord Byrons letters is said to be so altered after an illness that his *oldest Creditors* would not know him)[119]

This was much the longest period of illness that Charles recorded having had during the voyage. Its precise cause has not been established, but it has been suggested that six weeks' illness without presenting features is most likely to have been typhoid fever.[120] He was very disappointed to be stuck in bed in this fashion, for he had been hoping to collect many animals, but the time was not wholly wasted for he was able to write a long series of notes about the birds that he had so far seen in Chile. Among them, the highlight was some more observations on condors:

Having an opportunity of seeing very many of these birds in a Garden, I observe that all the females have bright red eyes; but the male yellowish brown. I however found that a young female (known by dissection, as this was in the Spring [of the southern hemisphere] the bird must at least be one year old), whose back was brownish & ruff scarcely as yet at all white, has her eyes dark brown. The young male has also its back & ruff brown, & the comb simple. These were fed only once a week. The Guassos state they can well live 6 weeks without food. They are caught on Corralitos [small corrals] or when roosting 5 or 6 together in a tree. They are very heavy sleepers (as I have seen) & hence a person easily climbs up a tree & lazoes them. They are only taken in winter & Spring; in the summer are said to retire into Andes. There are so many brought in that a live Condor has been sold for 6^{d}. Common price 2 or 3 dollars. They are wonderfully ravenous. One brought in lashed with rope & much injured, & surrounded by people, the instant the line was loosed which secured the beak, began to tear a piece of carrion.

The condors appear suddenly in numbers where an animal dies, in the same unaccountable manner in which all Carrion Vultures are well known to do. Tying a piece of meat in a paper, I passed by a whole row of them within 3 yards & they took no notice. I threw it on the ground within one yard, an old male condor looked at it & took no further notice: placing it still closer, the Condor touched it with his beak & *then* tore the paper off with fury. In an instant the whole row of Condors were jumping & flapping their wings. I think it is certain a Condor does not smell at a greater distance than a few inches. Mem: M. Audubon in Wern: transactions, similar observ. [see p.5]

[Three notes added later] The country people inform me the Condor lays two large white eggs in November or December; they make no nest but place the eggs on any small ledge. I am assured the young Condors cannot fly for a whole year. At Concepcion on *March 5th* [1835] I saw a young Condor, it was nearly full grown, but covered with a blackish down, precisely like a Gosling. I am sure this bird would not have been able to fly for many months. After the young birds can fly apparently as well as the old ones, from what I saw on coast of Patagonia, they appear to remain for some time with their parents. They hunt separately, before the ring around the neck is changed white. When at the S. Cruz river in months of April & May, two old birds were generally perched on the ledges or sailing about with a *full* fledged young bird, not white collar. Now I think it certain that this could not have been hatched during the same summer: if so the Condor probably lays only once in two years. It is rather singular that the name Condor is *only* applied to the young ones, before the white feathers appear; the old birds being called 'El Buitre', the Spanish of 'the Vulture'.

The Condors attack young goats & sheep, I have seen dogs trained to chace [*sic*] them away. It is beautiful to watch several Condors wheeling over any spot. Although you may never take your eyes off any one bird, for a quarter or half hour you can never see the slightest motion of their wings. I believe a Condor will go on flying in curves ascending & descending for any length of time without flapping its wings. When the bird wishes to descend rapidly, the wings are collapsed for a second. When soaring close above the beholder no tremulous motion or indistinct appearance can be observed in the separate feathers which terminate the wing. The head & neck are moved frequently & apparently with force, as a rudder of a ship, but perpendicularly as well as laterally. By the former motion, the whole body seems to alter its inclination with the horizon & by action of contrary current of wind to rise. The bird critically *views* the ground.

Shortly before any one of the Condors dies, all the lice which infest it crawl to the outside of the feathers.[121]

In the meantime, FitzRoy had been having a difficult time after the Lords of the Admiralty in London, holding strong and wholly unfair political prejudices against him, had flatly refused to give official approval for his purchase of the *Adventure* out of his own pocket, and funding for the extra twenty men that he needed to man her. In private letters dated 26 and 28 September to the ever-sympathetic but unfortunately not all-powerful Captain Beaufort at the Hydrographic Department, he wrote:

My Schooner is *sold*. Our painting man Mr Martens is *gone*. The Charts &c are progressing slowly – They are not ready to send away yet – I am in the dumps. It is heavy work – all work and no play – like *your* Office, something – though not half so bad probably . . .

September 1834

The *Beagle* and the *Mary Walker* of Glasgow at anchor off Valparaiso, by Martens

Troubles and difficulties harass and oppress me so much that I find it impossible either to say or do what I wish. Excuse me then I beg of you if my letters are at present short and unsatisfactory – My mind will soon be more at ease. Letters from my friends – Having been obliged to sell my Schooner, and crowd everything again on board the Beagle – Disappointment with respect to Mr Stokes – also the acting Surgeon – and the acting Boatswain – Continual hard work – and heavy expense – – These and many other things have made me ill and very unhappy.

The Beagle has been refitting, while paper work has been going on steadily. When I look back at the time we have been in Valparaiso, I am annoyed at it's length and yet I cannot see any way by which it could have been made shorter. Much material had been collected, and must have been put together somewhere. I see no hope of finishing before the middle or latter part of October. I have affronted and half quarrelled with most people by shutting myself up and refusing to visit or be visited. As Captain of a Ship in a bustling sea port it is a difficult matter to keep sufficiently quiet to make such progress as one would wish. Yet to this port a vessel *must* come for supplies. Besides after a long cruize upon salt meat, it is absolutely necessary that the Crew should have fresh meat and Vegetables for sufficient time to do away with all scorbutic* inclinations.

I sold my Schooner a fortnight ago, for more than I first gave for her, but not near enough to cover what has been laid out upon her, or what her crew and provisions have cost.[122]

On returning from Corfield's house to the *Beagle* while she was still refitting in Valparaiso, Charles found FitzRoy in an unhappy state of mind:

Poor FitzRoy was sadly overworked and in very low spirits; he complained bitterly to me that he must give a great party to all the inhabitants of the place. I remonstrated and said that I could see no such necessity on his part under the circumstances. He then burst out into a fury, declaring that I was the sort of man who would receive any favours and make no return. I got up and left the cabin without saying a word, and returned to Conception where I was then lodging [Valparaiso was meant rather than Concepción]. After a few days I came back to the ship and was received by the Captain as cordially as ever, for the storm had by that time quite blown over. The first Lieutenant, however, said to me: 'Confound you, philosopher, I wish you would not quarrel with the skipper; the day you left the ship I was dead-tired (the ship was refitting) and he kept me walking the deck till midnight abusing you all the time.' The difficulty of living on good terms with a Captain of a Man-of-War is much increased by its being almost mutinous to answer him as one would answer anyone else; and by the awe in which he is held – or was held in my time, by all on board.[123]

However in Charles's next letter to his sister Catherine, written on 8 November, his news about the situation was better:

* That is, an outbreak of scurvy caused by insufficient vitamin C in the diet.

November 1834

My last letter was rather a gloomy one, for I was not very well when I wrote it – Now everything is as bright as sunshine. I am quite well again after being a second time in bed for a fortnight. Capt FitzRoy very generously has delayed the Ship 10 days on my account & without at the same time telling me for what reason. We have had some strange proceedings on board the Beagle, but which have ended most capitally for all hands. Capt FitzRoy has for the last two months, been working *extremely* hard & at the same time constantly annoyed by interruptions from officers of other ships: the selling the Schooner & its consequences were very vexatious: the cold manner the Admiralty (solely I believe because he is a Tory) have treated him, & a thousand other &c &c has made him very thin & unwell. This was accompanied by a morbid depression of spirits, & a loss of all decision & resolution. The Captain was afraid that his mind was becoming deranged (being aware of his hereditary disposition*), all that Bynoe could say, that it was merely the effect of bodily health & exhaustion after such application, would not do; he invalided & Wickham was appointed to the command. By the instructions Wickham could only finish the survey of the Southern part & would then have been obliged to return direct to England. The grief on board the Beagle about the Captains decision was universal & deeply felt. One great source of his annoyment, was the feeling it impossible to fulfil the whole instructions; from his state of mind, it never occurred to him that the very instructions order him to do as much of West coast, as *he has time* for, & then proceed across the Pacific. Wickham (very disinterestedly giving up his own promotion) urged this most strongly, stating that when he took the command, nothing should induce him to go to T. del Fuego again; & then asked the Captain, what would be gained by his resignation. Why not do the more useful part & return as commanded by the Pacific. The Captain, at last, to every ones joy consented & the resignation was withdrawn.

* Lord Castlereagh, his uncle on his mother's side, had committed suicide a few years earlier.

November 1834

Buildings on the quayside at Valparaiso, by Martens

Hurra Hurra it is fixed the Beagle shall not go one mile South of C. Tres Montes (about 200 miles South of Chiloe) & from that point to Valparaiso will be finished in about five months. We shall examine the Chonos archipelago, entirely unknown & the curious inland sea behind Chiloe. For me it is glorious. C.T. Montes is the most southern point where there is much geological interest, as there the modern beds end. The Captain then talks of crossing the Pacific; but I think we shall persuade him to finish the coast of Peru: where the climate is delightful, the country hideously sterile but abounding with the highest interest to a Geologist. For the first time since leaving England I now see a clear & not so distant prospect of returning to you all; crossing the Pacific & from Sydney home will not take much time.[124]

CHAPTER 20

Chiloe and the Chonos Archipelago

The *Beagle* duly set sail for Chiloe on 10 November, and arrived in the harbour of San Carlos ten days later. The yawl and whaleboat were at once sent out under the command of Lieutenant Sulivan to check the correctness of the charts of the east coast of Chiloe. Charles hired horses to join them at Chacao, once the principal port of the island located at its north-easternmost corner, but abandoned because of the dangerous rocks and currents around it. Some of the inhabitants seemed to hope that the two boats belonged to a Spanish fleet coming to rescue the island from the patriot forces that had liberated Chile from the Spaniards in 1810. However, the authorities had been informed about the *Beagle's* visit, and the Governor of Chacao came to call politely on Sulivan's little party while they were eating their supper. He had once been a Lieutenant Colonel in the Spanish army, but was now miserably poor. He gave them two sheep, and accepted two cotton handkerchiefs, some brass trinkets and a pinch of tobacco in return.

In the usual torrents of rain, they sailed down the coast, and when the clouds cleared for a while had a splendid view of the snow-covered cone of the volcano Osorno to their north spouting huge volumes of smoke, with two other seven-thousand-foot volcanoes, Huequi to the east, and Corcovado to the south. After five days they arrived at Castro, the former capital of the island, where they found a typical Spanish town neatly laid out in a rectangular fashion with its plaza and large church, but totally deserted, with sheep grazing on the fine green turf that covered the streets and the plaza. The majority of the population, numbering over forty thousand, were little

The volcano Osorno as seen from Chiloe, by Conrad Martens

copper-coloured men of mixed blood, but coming from different Indian tribes and speaking a variety of languages. Many were descended from Indians who had been sent to the north '*en nomiendas commendo*', that is to be taught the Christian religion and in return to work as slaves for the first Spanish settlers before returning to their homes. When, however, they had then cleared some land, it would immediately be seized by the government. They retained their own *caciques* (chieftains), who had very little power, although the Chilean authorities had made some retribution for their treatment by the Spaniards, by distributing to the *caciques* and their families, to any man who had served in the militia, and to the aged, land that could legally be held. They lived chiefly on fish, shellfish and potatoes, and a few of them possessed fowls, sheep, goats, and less often pigs, horses and cattle. But poverty was universal, and even the Governor was a quiet old man who in his appearance and manner of life was scarcely superior to an English cottager.

On 1 December they arrived at the island of Lemuy, where Charles wished to examine a reported coalmine that turned out to be lignite of little value in the Tertiary sandstones of which the islands were composed. However, in his pocketbook he recorded a finding of significant geological interest:

> *At last* I found in the yellow sandstone a great trunk (structure beautifully clear) throwing off branches: main stem much thicker than my body and standing out from weathering 2 feet – central parts generally black and vascular and structure not visible. It is curious chemical action – such a sandstone in sea – holding such silex in solution: vessels transparent quartz: This observation most important as proof of general facts of petrified wood; for here the inhabitants firmly believe the process is now going on.[125]

He was next to find fossilised trees four months later, high up in the Andes.

The party was soon surrounded by a large group of the nearly pure Indian inhabitants, who were much surprised at their arrival, and said to one another, 'This is the reason we have seen so many parrots lately; the Cheucau has not cried "Beware" for nothing.' The cheucau was a very strange little bird, *Pteroptochos rubecula*, that Charles described as follows:

> It frequents the most gloomy and retired spots within the damp forests. Sometimes, although its cry may be heard close at hand, let a person watch ever so attentively, he will not see the cheucau; at other times, let him stand motionless, and the red-breasted little bird will approach within a few feet, in a most familiar manner. It then busily hops about the entangled mass of rotting canes and branches, with its little tail cocked upwards. I opened the gizzard of some specimens: it was very muscular, and contained hard seeds, buds of plants, and vegetable fibres, mixed with small stones. The cheucau is held in superstitious fear by the Chilotans, on account of its strange and varied cries. There are three very distinct kinds, – one is called "chiduco," and is an omen of good; another "huitreu," which is extremely unfavourable; and a third, which I have forgotten. These words are given in imitation of its cries, and the natives are in some things absolutely governed by

them. The Chilotans assuredly have chosen a most comical little creature for their prophet.[126]

Sulivan and his party continued to work to the south, and reached Caylen, Puerto Quellón on a modern map, the so-called '*fin del Christianitad*', though Charles thought it was rather better inhabited than that. The next day they stopped for a few minutes at a house at the northern tip of the island of Laylec, a miserable hovel that really was the southern end of South American Christianity. In the evening they arrived at the island of San Pedro, the south-eastern extremity of Chiloe. When doubling the point of the harbour, Messrs Stuart and Usborne landed to take a round of angles. A Chilotan fox sat on the point, and was so absorbed in watching their manoeuvres that he allowed Charles to walk up from behind and kill him with his geological hammer. He thereby earned an epitaph which ran: 'This fox, more curious or more scientific, but less wise than the generality of his brethren, is now mounted in the Museum of the Zoological Society.'[127]

Forest scene on Chiloe, by Martens

December 1834

The *Beagle* had arrived the previous day, having failed thanks to bad weather to complete the survey of the outer coast of Chiloe, but having established that existing surveys had made Chiloe too long by thirty miles, so that the main island would have to be cut down by a quarter. On 8 December, FitzRoy led a party to try to reach the summit of San Pedro. Charles wrote a graphic description of the difficulties of penetrating the virgin temperate forest:

> The woods here have a different aspect from those in the North, there is a much larger proportion of trees with deciduous leaves. The rock also being primitive Micaceous slate, there is no beach, but the steep sides of the hills dip directly down into the sea; the whole appearance is in consequence much more that of T. del Fuego than of Chiloe. In vain we tried to gain the summit; the wood is so intricate that a person who has never seen it will not be able to imagine such a confused mass of dead & dying trunks. I am sure oftentimes for quarter of an hour our feet never touched the ground, being generally from 10 to 20 feet above it; at other times, like foxes, one after the other we crept on our hands & knees under the rotten trunks. In the lower parts of the hills, noble trees of Winters bark, & the Laurus sassafras (?) with fragrant leaves, & others the names of which I do not know, were matted together by Bamboos or Canes. Here our party were more like fish struggling in a net than any other animal. On the higher parts brushwood took the place of higher trees, with here & there a red Cypress or an Alerce. I was also much interested by finding our old friend the T. del F. Beech, Fagus antarcticus; they were poor stunted little trees, & at an elevation of little less than a thousand feet. This must be, I should apprehend from their appearance, nearly their Northern limit. We ultimately gave up the ascent in despair.[128]

Leaving parties under Sulivan in the yawl and under Stokes in a whale-boat to survey the western coasts of Chiloe and of the Chonos Group, FitzRoy took the *Beagle* down to Cape Tres Montes at the southern end of the Chonos Archipelago. Here they found a harbour near the most perfectly conical hill that Charles had ever seen,

Forest scene on Chiloe, by Martens

rivalling the Sugar Loaf at Rio de Janeiro, so he naturally set out to climb to its summit. The ascent was so steep that he had to use the trees and the great thickets of what he called 'Fushza' – presumably *Fuchsia magellicana* – as ladders, to be rewarded at last by a magnificent view that he felt for a short while he was the first man ever to have seen. But nearby was a small piece of wood with a nail in it, and his thoughts turned to the poor shipwrecked sailor trying to travel up the coast who might have left it there, only in the end to have laid himself down to die.

The *Beagle* spent Christmas trapped at this melancholy place, swampy with rain, tormented with gales, without the interest even of population, for they had neither heard the voices nor found any traces of any natives. Then on 28 December the wind dropped, and they ran along the coast to the Bay of San Estevan. While they were furling sails, they saw some men on a point of land near the ship signalling to them 'in a very earnest manner'. Having been taken on board, they turned out to be American sailors who had, off Cape Tres Montes in October 1833, deserted in a small boat from a New Bedford whaler with the intention of working their way up the coast to Chiloe. But during their first landing they stove in their boat, and quickly discovered how impossible it was to make any effective progress on foot along the precipitous shores thickly coated with trees. They therefore had no alternative but to camp out near San Estevan for the next fourteen months, living on seals' flesh, shellfish and wild celery, a diet that kept them in surprisingly good health. All the ships sailing past off the coast kept well away from the rocky and dangerous lee shore, and the *Beagle* was the first whose attention they had succeeded in attracting. They had found traces of previous castaways, but had seen no natives.

Lieutenant Sulivan's party had spent a more cheerful Christmas on the west coast of Chiloe. From San Carlos he wrote home on 9 January 1835:

> I shall amuse you with a few stories. For instance, our foraging on a small island inhabited by Indians, on Christmas morning, from nine

to twelve, in a heavy gale of wind and tremendous rain, before we could get eggs enough to make our plum-pudding or a sheep to eat. However, we got into the padre's house attached to the church, as our tents, clothes, and blankets were wet through, and by 4 p.m. had one side of a sheep roasted, another side boiled, twelve pounds of English fresh roast beef heated, and two immense plum-puddings made. No bad quantum for twelve men! It would have amused you if you could have seen us in a dirty room with a tremendous fire in the middle, and all our blankets and clothes hung round the top on lines, getting smoked as well as dry, while all hands were busily employed for four hours killing a sheep, picking raisins, beating eggs, mixing puddings which were so large that, in spite of two-thirds of the party being west-country men, we had enough for supper also. However, we passed a pleasant day in spite of wind and weather, and it was a holiday to us, as we could afford to knock off work when it rained too hard constantly to be able to move, which happened on Christmas Day and New Year's Day. Every other day for eight weeks we were hard at work. It is very curious that I am always in better health in a boat, for I never have enjoyed such perfect good health for two months since leaving England.[129]

The *Beagle* sailed on up the archipelago. Charles did not see a great deal of geological interest, though at Yuche Island he indulged himself by reflecting on the importance of granite:

The chief part of the range is composed of grand solid abrupt masses of granite, which look as if they had been coeval with the very beginning of the world. The granite is capped with slaty gneiss, & this in the lapse of ages of time has been worn into strange finger-shaped points. These two formations, thus differing in their outlines, agree in being almost destitute of vegetation; and this barrenness had to our eyes a more strange appearance, from being accustomed to the sight of an almost universal forest of dark green trees. I took much delight in examining the structure of these mountains. The complicated & lofty ranges bore a noble aspect of durability – equally profitless however to man & to all other animals. Granite to the Geologist is a classic ground: from its wide-spread limits, its beautiful & compact texture, few rocks have been more anciently recognised. Granite has given rise perhaps to more discussion concerning its origin than any other formation. We see it generally the fundamental rock, & however formed we know it to be the deepest layer in the crust of this globe to which man is able to penetrate. The limit of mans knowledge in every subject possesses a high interest, which is perhaps increased by its close neighbourhood to the realms of imagination.[130]

On 4 January, driven by a heavy north-westerly gale, the *Beagle* anchored in the bay where the small merchantman or 'pink' *Anna* had taken refuge in April 1741 to carry out repairs after being severely damaged in a storm while rounding Cape Horn with Sir George Anson's squadron in which two ships were lost. Anson went on to capture the Spanish galleons laden with treasure from Manila, and returned to England in 1744 after circumnavigating the globe. A boat with the Captain took Charles to the head of the bay, of which he wrote:

> The number of the seals was quite astonishing; every bit of flat rock or beach was covered with them. They appear to be of a loving disposition & lie huddled together fast asleep like pigs; but even pigs would be ashamed of the dirt & foul smell which surrounded them. Often times in the midst of the herd, a flock of gulls were peaceably standing: & they were watched by the patient but inauspicious eyes of the Turkey Buzzard. This disgusting bird, with its bald scarlet head formed to wallow in putridity, is very common on this West Coast. Their attendance on the seals shows on the mortality of what animal they depend.

Since the Indians had certainly lived at one time in the region, and there was still an abundance of their favourite food to be had there, Charles was puzzled to find that there was no longer any trace of them. He feared that like the rest of their tribes the natives were doomed in the end to extinction in South America.

January 1835

The *Beagle* had a rendezvous with Mr Stokes in the yawl at a harbour near the northern end of the Chonos Archipelago often used by their friend William Low. Charles noted grumpily that the next week passed 'rather heavily', but in fact it was the occasion when he crowned his collections in Chiloe and the Chonos Archipelago of frogs, turbellarian flatworms, siphonophores and other hydrozoans, some of which turned out to be new species, such as a prettily coloured sea slug that has recently been renamed *Thecacera darwinii*, with an important discovery. This was his finding on the beach at Low's Harbour an exceptionally small variety of barnacle that riddled the thick shells of the mollusc *Concholepas peruviana* with holes, and lived inside these unfortunate creatures as parasites.

The barnacle[131] was contained in a flask-shaped and orange-coloured sack just under one-tenth of an inch in length (Fig. 2), which was held in its cavity in the shell of its host by longitudinal stony bands. The body lay in the sack with three double cirrhi protruding at its hinder end (Fig. 3). In the final state of its pupal development, Charles wondered, 'Who would recognise a young Balanus in this illformed little monster? Are the two *strong* legs (with spiny plate capable of rotatory & other motion) for boring holes in the shell?'*

Always frustrated by the incessant rain, the *Beagle* made her way back up the west coast of Chiloe, and on 19 January anchored again at Punta Arenas near San Carlos. During that night the volcano of Osorno began to erupt, and at three o'clock in the morning all the officers were on deck watching its glowing cone casting a long red streak on the water while a constant succession of dark objects were

* When ten years later, Charles paid heed to his friend Joseph Hooker's advice that in order to make his mark as a biologist he needed to do some serious taxonomic work on the classification of unknown species, he embarked on a series of intensive studies at Down House in Kent on barnacles, which were described in his monograph on the Cirripedia published in 1854. One of his starting points was this smallest of all cirripedes, that he christened *Cryptophialus minutus*, and that was in due course classified formally as a crustacean of order Acrothoracica, a naked, boring barnacle, living parasitically in shells and corals. Charles's

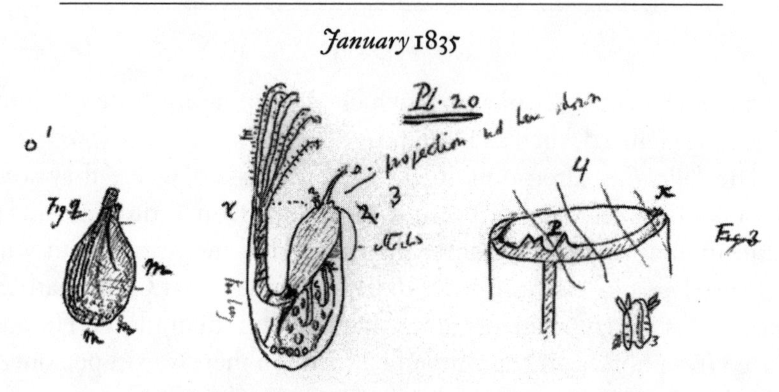

The anatomy of the smallest barnacle, *Cryptophialus*

thrown up and cascaded down. But by the morning the volcano seemed to have regained its composure.

FitzRoy was anxious that some further bearings should be taken on the outer coast of Chiloe, and dispatched Philip Gidley King to ride with Charles to Castro at the centre of the island, and thence to the Capella de Cucao on the west coast. Having hired horses and a guide, they were joined by a woman with two boys, for on this road Charles said that the rule was for everyone to act in a 'hail fellow well met' fashion and enjoy the privilege so rare in South America of travelling without any firearms. Although she belonged to one of the best families in Castro, the woman rode cross-legged without shoes or stockings, and impressed Charles by her lack of social pride. The road was constructed of broad slabs of wood laid longitudinally or of smaller ones laid transversely, and since it was summer and not too slippery was not hard to ride. 'The number of trees which were in full flower perfumed the air; yet even this could scarcely dissipate the gloomy dampness of the forest. The number of dead trunks, which stand like great white skeletons, never fails to give these primeval

researches at Down House revealed that the adults he had dissected on the *Beagle* were all females that possessed neither testicles nor seminal vesicles, whereas the males were minute creatures not much more than one-hundredth of an inch in length, lacking all organs except for testicles and seminal vesicles, and 'a wonderfully elongated probosciformed penis'. Four or five of these tiny males were attached to each female. This unique reproductive system gave him much valuable food for thought.

woods a character of solemnity which is wanting in those of countries long civilized, such as England.'

The following afternoon, they arrived at Castro, where they were most kindly received by the new Governor, Don Pedro. He was a Chilean now very much poorer than he had formerly been, but who displayed a degree of hospitable disinterestedness that Charles attributed to his purely Spanish rather than partly Indian blood. He produced fresh horses and accompanied them on their way in person. At Vilèpille, Don Pedro asked the military *Comandante* of the region for a guide to Cucao, but although the old gentleman at once offered his own services, he found it very hard to understand why the two Englishmen could possibly want to go to such an out of the way place. 'But where are you *really* going?' he asked repeatedly; and when Don Pedro answered, 'To Cucao,' the *Comandante* said, '*A los infiernos, hombre*; – what is the good of deceiving me?' And thus escorted by the two highest men in the land, as was clear from the grovelling respect with which the party was greeted by all the Indians, they rode through the forest and occasional clearings in which corn and potatoes were cultivated until they reached the great lake of Cucao, stretching twelve miles to the coast. Here the *Comandante* loaded them into a roughly-built little boat called a '*periagua*', paddled by six of the ugliest Indians that Charles had ever seen, who despite their unpromising appearance propelled the boat against a light breeze at fully three miles an hour. A cow was neatly levered with oars under its belly over the side of the boat to accommodate them. In Cucao they were housed in reasonable comfort in a hut used by the priest when he visited his *capella*.

Cucao was at that time the only inhabited part of the whole west coast of Chiloe, with thirty or forty Indians living there, and not a single Spaniard. Charles was distressed to find that these people were treated by their rulers, who were so polite to himself and Philip King, as if they were slaves rather than free men. Having made friends with the Indians by giving them some cigars, maté leaves and sugar, he noted that they ended all their complaints by saying, 'and it is only because we are poor Indians and know nothing, but it was not so when we had a King'. He concluded that a boat's crew flying the Spanish flag might be able to take the island of Chiloe at a stroke.

After breakfast he and King rode northwards a short way along the beach, on which even after days of good weather a tremendous surf was breaking. They would have liked to have returned to Punta Arenas along the coast, but were assured by the Indians that no one had ever got through that way. So they re-embarked on the *periagua*, crossed the lake, recovered their horses, and after dining with the friendly *Comandante*, reached Castro that evening and Punta Arenas on the next one. Heavy rain kept the *Beagle* at anchor for yet another week, until at last they could set sail for Valdivia. Everyone was glad to say goodbye to Chiloe, though without the gloom and ceaseless rain of winter it might have passed for a charming island.

CHAPTER 21

The Great Earthquake of 1835 Hits Valdivia and Concepcion

The *Beagle* anchored at the port of Valdivia on 8 January, and the next morning Charles went up the river to the quiet little town with houses entirely built of alerce planks. Like Chiloe the country was everywhere heavily forested, though with fewer dark evergreens, and therefore of a lighter green appearance. A form of bamboo some twenty feet high furnished the long tapering *chusas* or spears used by the Araucanian Indians from further north who were fighting against General Rosas in Argentina. But the twenty-six tribes living around Valdivia were 'reducidos & Christianos' under Spanish residents, and the *caciques* of several of them received pensions of thirty dollars a year to keep them quiet. The Friar of the district said that they respected the Catholic religion, but did not much like to come to Mass, and were strongly resistant to the marriage service, perhaps because the most sought-after position was to be one of the *cacique*'s ten wives who took turns to live with him for a week. Apple trees flourished well in the country, yielding cider and a stronger spirit, and drunkenness was regrettably the besetting sin of the Indians.

The people in the town were mainly Spaniards, and during the fortnight spent there by the *Beagle* the gaiety on board rose appreciably. The Mayor brought a whole boatload of ladies to visit the ship one day, who were obliged by bad weather to stay overnight. In return a ball was given, attended and greatly enjoyed by almost all of the crew. Charles noted that 'The Signoritas are pronounced very charming; & what is still more surprising, they have not forgotten how to blush, an art which is at present quite forgotten in Chiloe.'

February 1835

Charles crossed over the bay to the side opposite the *Beagle*'s anchorage to visit the fort called Niebla. It was in a ruinous state, and when John Wickham said to the commanding officer that the carriages of the guns would collapse after one shot, he replied, 'No, I am sure, Sir, they would stand two!' The Spaniards must at one time have intended to make the fort impregnable, but when in December 1819 Lord Cochrane was commanding the Chilean navy, it was at Valdivia that he landed with only three hundred men and overcame the base with very few casualties of his own.

On 20 February 1835 Charles wrote in his commonplace journal:

> This day has been remarkable in the annals of Valdivia for the most severe earthquake which the oldest inhabitants remember. Some who were at Valparaiso during the dreadful one of 1822, say this was as powerful. I can hardly credit this, & must think that in Earthquakes as in gales of wind, the last is always the worst. I was on shore & lying down in the wood to rest myself. It came on suddenly & lasted two minutes (but appeared much longer). The rocking was most sensible; the undulation appeared both to me & to my servant to travel from due East. There was no difficulty in standing upright; but the motion made me giddy. I can compare it to skating on very thin ice or to the motion of a ship in a little cross ripple.
>
> An earthquake like this at once destroys the oldest associations; the world, the very emblem of all that is solid, moves beneath our feet like a crust over a fluid; one second of time conveys to the mind a strange idea of insecurity, which hours of reflection would never create. In the forest, a breeze moved the trees, I felt the earth tremble, but saw no consequence from it. At the town where nearly all the officers were, the scene was more awful; all the houses being built of wood, none actually fell & but few were injured. Every one expected to see the Church a heap of ruins. The houses were shaken violently & creaked much, the nails being partially drawn. I feel sure it is these accompaniments & the horror pictured in the faces of all the inhabitants, which communicates the dread that every one feels who has *thus seen* as well as felt an earthquake. In the forest it was a highly interesting but by no

means an awe-exciting phenomenon. The effect on the tides was very curious; the great shock took place at the time of low water; an old woman who was on the beach told me that the water flowed quickly but not in big waves to the high water mark, & as quickly returned to its proper level; this was also evident by the wet sand. She said it flowed like an ordinary tide, only a good deal quicker. This very kind of irregularity in the tide happened two or three years since during an Earthquake at Chiloe & caused a great deal of groundless alarm. In the course of the evening there were other weaker shocks; all of which seemed to produce the most complicated currents, & some of great strength in the Bay.[132]

The *Beagle* sailed on northwards up the coast, continuing her survey. On 4 March she entered the harbour of Concepción, to be greeted with the news, as they could see all too well for themselves, that the great earthquake had left scarcely a house standing in either Concepción itself or at the port, Talcahuano. Nearly all the inhabitants had escaped the main shock uninjured, but immediately after it an alarm was given that the sea was retiring, just as was remembered to have happened in 1751 at the former port of Concepción, Penco, when it was destroyed by an earthquake and a subsequent wave. In FitzRoy's words:

About half an hour after the shock, when the greater part of the population had reached the heights, the sea having retired so much that all the vessels at anchor, even those which had been lying in seven fathoms water were aground, and every rock and shoal in the bay was visible, an enormous wave was seen forcing its way through the western passage which separates Quiriquina Island from the mainland. This terrific swell passed rapidly along the western side of the Bay of Concepcion, sweeping the steep shores of everything moveable within thirty feet (vertically) from high water mark. It broke over, dashed along, and whirled about the shipping as if they had been light boats; overflowed the greater part of the town, and then rushed back with such a torrent that every moveable which the earthquake had not buried under heaps of ruins was carried out to sea. In a few minutes, the vessels were again aground, and a second great wave was seen approaching, with more noise and impetuosity than the first; but although this was more powerful, its effects were not so considerable – simply because there was less to destroy. Again the sea fell, dragging away quantities of woodwork and the lighter materials of houses, and leaving the shipping aground.

After some minutes of awful suspense, a third enormous swell was seen between Quiriquina and the mainland, apparently larger than either of the two former. Roaring as it dashed against every obstacle with irresistible force, it rushed – destroying and overwhelming – along the shore. Quickly retiring, as if spurned by the foot of the hills, the retreating wave dragged away such quantities of household effects, fences, furniture, and other movables, that after the tumultuous rush was over, the sea appeared to be covered with wreck. Earth and water trembled: and exhaustion appeared to follow these mighty efforts.

Numbers of the inhabitants then hastened to the ruins, anxious to ascertain the extent of their losses, and to save some money, or a few valuable articles, which, having escaped the sweep of the sea, were exposed to depredators.

Remains of the Cathedral at Concepcion, by J.C. Wickham

March 1835

During the remainder of the day, and the following night, the earth was not quiet many minutes at a time. Frequent, almost incessant tremors, occasional shocks more or less severe, and distant subterranean noises, kept everyone in anxious suspense. Some thought the crisis had not arrived, and would not descend from the hills into the ruined town. Those who were searching among the ruins, started at every shock, however slight, and almost doubted that the sea was not actually rushing in again to overwhelm them.[133]

Charles was deeply impressed to see the ruins of Talcahuano and Concepción, which he described as:

The most awful yet interesting spectacles I ever beheld. In Concepcion each house or row of houses stood by itself a heap or line of ruins: in Talcuhano, owing to the great wave little more was left than *one* layer of bricks, tiles & timber, with here & there part of a wall still standing up. The town of Concepcion is built, as is usual, with all its streets at rt angles; one set runs (SW by W & NE by E) & the other (NW by N & SE by S). The walls which have the former direction certainly have stood better than those at right angles to them. If, as would seem probable Antuco may be considered as the centre, it lying rather to the Northward of Concepcion, the concentric lines of undulation would not be far from coincident with NW by N & SE by S walls: this being the case the whole line would be thrown out of its centre of gravity at the same time & would be more likely to fall, than those which presented their ends to the shock. The different resistance offered by the two sets of walls is well seen in the great Church. This fine building stood on one side of the Plaza; it was of considerable size & the walls very thick, 4 to 6 ft & built entirely of brick: the front which faced the NE forms the grandest pile of ruins I ever saw; great masses of brickwork being rolled into the square as fragments of rock are seen at the base of mountains. Neither of the side walls are entirely down, but exceedingly fractured; they are supported by immense buttresses, the inutility of which is exemplified by their having been cut off smooth from the walls, as if done by a chisel, whilst the walls themselves remain standing. There must have been a rotatory motion in the earth, for square objects placed on the coping of this wall are now seated edgeways.

What had been experienced was, however, not simply an earthquake whose epicentre was somewhere to the east of Valdivia, as Charles imagined. The retreat of the water after the shock, followed in Concepcion by the arrival a short while afterwards of a train of three huge waves of the type christened a '*tsunami*' in Japan, were typical of an underwater 'tsunamigenic' earthquake and landslides taking place some distance away beneath the sea. Juan Fernandez Island – Robinson Crusoe's legendary island located in the Pacific 360 English miles north-west of Concepcion – had been reported by its Governor to have been even more violently shaken on 20 February than the opposite shore of the mainland, when at around the same time a submarine volcano close to the island at a depth of sixty-nine fathoms had started to erupt. When Charles presented to the Geological Society in London on 7 March 1838 his paper entitled 'On the connexion of certain volcanic phenomena in South America; and on the formation of mountain chains and volcanoes, as the effect of the same power by which continents are elevated', he listed three groups of earthquakes and eruptions that were apparently connected by having taken place in the same part of North or South America over relatively short periods of time. His third group included the earthquakes both in Concepción and at Juan Fernandez Island on 20 February 1835, though he did not appreciate that the shocks at the two places must have been felt at nearly the same time, after which the big waves at Concepción had followed with an appreciable delay. In the 1845 edition of the *Journal of Researches* he noted correctly that also in the earthquake of 1751 both Juan Fernandez Island and Concepción had been involved, which 'seems to show some subterranean connection between these two points'. Hence he and FitzRoy had in fact been among the first Europeans to have unknowingly witnessed one of the highly destructive *tsunami* originating along the eastern edge of the Nazca crustal plate that are today recognised to inundate the coastlines of Chile and Peru at intervals of roughly thirty years.*

A further issue that greatly interested both FitzRoy and Charles was to know how far the earthquake had resulted in any raising of the ground level at the coast. FitzRoy recorded that for some days afterwards the sea failed by four or five feet to rise to its usual high tide marks, but two months later the difference had declined to no more than two feet. Only on the little island of Santa Maria a few miles south of Concepción did his careful soundings and measurements of the heights of various marks that he made with the help of an intelligent Hanoverian, Anthony Vogelborg, who had lived on the

* As explained by Edward Bryant in *Tsunami: The Underrated Hazard* (Cambridge University Press, 2001), the Peru–Chile trench in the eastern Pacific is an important faultline six hundred miles in length where the small Nazca Plate undergoes subduction beneath the large South American Plate, often resulting in the seismic generation of tsunami. In a major event of this kind which occurred on 21 and 22 May 1960, the first submarine earthquake took place early in the morning of 21 May in the sea to the west of Valdivia, yielding a tsunami that destroyed the area around Concepción. Large aftershocks accompanied by submarine landslides followed during the next thirty-three hours, culminating in a very strong submarine earthquake with an epicentre in roughly the same place as the first one, setting up tsunami that laid waste all the coastal towns down to the south of Chiloe. The initial wave was recorded at tidal gauges across the Pacific as far west as Japan, showing that it travelled over the deep ocean with a height of only 16 inches and at a speed of up to 460 miles per hour. On arrival at shallower coastal water, the wave front was greatly slowed down and greatly increased in height, undergoing refraction and interference to an extent that depended on the precise profile of the shore, with further complications introduced by submarine landslides. At Concepción the presence of the island of Quiriquina in the mouth of the bay helped to set up the train of waves with increasingly large crests reported by FitzRoy, although at Valdivia the profile of the coast with a steeply sloping shore was such that no large waves arose. The velocity of propagation of a shock-wave through the earth's crust is of the order of seven thousand miles per hour, so that for the relatively minor tsunami of 1835, whose epicentre was near Juan Fernandez Island, the expected delay between the arrival of the shock at Concepción and the arrival of the peak of the first tsunami wave would be slightly over forty minutes. The estimate of FitzRoy's informant of 'about half an hour' was therefore not far out.

island for two years, together with evidence of dead shellfish stranded well above sea level, satisfy him that the southern end of the island had been permanently raised by some eight feet, the middle by nine, and the northern end by upwards of ten feet.

When the *Beagle* had arrived at the Bay of Concepción, she had only one large anchor left, having lost or broken all the others. None of suitable size were available at Talcahuano, so on 7 March she left the melancholy ruins and their disconsolate tenants, and sailed back to Valparaiso.

CHAPTER 22

On Horseback from Santiago to Mendoza, and Back Over the Uspallata Pass

After staying with Richard Corfield in the Almendral at Valparaiso for a night or two, Charles took a *birloche* (stage coach) to Santiago, where another English businessman, mine-owner, plant-collector and Fellow of the Royal Society Alexander Caldcleugh, had offered to help him with preparations for a long ride to examine the geology of the Cordilleras. He took with him on this trip his former companion Mariano Gonzales, and a muleteer with his *madrina* and ten mules, of which six were for riding and four to carry food and baggage. The *madrina* was a mare with a bell round her neck whose function was to hold together the little troop of mules and to act most effectively as their collective godmother.

Leaving Santiago on the morning of 18 March, they rode over the dry plains around the city until they arrived at the mouth of the narrow valley of the River Maípo, bounded by the high mountains of the first Cordilleras. There the cottages were surrounded by vineyards and fruit trees heavily laden with ripe apples, nectarines and peaches. In the evening Charles's boxes were examined at the Custom House for departure from Chile, where the officers were extremely civil, no doubt partly because Charles carried a strong passport from the President of Chile. But Charles was also highly impressed by the politeness of every Chileno, which compared very favourably with that of the corresponding officials in England. In the inhabited districts, the party slept at a cottage where they hired pasture for the animals, bought a little firewood, set up their cooking apparatus, and

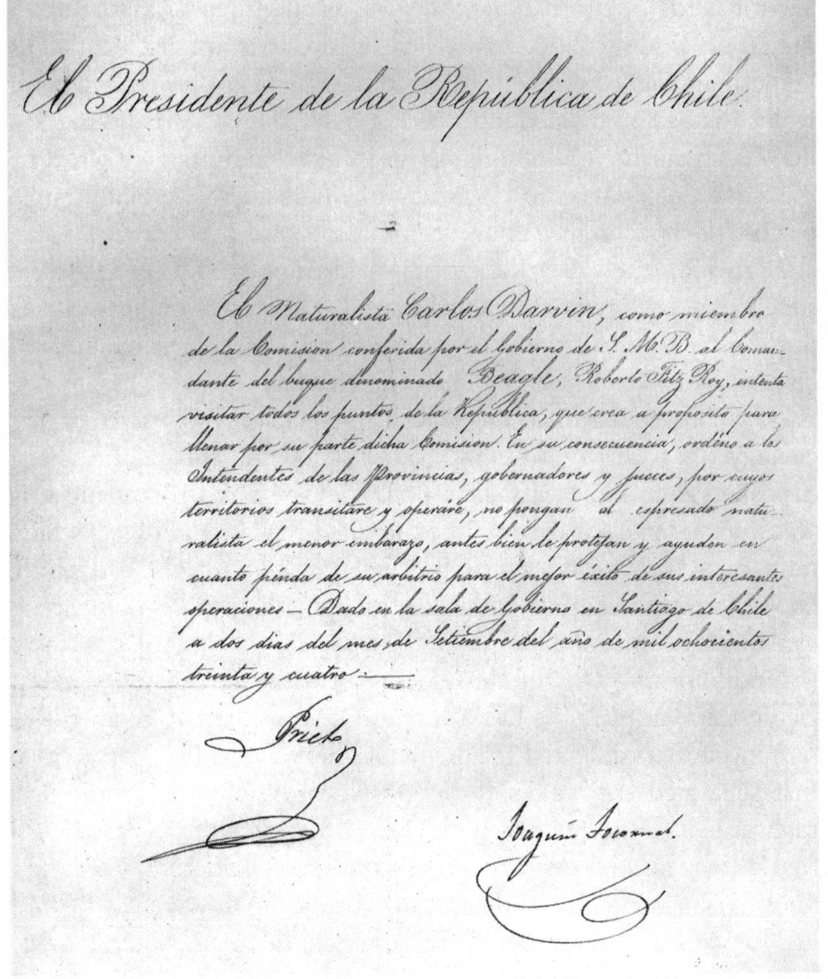

El Presidente de la República de Chile.

El Naturalista Carlos Darvin, como miembro de la Comision conferida por el Gobierno de S. M. B. al Comandante del buque denominado Beagle, Roberto Fitz Roy, intenta visitar todos los puntos de la República, que crea a proposito para llenar por su parte dicha Comision. En su consecuencia, ordeno a los Intendentes de las Provincias, gobernadores y jueces, por cuyos territorios transitare y operare, no pongan al espresado naturalista el menor embarazo, antes bien le protejan y ayuden en cuanto penda de su arbitrio para el mejor éxito de sus interesantes operaciones.— Dado en la sala de Gobierno en Santiago de Chile a dos dias del mes de Setiembre del año de mil ochocientos treinta y cuatro.—

Prieto

Joaquin Tocornal.

Charles's passport from the President of Chile

bivouacked in the corner of a field where they ate their supper under a cloudless sky and knew no troubles.

The next morning they reached the last and highest house in the valley. Here the Maípo rushed furiously down across a narrow plain of shingle, with hills rising three to five thousand feet on either side, their steep faces mainly purple in colour, and strikingly stratified.

Several herds of cattle driven down from the higher valleys passed the group, a sign of the approaching winter that hurried their steps 'more than was convenient for geology'. The next night was passed where the Maípo divided into the Rio del Valle del Yeso and the Rio del Volcan, at the foot of a ten-thousand-foot mountain crowned by the mines of San Pedro de Nolasko, where there were still some patches of snow on the summit.

Higher up the valley the vegetation became scanty, though there were a few very pretty alpine plants; and scarcely any birds or insects were to be seen. All the valleys were filled with huge quantities of brightly coloured gravel, through which channels were cut by the rivers. Charles reckoned from what he had already seen of the geology that this gravel and other alluvium must have been deposited from the ocean during a previous era when the land was all underwater. That evening they reached the Valle del Yeso, which must once have been a large lake, but now contained a bed nearly two thousand feet thick of white and in many parts quite pure gypsum (calcium sulphate).

Next morning the hitherto good road degenerated into a steep zigzag track up the side of the westernmost of the two main ridges forming the Cordillera at this latitude. This was the Peuquenes ridge that separated the waters flowing westwards into the Pacific and eastwards into the Atlantic, and which was crossed by a pass 13,200 feet above sea level. At this altitude the mules had some difficulty, known as '*puna*' by the Chilenos, with breathing the rarefied air, and had to take a short rest every fifty yards. Charles made light of it, but was constrained to admit that it was nice when the excitement of finding abundant impressions of fossil seashells on the highest part of the ridge made him quite forget about the discomfort of the *puna*. In his notebook he wrote:

> When we reached the crest & looked backwards, a glorious view was presented. The atmosphere so resplendently clear, the sky an immense blue, the profound valleys, the wild broken forms, the heaps of ruins piled up during the lapse of ages, the bright colored rocks, contrasted

> with the quiet mountains of Snow, together produced a scene I never could have imagined. Neither plant or bird, excepting a few condors wheeling around the higher pinnacles, distracted the attention from the inanimate mass. – I felt glad I was by myself, it was like watching a thunderstorm, or hearing in the full Orchestra a Chorus of the Messiah.[134]

They descended the further side of the ridge to spend the night at an altitude of no more than ten thousand feet. During the night the sky clouded over in what Charles thought might be a threatening fashion, but he was relieved to be assured by the muleteer that without any thunder and lightning there was no danger of a bad snowstorm. Alexander Caldcleugh, when crossing the pass at the same time of year, had been caught in such a storm, and was trapped uncomfortably for some time in a cave that was the only available refuge.

In the morning, when the potatoes were still hard after boiling all night at the reduced pressure and temperature, Charles was amused to hear his companions conclude that 'the cursed pot (which was a new one) did not choose to boil potatoes'. After their potatoless breakfast, they crossed the intermediate area where cattle were brought to graze in the middle of the summer, but from which with the approach of winter they had been removed for fear of early snowstorms, just as the flocks of guanaco had already decamped of their own volition. They were then faced with the long climb across areas of permanent snow, with bold conical hills of red granite on either side, to reach the Portillo Pass at a height of 14,305 feet. After crossing the pass, they descended to the level of the first vegetation, and spent the night in the comforting shelter of some large rocks. Two days and another ridge later, and seven thousand feet lower, Charles had his first extended view of the Pampas extending to the east, but was rather disappointed to find it in no way superior to that from the crest of the Sierra de la Ventana that he had seen two years before on the far side of Argentina.

He was interested to find that the geological structures of the two main ridges he had now crossed differed in some significant respects. That of the Peuquenes ridge, and of the lower ridges flanking it to the west, was composed of a vast pile many thousands of feet thick of porphyries that had flowed as submarine lavas, mixed with angular and rounded fragments of the same rocks, thrown out of the submarine craters. The centres of these masses had then been covered by a considerable thickness of sedimentary material comprising red sandstone, porphyries fused by heat and pressure, and a chalky clay-slate mixture, passing into huge beds of gypsum formed by the evaporation of sea water. In the upper beds, fossil seashells that had once crawled on the bottom of the sea were not infrequent, so that the land had been upraised fourteen thousand feet since the Secondary period some three hundred million years ago. What is more, the shells in question had been accustomed to live in fairly deep water, so that the mountain must then have subsided several thousand feet to allow the necessary submarine strata to accumulate above the shells, followed by a final re-elevation.

The formation of the Portillo ridge was quite different. It consisted mainly of grand bare pinnacles of a red potash granite, covered low down on its western flank by a sandstone that had been heated enough to convert it into quartz. On top of the quartz there was a conglomerate consisting of a mixture of the pebbles and seashells from the Peuquenes ridge with the red potash granite of the Portillo ridge. It was evident that both ridges had been upheaved when the conglomerate was forming, but since the conglomerate had been thrown off at an angle of 45° by the red potash granite, the bulk of the injection had taken place after the accumulation of the conglomerate, and long after the elevation of the Peuquenes ridge. Hence the loftiest ridge in this part of the Cordillera was actually younger than the less lofty Peuquenes ridge, and the waters flowing to the east must have continued to break through it during the later stages of its growth. From this kind of argument Charles concluded that in most if not all of the Cordillera each of the parallel lines had been formed by repeated upheavals and injections at different times. 'Daily,' he wrote, 'it is forced home on the mind of the geologist that nothing, not even the wind that blows, is so unstable as the level of the crust of this earth.'

Although during this most geologically fruitful of all Charles's long rides in South America his attention was focused mainly on the rocks that he found, biology was not wholly neglected. On the patches of snow on both high ridges he noticed that the footsteps of the mules were stained a pale red, as though their hoofs were bleeding slightly. This was not the explanation, nor was the possibility that a red dust of porphyry was blowing from the mountains. On rubbing the snow on paper, and applying his lens he found that the colour was contained in groups of little circular balls, each a thousandth of an inch in diameter, which appeared to be 'the eggs of small molluscous animals'. He posted samples to Henslow, and the red snow was in due course identified as arising from an alga familiar to Arctic explorers that Charles called *Protococcus nivalis*.

More significant was a comment made after his descent from the Portillo Pass about the role of the Cordilleras in the geographical distribution of species, which was a subject about which he had been thinking for some while (see p.241). Its importance had been stressed by the geologist Charles Lyell – in due course to become one of Charles's two closest confidants – in the second of his three volumes on Principles of Geology, which Charles had with him on the *Beagle*.[135] In the *Journal of Researches*, Charles wrote:

> I was much struck with the marked difference between the vegetation of these eastern valleys and those on the Chilian side: yet the climate, as well as the kind of soil, is nearly the same, and the difference of longitude very trifling. The same remark holds good with the quadrupeds, and in a lesser degree with the birds and insects. I may instance the mice, of which I obtained thirteen species on the shores of the Atlantic, and five on the Pacific, and not one of them is identical. We must except all those species, which habitually or occasionally frequent elevated mountains; and certain birds, which range as far south as the Strait of Magellan. This fact is in perfect accordance with the geological history of the Andes; for these mountains have existed

as a great barrier since the present races of animals have appeared; and therefore, unless we suppose the same species to have been created in two different places, we ought not to expect any closer similarity between the organic beings on the opposite sides of the Andes, than on the opposite shores of the ocean. In both cases, we must leave out of the question those kinds which have been able to cross the barrier, whether of solid rock or salt-water.

Footnote. This is merely an illustration of the admirable laws, first laid down by Mr. Lyell, on the geographical distribution of animals, as influenced by geological changes. The whole reasoning, of course, is founded on the assumption of the immutability of species; otherwise the difference in the species in the two regions might be considered as superinduced during a length of time.

A great number of the plants and animals were absolutely the same as, or most closely allied to those of Patagonia. We here have the agouti, bizcacha, three species of armadillo, the ostrich, certain kinds of partridges and other birds, none of which are ever seen in Chile, but are the characteristic animals of the desert plains of Patagonia. We have likewise many of the same (in the eyes of a person who is not a botanist) thorny stunted bushes, withered grass, and dwarf plants. Even the black slowly-crawling beetles are closely similar, and some, I believe, on rigorous examination, absolutely identical. It had always been to me a subject of regret, that we were unavoidably compelled to give up the ascent of the S. Cruz river, before reaching the mountains: I always had the latent hope of meeting with some great change in the features of the country; but I now feel sure, that it would only have been following the plains of Patagonia up a mountainous ascent.[136]

The footnote explaining that the argument assumed that species were immutable was of course added some while after Charles had opened his first and strictly private notebook on the Transmutation of Species in July 1837, but in 1845 he was still careful to give no hint of the drastic change of his views that would not be publicly unveiled until the publication of *On the Origin of Species* in 1859.

Charles and his companion rode on with their mules, first eastwards and then north to the village of Luxan, where Charles had his first sight of a swarm of locusts from the sterile plains of the south descending to strip the trees of their leaves. And that night he wrote in his pocket-book:

> at night good to experience everything once – Chinches, the giant bugs of the Pampas; horribly disgusting to feel numerous creatures nearly an inch long & black & soft crawling in all parts of your person – gorged with your blood.[137]

The word '*chinches*' was the Spanish term in general use for bedbugs, though in his diary Charles referred to them as 'Benchuca' bugs. The insect *Triatoma infestans*, known today as the 'Vinchuca' bug, is the vector for *Trypanosoma cruzi*, the causative agent of Chagas's disease, which then as now was endemic in this and neighbouring parts of South America.

Three and a half months later, Charles wrote in his diary:

> They [Benchuca bugs] are also found in the Northern part of Chili & in Peru: one which I caught at Iquiqui was very empty; being placed on the table & though surrounded by people, if a finger was presented its sucker was withdrawn, & the bold insect began to draw blood. It was curious to watch the change in the size of the insects body in less than ten minutes. There was no pain felt. This one meal kept the insect fat for four months; In a fortnight, however, it was ready, if allowed, to suck more blood.[138]

The person who was bitten on this second occasion was not Charles himself, but one of his companions.

It has been suggested that the chronic illness involving repeated digestive upsets from which Charles suffered so severely from 1839 onwards, after he was married and lived at Down House, was Chagas's disease, with which he might have been infected at Luxan. However, this diagnosis does not fit with his failure to record any period of the fever that characteristically accompanies the initial infection. Moreover a strong reason for rejecting Chagas's disease as the cause of his illness is that during the last years of his life his symptoms decreased appreciably in severity, rather than increasing as they would had he been suffering from the trypanosomiasis. A variety of suggestions have been put forward as the cause of Charles's illness, of which the most plausible seems to be that some form of psychoneurosis was involved, perhaps triggered by a food allergy, though the evidence does not totally rule out other possibilities.[139] The contrast between the superbly energetic young man on the *Beagle*, and the semi-invalid at Down House who achieved so much despite his persistent sickness, remains something of an enigma.

From Luxan, Charles and his party rode on to Mendoza, where grazing land for cattle and orchards filled with fruit flourished thanks to artificial irrigation, but the inhabitants were 'sad drunken raggedmuffins' who had nothing to do except 'to eat, sleep & be idle'. They continued over a very dry and dusty plain to 'the solitary hovel which bears the imposing name of Villa Vicencio, and has been mentioned by every traveller'. A mile from the house there were streams containing tepid and slightly mineral water in which some water beetles were swimming. The next night was spent at the miserable village of Hornillos, and then there followed a memorable day.

Although the height of the Uspallata range was only half that of the main eastern chain of the mountains, it was geologically very similar with its nucleus of granite and great beds of crystalline rocks, which exactly resembled those that Charles had seen in the upper Tertiary strata of Patagonia, Chiloe and Concepción. He could therefore not expect to see any of the shells that never occurred in this formation, but ought to find Lignite or Carbonaceous shale, and particularly the silicified wood that he had discovered in Chiloe (see p.257). So with a conviction that he was again on the Tertiary strata, he set out on an apparently forlorn hunt, and as he explained in his next letter to Henslow, was splendidly rewarded:

> How do you think I succeeded? In an escarpment of compact greenish Sandstone I found a small wood of petrified trees in a vertical position,

or rather the strata were inclined about 20–30° to one point & the trees 70° to the opposite one. That is they were, before the tilt, truly vertical. The Sandstone consists of many layers & is marked by the concentric lines of the bark (I have specimens).* 11 are perfectly silicified, & resemble the dicotyledonous wood which I have found at Chiloe & Concepcion: the other 30–40 I only know to be trees from the analogy of form & position; they consist of snow white columns like Lots Wife† of coarsely crystall. Carb. of Lime. The longest shaft is 7 feet. They are all close together within 100 yd & about same level; no where else could I find any. It cannot be doubted that the *layers* of fine Sandstone have quietly been deposited between a clump of trees, which were fixed by their roots. The Sandstone rests on Lavas, & is covered by a great bed, apparently about 1000 ft thick, of black Augitic lava,§ & over this there are at least 5 grand alternations of such rocks & aqueous sedimentary deposits, amounting in thickness to several thousand feet. I am quite afraid of the only conclusion which I can draw from this fact, namely that there must have been a depression in the surface of the land to that amount. But neglecting this consideration it was a most satisfactory support of my presumption of the Tertiary age of this Eastern Chain (I mean by Tertiary that the shells of the period were closely allied, or some identical, to those which now live as in lower beds of Patagonia).[140]

In the *Journal of Researches* he wrote:

It required little geological practice to interpret the marvellous story which this scene at once unfolded; though I confess I was at first so

* The botanist Robert Brown was much impressed by the specimens, which he identified as coniferous trees with characteristics of the Araucaria (monkey puzzle) trees now common in Chile, and some affinities with yews.

† According to the book of Genesis, when Lot and his wife were escaping from the brimstone and fire rained by the Lord on Sodom and Gomorrah, Lot's wife disobeyed instructions not to look back behind them, and was changed into a pillar of salt.

§ Consisting chiefly of silica, magnesia, iron and lime.

much astonished that I could scarcely believe the plainest evidence of it. I saw the spot where a cluster of fine trees had once waved their branches on the shores of the Atlantic, when that ocean (now driven back 700 miles) approached the base of the Andes. I saw that they had sprung from a volcanic soil which had been raised above the level of the sea, and that this dry land, with its upright trees, had subsequently been let down to the depths of the ocean. There it was covered by sedimentary matter, and this again by enormous streams of submarine lava – one such mass alone attaining the thickness of a thousand feet; and these deluges of melted stone and aqueous deposits had been five times spread out alternately. The ocean which received such masses must have been deep; but again the subterranean forces exerted their power, and I now beheld the bed of that sea forming a chain of mountains more than seven thousand feet in altitude. Nor had those antagonist forces been dormant, which are always at work to wear down the surface of the land to one level: the great piles of strata had been intersected by many wide valleys; and the trees now changed into silex were exposed projecting from the volcanic soil now changed into rock, whence formerly in a green and budding state they had raised their lofty heads. Now, all is utterly irreclaimable and desert; even the lichen cannot adhere to the stony casts of former trees. Vast, and scarcely comprehensible as such changes must ever appear, yet they have all occurred within a period recent when compared with the history of the Cordillera; and that Cordillera itself is modern as compared with some other of the fossiliferous strata of South America.[141]

Today the remains of the silicified trees may still be seen near the Villa Vicencio at the head of the Uspallata Pass beside the main road from Mendoza to Santiago, although only the bases of the sandstone shells surrounding the trunks themselves have survived the onslaught of souvenir hunters over the last hundred years. A plaque erected in 1959 commemorates the visit of Carlos R. Darwin in 1835. About fifty years ago when the modern road was being rebuilt, the engineers found it necessary to do some dynamiting near the site, and had to destroy one of the trees. This yielded a large piece of silici-

The base of one of the fossilised trees as it may be seen today at the head of the Uspallata Pass

fied wood which has now joined the much smaller specimen brought back by Charles on the *Beagle*, for preservation at the Sedgwick Museum in Cambridge.

After crossing the Uspallata range and spending a night at the *estancia* of Uspallata, where the Custom House was the last inhabited place on the Argentinian side of the Cordilleras, Charles rode on across passes reputed to be some of the worst in the Cordilleras. They crossed the famous Puente del Inca, where soil and stones falling into a narrow channel had formed a natural bridge on meeting an overhanging part of the opposite bank. Charles was unimpressed, and thought it unworthy of the great monarchs whose names it bore. Nearby were the ruins of Tambillos, the southern limit of the huge area ruled by the Incas before their conquest by the Spaniards in 1533, where the remains of a paved Inca path can still be seen, and where the buildings of three Inca staging posts have recently been excavated. Charles had been told of similar buildings in other parts of the Cordilleras, often in places far from any supply of water, so that they could not have been in regular occupation. He was led to speculate whether there might not have been a reduction in rainfall during the past three hundred years, but after talking in Lima to a civil engineer about the number of abandoned ruins and areas of agricultural land to be seen in Peru was persuaded that it was more likely that the extensive irrigation systems often set up by the Incas had fallen out of use.

They continued down the valley of the Rio Aconcagua, and then turned south across a fertile plain where 'the Autumn being well advanced, the leaves of many of the fruit trees were falling, & the labourers were all busy in drying on the roofs of their cottages, figs & peaches; while others were gathering the grapes from the vineyards. It was a pretty scene, but there was absent that pensive stillness & the song of the robin at dusk which makes the autumn in England indeed the evening of the year'. Charles's homesickness was beginning to make itself felt after nearly three and a half years absence. They reached Santiago on 10 April after a twenty-four-day journey that well repaid their trouble.

Charles returned to Richard Corfield's house in Valparaiso on 17 April. Six days later the *Beagle* anchored in the harbour, having completed her survey of the coast south of Concepción, and FitzRoy sent boats ashore. Charles came on board extra well-armed against finding the Captain in one of his difficult moods, for he could report that FitzRoy had at last been promoted to become a full Post-Captain by rank as well as title. But FitzRoy's pleasure at the news was tempered by his continued failure to have obtained from headquarters any reward for Stokes or Wickham for their manifest exertions.

CHAPTER 23

A Last Ride in the Andes, from Valparaiso to Copiapó

The *Beagle* sailed on up the coast to the north, while Charles set off on 27 April 1835 for the longest of his rides through the Andes, 220 miles in a straight line to Coquimbo, though much further on the actual road, and then on to Copiapó where FitzRoy had promised to call for him. He was again accompanied by Mariano Gonzales, but this time he bought four horses for riding, and two mules which would carry the baggage on alternate days. The six animals only cost him twenty-four pounds, and he sold them again at Copiapó for twenty-three. They travelled independently as before, cooking their own meals and sleeping in the open air, though the nights were sometimes frosty.

For geological purposes, they made an initial detour to the foot of the Bell Mountain, whose height had been determined by triangulation from the *Beagle* as 6200 feet, similar to others in the truly alpine country, though all were dwarfed by Acongagua's 23,000 feet in the distance. Charles rated the view up the Quillota valley of the Cordilleras thickly covered with new snow as one of the finest in Chile.

On 3 May they passed through the port of Quilimar on the coast, where Charles heard that the *Beagle* had conducted a survey, and all the inhabitants were convinced that she was a smuggler. They complained of the entire want of confidence shown by the Captain in not coming to any terms, each man thinking wrongly that his neighbour was in on the secret. They proved very hard to undeceive, and Charles commented on how little even the upper classes understood the wide distinction of manners, for as he said, 'A person who could possibly mistake Capt. FitzRoy for a smuggler, would never perceive any difference between a Lord Chesterfield & his valet.'

Finding the country along the coast of little geological interest, they struck inland into a *mineral* or mining district called Los Hornos, where the principal hill was so thickly drilled with excavations that it resembled a hugely magnified anthill. The miners were a race of men not unlike men-of-war sailors in some of their habits, for after working for weeks on end in the most desolate places, they would come down to the villages on feast days, and squander their sometimes considerable earnings as rapidly as possible on drink and extravagant clothes before returning to their miserable abodes. At the *mineral* of Punitague much copper and gold had been extracted, and there were mercury mines that had not been worked.

On 12 May they visited the copper mines of Panuncillo, which belonged to Charles's friend Alexander Caldcleugh of Santiago. They had been bought for a song by him, the ore being the common yellow copper pyrites, a sulphide which was not the most readily reducible form. Charles had read, and found hard to believe, an account of the huge size of the loads of ore that the Chilean miners known as '*apires*' could carry up daily from a deep mine. But he confirmed for himself that twelve loads, each averaging two hundred pounds, could be brought up every day from a depth of 240 feet. It was 'a wonderful instance of the amount of labour which habit, for it can be nothing else, will teach a man to endure'.

He was interested to be told by the Mayor-domo of the mines, Don Joaquin Edwards, a young man and himself the son of an Englishman, how the reputation of the English buccaneers for heresy, contamination and evil still persisted in some quarters in Coquimbo. It had not yet been forgotten how one of them had removed the Virgin Mary from the church, and returned the following year for St Joseph, for he thought it a pity that the Lady should not have a husband. And Mr Caldcleugh related how an old lady at a dinner in Coquimbo had remarked how strange it was that she should have lived to dine in the same room with an Englishman, when twice as a girl, the cry of '*Los*

Ingleses' had caused everyone to take to the mountains with their valuables.

Two days' ride to the north they found the *Beagle* in the little harbour of Herradura, a few miles south of Coquimbo, with the crew living in tents on the shore while the ship was thoroughly refitted for her long journey home across the Pacific. Charles found some lodgings in Coquimbo that he shared with FitzRoy, and spent a few days exploring the neighbourhood. When he arrived there, the ground was totally dry, but the next morning the first rain of the season fell, and he was impressed to find that only ten days later all the hills were covered with an inch of grass. Dining one evening with Don Joaquin there was the initial rumble of an earthquake, but thanks to the screams of the ladies, the running of servants and the rush of the gentlemen to the doorway, he could not distinguish the actual movement. He took a walk up the valley to inspect the step-like plains of shingle which had been described in the third volume of Lyell's *Principles* that had just reached him,[142] and was happy to find himself confirming that they were most likely to have been caused by an elevation of the shore in successive steps of the kind that he and FitzRoy had observed at Concepción. He went with Don José Maria Edwards, a pleasant young Anglo-Chilean, up the valley of the Elque to visit his father's famous silver mine, this being a specially enjoyable trip because at its altitude of three thousand feet there were no fleas in the rooms. Leaving Don José behind at a pleasant hacienda belonging to a relative, Charles rode further up the Rio Elque to the point where it was joined by the Rio Claro in a beautiful valley famous for its figs and grapes. He had been told he would find petrified shells there, which turned out to be true, and petrified beans, which turned out to be small white quartz pebbles. Having seen all he wanted, he returned to the hacienda to pick up Don José, and on the evening of 27 May they arrived back in flea-ridden Coquimbo.

The *Beagle* was due to sail back to Valparaiso in a few days' time, and then up to Copiapó to pick up Charles and proceed to Lima. FitzRoy had also hired a small vessel which would survey the north coast of Chile under Sulivan's command before rejoining the *Beagle* at Lima. Charles set out for the valley of Guasco (spelled Huasco on a modern map), choosing the road nearest the coast because in the interior there would be no fodder at all for the animals. The recent shower had reached about halfway to Guasco, giving the country a slight tinge of green. He was once more reminded nostalgically of England, and wrote that 'travelling in these countries, like a prisoner shut up in gloomy courts, produces a constant longing for such scenes'. The northern part of Chile was indeed not easy on travellers. A mountainous rocky desert extended inland to the Cordilleras, generally uninhabited except by multitudes of small snails that appeared only at dawn when the ground was damp with dew from the clouds that hung low over most of the coast of Chile and Peru during the winter months. In some places there were herds of guanaco, which as Charles had seen on the great pebble-coated plains of southern Patagonia, were the only large mammals capable of surviving in such an arid environment. Even where the desert plain was crossed by a river, the watered area on its banks was very narrow, and provided no pasture for animals. The great majority of the towns and villages through which they passed depended on a mine nearby for their existence, and the most agriculturally productive area of the whole region was Guasco Alto at the head of the valley near the Cordilleras, famous for its dried fruit. Between Coquimbo and Guasco, they could buy for their animals only a little corn and straw at a small village on the first day, and on the second an armful of dirty straw from a civil old gentleman looking after a copper-smelting furnace.

In the Guasco valley Charles stayed for two days 'on account of my animals' with Mr Hardy, an owner of copper mines at Freyrina. The view up the valley had been described as outstandingly beautiful, but it seemed to Charles that a little washing with a neutral tint would have improved its accuracy, for the hills were as bare as a bone. Rain had last fallen over a year ago, and the inhabitants heard with great envy of the shower at Coquimbo. From the looks of the weather, however, they thought that luck might be coming their way, and soon it did. Charles was at Copiapó by then, and the people spoke with equal envy of the abundant rain at Guasco.

He called one evening at the house of the '*Gobernador*' of Freyrina. The signora was from Lima, and affected 'blue-stockingism & superiority' over her neighbours. Yet this learned lady could never have seen a map, for when one day she saw a coloured atlas lying on a piano at Mr Hardy's house, she exclaimed, '*Esta es contradanca.*' (This is a country dance.) '*Que bonita*!' (How pretty!)

After spending a couple of days at Ballenar (now Vallenar), a handsome town of appreciable size that owed its prosperity entirely to some silver mines, Charles decided not to take the direct road to Copiapó, today part of the Pan American Highway, but to cross the hills in order to get into the valley of the Rio Copiapó, running back in a south-easterly direction. As before, the country was barren in the extreme, and the supply of fodder for the animals was extremely limited. On the first evening they found a valley in which a small stream flowed further at night than in the day, thanks to evaporation. At its source there was plenty of firewood, so that it was a good place to bivouac, but there was not a mouthful for their animals to eat. There were all the same two cottages of Indians there with a troop of donkeys employed on carrying firewood and other goods to the mines, whose sole food was evidently the stumps of the dry twigs of the bushes.

The next night they found a site where there had once been a smelting furnace, and where water and firewood were available, but once more no food for the animals. At noon on 12 June they at last arrived at the hacienda of Potrero Seco in the Copiapó valley, for which Charles was heartily grateful, for 'it is most disagreeable to hear whilst you are eating a good supper, your horse gnawing the post to which he is tied & to know that you cannot relieve his hunger'. But to all appearance the horses were quite fresh, and showed no sign that they had eaten nothing for two and a half days.

The hacienda belonged to some English merchants, whose agent Mr Bingley was there at the time. His main business was the shipment of copper ores, and its success depended wholly on the cultivation of sufficient clover every year for the pasturage of mules, on the part of the estate twenty-five miles in length on either side of the river that could be properly irrigated. The water this year reached up to a horse's belly at the hacienda, where the stream was fifteen yards wide and rapid, and a little water was actually reaching the sea, for there had been a good fall of snow in the Andes. This could not be relied upon, and thirty years had sometimes passed without a single drop entering the Pacific. Rain in the lower country might save the situation by enabling the mules and cattle to be pastured for a while in the mountains, but this too often failed. Mining companies that did not own enough land to pasture their mules when the river was very low were then ruined.

Proceeding up the valley towards the Cordilleras, Charles lunched with Don Eugenio Matta – 'the pleasantest gentleman I have met in Chili' – and met General Aldunate, who was the first Governor of Chiloe after its liberation from the Spaniards, and was well known to Captain P.P. King and his officers on the previous voyage of the *Adventure* and *Beagle*. That evening Charles stayed with Don Benito Cruz, yet another most hospitable and kind gentleman – 'indeed I defy a traveller to do justice to the goodnature with which strangers are received in this country' – at the hacienda of Las Amolanos.

Next morning he hired mules and a guide to penetrate a little further into the desolate and barren country at the head of one of the branches of the Copiapó valley, but finding little of interest retraced his steps to Don Benito's hacienda, where the geology proved to be more to his liking. First, as he had been doing systematically wherever he went, he collected the fossil shells which abounded in the strata of yellowish limestone or equivalent chalky slate-rock, and the identification of whose species provided valuable information for dating their source. Second, he looked at the great pile above it, two or three thousand feet thick, of a coarse bright-red conglomerate mixed with beds of red sandstone and layers of green and other-coloured porphyries, embedded in which were thousands of huge silicified trunks of coniferous trees. One that he found was eight feet long, and another was eighteen feet in circumference. He felt it 'marvellous' that every vessel in so thick a mass of wood could have been converted into silicate. Above this conglomerate there was red sandstone two or three hundred feet thick, and more chalky slate-rock, containing shells which he identified as the species *Gryphæa darwinii* and *Turritella andii*, whose presence established that the whole vast pile of strata, not less than eight thousand feet thick, was of the same age. The shells would probably have lived at a depth of not more than two hundred feet, so that the bottom of the sea in their day must have subsided many thousands of feet to receive the submarine strata that eventually lay above them. Charles thus reached a conclusion about the manner in which the land had risen and fallen several times in previous ages that nicely confirmed what he had found at the Uspallata Pass.

On 22 June Charles arrived back in Copiapó, where he stayed at Mr Bingley's house. The *Beagle* had not yet arrived, so he hired a muleteer and eight mules for another brief trip to the Cordilleras, where he climbed near enough to the crest of the first ridge of mountains to suffer from the *puna*. But the extreme dryness at this southern end of the Atacama Desert with a cloudless sky, a very low temperature, and a furious gale of wind, made conditions less than comfortable, and he returned with relief to the smell of the fresh clover at Copiapó, where the *Beagle* had berthed on 3 July, under the temporary command of Wickham. On the evening of 5 July he gave his '*adios*' with hearty goodwill to Mariano Gonzales, with whom he had ridden so many leagues in the Andes. Next morning the *Beagle* sailed for Iquique in Peru.

CHAPTER 24

The Wreck of HMS *Challenger*

Later that month, Charles wrote to his sister Caroline:

> It is very hard & wearisome labor riding so much through such countries as Chili, & I was quite glad when my trip came to a close. Excluding the interest arising from Geology, such travelling would be down right Martyrdom. But with this subject in your mind, there is food in the grand surrounding scenes for constant meditation. When I reached the port of Copiapo, I found the Beagle there, but with Wickham as temporary Captain. Shortly after the Beagle got into Valparaiso, news arrived that H.M.S. Challenger was lost at Arauco, & that Capt Seymour a great friend of FitzRoy & crew were badly off amongst the Indians. The old Commodore in the Blonde was very slack in his motions, in short afraid of getting on that lee-shore in the winter; so that Capt FitzRoy had to bully him & at last offered to go as pilot. We hear that they have succeeded in saving nearly all hands, but that the Captain & Commodore have had a tremendous quarrel; the former having hinted some thing about a Court-Martial to the old Commodore for his slowness. We suspect that such a taught [*sic*] hand, as the Captain is, has opened the eyes of every one fore & aft in the Blonde to a most surprising degree. We expect the Blonde will arrive here in a very few days & all are very curious to hear the news; no change in state politicks ever caused in its circle more conversation, than this wonderful quarrel between the Captain & the Commodore has with us.[143]

According to FitzRoy[144] the *Challenger*, commanded by Captain Michael Seymour, who served at the South American Station of the

Royal Navy in 1827–29 and 1833–35, had sailed from Rio de Janeiro bound for Chile on 3 April 1835. She had much bad weather off Cape Horn, which appreciably lengthened her passage, and on 19 May in heavy rain, with a strong north-west wind and poor visibility, her position was thought by dead reckoning to be well out to sea south of Concepción. At 5 p.m. a precautionary sounding was taken, and no bottom was found with 210 fathoms of line, though in fact they were only about twelve miles north of the island of Mocha, not far from the coast of the mainland. On the insistence of the Master, though against the Captain's judgement, a course north by east to the entrance of Concepción Bay was held, until soon after 8 p.m. the water suddenly shallowed and breakers were seen by the lookout men. The orders 'Helm down,' 'About ship' and 'Mainsail haul' were immediately given, but it was too late. The ship was rising to a heavy rolling breaker, and struck heavily astern on the rocks not far from the mouth of the Rio Leübu, about twenty miles south of Concepción. Here the *Challenger* was soon reduced by the breakers to a helpless wreck that would never float again.

In the following days, every transportable article of value was removed from the ship to an encampment on the shore. A great many Indians, many on horseback, assembled and gave valuable assistance by hauling the rafts ashore, while the women rode into the furious surf and brought the boys out behind them on their saddles. What a lesson this was, FitzRoy reflected, to the 'wreckers' of some other coasts, whose inhabitants are called civilised.

On 21 May, Lieutenant Collins and the Assistant Surgeon of the *Challenger* managed to get through to Concepción for help. The excellent British Consul Mr Rouse immediately responded by setting out for the wreck with horses and mules carrying the few useful things he could muster on the spot, which included two small tents that had been lent to him by the *Beagle* after February's earthquake. From Concepción, Lieutenant Collins went on to Talcahuano to hire a vessel, but the owner of the only suitable one at the port, the American schooner *Carmen*, asked such an unreasonable price for her services that Collins was obliged to return empty-handed to the wreck. Here he found that Captain Seymour had decided that the encampment should be moved from the site of the wreck to a more convenient spot under the Tucapel Heights at the mouth of the river Leübu. By 8 June everything had been transferred to the new camp, a flagstaff had been erected and the flag was flying. Here the officers and crew of the *Challenger* settled down to wait, suffering from a veritable plague of mice, but with an excellent supply of the most delicious potatoes that they had ever eaten.

In the meantime, the *Beagle* had sailed back from Herradura, arriving at Valparaiso on 14 June. On 16 June an English merchantman received a letter from Santiago briefly reporting the total loss of HMS *Challenger*. FitzRoy at once sprang into action, and located a Swedish vessel which on 20 May had seen a ship in difficulties on the coast south of Concepción that might well have been the *Challenger*. At the post office next day he intercepted a letter for Commodore Mason labelled 'Despatches by *Challenger*', which he delivered in person to the *Blonde*. His offer of all the help that he could give was immediately accepted by the Commodore. After instructing Wickham to take charge of the *Beagle*, and call at Copiapó to pick up Charles, before proceeding to Iquique and Callao where he would rejoin the ship, FitzRoy embarked on the *Blonde* with the Master's Assistant Mr Usborne, his Coxswain James Bennett, and a whaleboat. Arriving at Talcahuano in the bay of Concepción on 21 June, he took a boat and went ashore to obtain a guide and horses in order to follow the Commodore's instructions to go overland to Captain Seymour's camp, and concert measures for rescuing the crew and the remaining ship's stores. For this purpose he recruited an additional assistant familiar with the half-Indian natives of the frontier, in the shape of the Hanoverian Anthony Vogelborg who had helped him to measure the elevation of the land at Santa Maria after the earthquake.

FitzRoy and his party were unable that evening to cross the first obstacle on their journey, the Bio Bio River, half a mile wide, because the crew of the ferry-boat were not to be found. But at dawn on 22 June they were soon across the river, and galloped on over the hills at

Point Coronel and the sandy beaches called Playa Negra and Playa Blanca, with Santa Maria in the distance, until they arrived at Arauco, where they stayed with the local *Comandante*, Colonel Valenzuela. Provided by him with fresh horses, they hurried on next day across well wooded land with an endless series of steep and muddy ravines, swelled by heavy rain. In his published account of their desperate ride, FitzRoy hoped that a story from the history of Arauco might 'shorten our journey, and divert us for a time from mud, and rain, and wind', but the legend of the exploits of the great Araucanian chief Colocolo at the end of the sixteenth century does not have exactly the effect on the reader that he intended. Late that evening they reached the banks of the Leübu, where FitzRoy hailed as loudly as he could, 'Challengers ahoy': there was no reply. He tried again, and a faint 'Hallo' repaid him for every difficulty. Soon he was embarked in the *Challenger*'s dinghy, the only boat saved, and was met by Captain Seymour at the landing place. At the encampment all had turned out to hear the welcome news that the *Blonde* was at Talcahuano, and coming to their relief. With the *Challenger*'s crew was the Consul Mr Rouse, who had responded so promptly to the shipwreck, and had been of great assistance in maintaining good relations with the Araucanians, and daily supplies of provisions.

Early next morning (the twenty-fourth), FitzRoy went with Captain Seymour to the heights of Tucapel, overlooking the river and commanding an extensive view of the sea. The orderly layout of the encampment was impressive, with its fourteen or fifteen tents surrounded by a palisade with a ditch, and seemed likely to cause even a large body of Indians to hesitate before attacking it. Anxious to return as quickly as possible to Talcahuano in order to tell the Commodore how easily the *Blonde* would be able to take off both people and stores at the mouth of the Leübu, FitzRoy recrossed the river, and rode off with two companions 'sparing neither whip nor spur' in the hope of reaching Arauco by midnight. But thanks to the incessant rain, and a mistake by the guide with the second set of horses that added two extra rivers to be crossed, they were obliged to spend the night at an old farmhouse where the fleas were distressingly rampant. The next morning they struggled on to Arauco to have breakfast with Colonel Valenzuela, whom they found surprised that they had succeeded in spending a night at the Leübu encampment, and worried by reports that about three thousand possibly hostile Indians were marching northwards. They rode away on the good horses that they had brought from Concepción three days earlier, but encountered further problems with ferries whose boatmen were reluctant to cross the swollen rivers, and roads in even worse condition than for the outward journey. It was not until ten o'clock on the morning of 26 June that FitzRoy was able to deliver Captain Seymour's letter to Commodore Mason on board the *Blonde*.

It turned out that the Commodore had engaged the American schooner *Carmen*, for which such exorbitant demands had been made a month previously, to go in search of the crew of the *Challenger*. Mr Usborne had been sent in her, with the second Master and three seamen from the *Blonde*, FitzRoy's Coxswain, and the *Beagle*'s whale-boat. But the *Carmen* was not a well-found vessel, and in the end became more of a hindrance than a help. The *Blonde* sailed from Concepción Bay on 27 June, but thick weather and half a gale from the north kept her offshore and prevented her crew for the whole of the next week from making any reliable sightings of the Tucapel heights that would guide them to the encampment at the mouth of the Leübu. At last on 5 July the weather cleared, and the *Blonde* found herself lying becalmed five miles from the land, but able to see the *Challenger*'s flags flying on the heights. Three boats were launched from the *Blonde*, but a current setting along the shore kept them from reaching the landing place at the mouth of the Leübu until the evening, when FitzRoy landed and agreed with Captain Seymour that further operations would have to be postponed to the next day. At nine o'clock in the morning the *Blonde* was duly anchored about a mile from the landing place, every boat was hoisted out, and the embarkation proceeded rapidly. At six in the evening, Captain Seymour came on board with the last party of his crew, and at eight the *Blonde* weighed and set sail for Talcahuano before a favourable southerly breeze.

14. Bay of Valparaiso looking towards Viña del Mar, by Conrad Martens.

15. Street with wooden houses in San Carlos, by Conrad Martens.

16. Walking dress of the Females of Lima, by Syms Covington.

17. Darwin's finch *Geospiza magnirostris*, by John Gould.

18. Cactus finch *Cactornis scandens*, by John Gould.

19. Galapagos mocking bird *Mimus parvulus*, by John Gould.

20. View of Jamieson Valley at the Wentworth Falls, by Conrad Martens.

August 1835

At eleven in the morning of 7 July, a dismasted vessel with an English blue ensign hoisted was sighted five miles to the north of the *Blonde*, which after being identified as the *Carmen*, was taken in tow. Mr Osborne explained that on 29 June, when she was investigating a fire that she had seen on Tucapel Head, four seamen had been furling the topsail, the schooner gave a sudden plunge into a high swell, bringing down the foremast head, fore-topmast and topsail yard, followed by the mainmast. The four men had been saved, though James Bennett was badly bruised, and the ship was left drifting helplessly southwards, with almost no tools for carrying out any repairs. At the latitude of Valdivia, the southerly wind had set in, and after Mr Usborne had used his knowledge of the coast from his recent surveys to save the ship from being blown on to the shore and wrecked, she had been carried north to where the *Blonde* had found her. The *Blonde* then worked steadily northwards with the *Carmen* in tow until she was able to anchor at Talcahuano shortly before midnight on 8 July.

After dealing with much official business, thanking the authorities for their assistance, and saying goodbye to Mr Rouse, FitzRoy sailed in the *Blonde* for Valparaiso where she arrived on 13 July. Four days later she proceeded to Coquimbo, where the *Conway* was at anchor, waiting to take the officers and two-thirds of the crew of the *Challenger* back to England. From Coquimbo the *Blonde* sailed north up the coasts of Chile and Peru, touching at Cobija, Arica and Ilo – 'hapless arid dwelling places for either man or beast, as I have ever seen' – until FitzRoy was able to rejoin the *Beagle* at Callao on 9 August.

So concludes FitzRoy's own account of his characteristically enthusiastic rescue of his old friend Michael Seymour and his crew after the wreck of the *Challenger*. What, however, remains unclear is precisely how the tremendous quarrel with the Commodore, described so graphically by Charles, actually came about. As FitzRoy told it, news of the shipwreck first reached Valparaiso from Santiago almost a month after it had taken place. When he himself took the *Challenger*'s dispatches in a letter from Santiago to the Commodore on 17 June, and no doubt asked very forcefully for something to be done, his request was immediately granted. And after that he had no cause for accusing the Commodore of anything worse than hiring a very unsuitable boat to help with the rescue. Nevertheless, the occurrence of the wreck must in fact have been known to the authorities in Concepción on 21 May, when Mr Rouse went to the scene, and the news was presumably passed on to the Commodore in Valparaiso soon afterwards. FitzRoy's main and not unreasonable grounds for complaint were therefore that the Commodore had taken no effective action between about 24 May and 17 June. However, for reasons of tact it seemed to him best to omit this detail from the story published in 1839, which he was able to do without making his story positively inaccurate.

At Seymour's court martial on board the *Victory* at Portsmouth three months later, he presented in his defence a document from FitzRoy suggesting that changes in the ocean currents due to the earthquake had been responsible for the error in the dead reckoning of her distance from the coast at the time of the wreck. He was honourably discharged, highly praised for his conduct, and given another ship.[145]

CHAPTER 25

From Copiapó to Lima

On 12 July the *Beagle* anchored at Iquique in Peru, built on a stretch of sand at the foot of a cliff of rock two thousand feet high. All the water and provisions for the thousand inhabitants had to be supplied from the neighbouring port at Arica, where there was a small river, or from elsewhere. In former times there had been two extremely rich silver mines in the vicinity which were now exhausted. With some difficulty, Charles hired a guide and two mules to take him to the saltpetre works that were the current support of Iquique, though their product was not pure potassium nitrate, an essential component in the manufacture of gunpowder, but was mixed with sodium nitrate and some common salt. After ascending to the top of the cliff up a zigzag track, the road strewn with the bones and skins of dead mules and jackasses led across a complete and utter desert covered by a thick crust of salt on a salty sandstone. The only living creatures to be seen were one or two vultures that fed on the carcases, and apart from them there were not even any insects. However, Charles was glad to have seen this typical example of much of the coast of Peru.

A week later the ship anchored in the harbour of Callao, the port serving Lima, described unflatteringly by Charles as a miserable, filthy, ill-built small seaport, whose inhabitants displayed every possible mixture of European, Negro and Indian blood. Its climate was famous for a proverbial absence of rain, though there was generally a thick drizzle or Scotch mist hanging over the coast, commonly called Peruvian dew, that kept everything slightly damp. Charles took a walk on the barren little island of San Lorenzo, where the side of the 1200-foot mountain fronting the harbour was worn into three terraces covered by many tons of shells of species still existing in the sea, providing proof of relatively recent elevation of the land. He was interested to find, embedded in the eighty-five-foot bed among the shells, a bit of cotton thread, a plaited rush, and the head of a stalk of Indian corn, which was the earliest recorded finding of kernels of maize at such a site.[146] Although he found no ceramics there, he did find some fragments of coarse red earthenware on the mainland plain opposite the island at about the same height, suggesting that the elevation to eighty-five feet above sea level had taken place within a period of the order of two thousand years.

Charles took a coach to Lima, where he spent five very pleasant days, noting that 'a residence of some years in contact with the polite & formal Spaniards certainly improves the manners of the English merchants'. Another of his comments was:

> There are two things in Lima which all travellers have discussed; the ladies "tapadas", or concealed in the saya y Manta, & a fruit called Chilimoya.* To my mind the former is as beautiful as the latter is delicious. The close elastic gown fits the figure very closely & obliges the ladies to walk with small steps which they do very elegantly & display very white stockings & very pretty feet. They wear a black silk veil, which is fixed round the waist behind, is brought over the head, & held by the hands before the face, allowing only one eye to remain uncovered. But then that one eye is so black & brilliant & has such powers of motion & expression that its effect is very powerful. Altogether the ladies are so metamorphised [*sic*] that I at first felt as much surprised as if I had been introduced amongst a number of nice round mermaids, or any other such beautiful animal. And certainly they are better worth looking at than all the churches & buildings in Lima. Secondly for the

* The *saya*, an overskirt with several stiff pleats, and a black silk *manto*, a sleeveless hood covering the head and bust and fastened at the waist, were worn uniquely by the women of Lima when they went into the town or to church. This costume had evolved in early colonial days around 1560, and led to the ladies being called *tapadas*, or women who hid their faces in a mantle. The chilimoya is *Anana cherimola*, the Peruvian custard apple.

> Chilimoya, which is a very delicious fruit, but the flavour is about as difficult to describe, as it would be to a Blind man some particular shade of colour; it is neither a nutritive fruit like the Banana, or a crude fruit like the Apple, or refreshing fruit like the Orange or Peach, but it is a very good & large fruit & that is all I have to say about it.[147]

Syms Covington painted a picture of one of the ladies of Lima (Plate 16) that fits Charles's description very nicely.

On 9 August, the *Blonde* arrived at Callao with FitzRoy, who then went to Lima for a while to examine some old charts and papers that interested him. The road from Callao to Lima was infested with gangs of mounted robbers, so Charles stayed on board the *Beagle* writing up his geology notes on Chile. He found the delay in departing, at long last bound west for home, for the Galapagos rather irksome, but FitzRoy wrote to reassure him: 'Growl not at all – Leeway will be made up. Good has been done unaccompanied by evil – ergo – I am happier than usual.'

At Herradura in May, FitzRoy had borrowed from Don Francisco Vascuñan a thirty-five-ton schooner called *Constitución*, in which Lieutenant Sulivan, Philip Gidley King, Mr Stewart and Mr Forsyth volunteered to complete the survey of the north coast of Chile, before rejoining the *Beagle* at Callao. With a small boat and crew from the *Beagle*, and a native with his balsa (a fishing raft built of reeds), they accomplished their survey in a most satisfactory fashion, and duly anchored at Callao on 30 July. Sulivan reported that the *Constitución* was a handy craft and an excellent sea boat. After some days' consideration FitzRoy then decided, notwithstanding his unhappy experience with the *Adventure*, to buy the *Constitución* for £400 from his own pocket as agreed with Don Francisco, and to fit her out at Callao to conduct a similar survey of the coast of Peru. Lieutenant Sulivan could not again be spared from the *Beagle*, but Mr Usborne and Mr Forsyth could do the job, and afterwards return in a merchant vessel to England. Although the Hydrographer fully supported FitzRoy's application to the Admiralty for permission to make this purchase, stating that the subsidiary craft would materially assist the survey, the Minutes written across his letter to their Lordships refer to 'former papers forbidding him to hire a tender', and state, 'Inform Capt. FitzRoy the Lords highly disapprove of this proceeding, especially after the orders which he previously received on the subject.'[148] However, thanks to FitzRoy's zeal to complete his survey as efficiently as possible, he was never deterred by the evident ill will of their Lordships towards him.

Before leaving for the Galapagos, Charles wrote a note to Alexander Usborne asking him to look out for any shells that he might find along the coast of Peru on the remnants of ancient beaches elevated to heights up to a few hundred feet above sea level. He was to 'mark on Paper the name of the Place, & estimate carefully the vertical height. It would be well always to state the amount of (& reasons for) conviction which you feel respecting their origin.'[149] There is no record of Usborne having ever followed Charles's careful instructions for collecting shells in Peru. When, however, he was asked many years later for his recollections of the *Beagle*, he recalled that Charles was invariably a most amiable person, and added: 'He was a dreadful sufferer from sea-sickness, and at times when I have been Officer of the Watch, and reduced the sails on the ship, making her moves easy, and relieving him; I have been pronounced by him to be "a good officer", and he would resume his microscopic observations in the Poop Cabin. He was a general favourite on board and dubbed the "Philosopher", but we never anticipated he would have become so distinguished a man as he afterwards proved to be.'[150]

CHAPTER 26

The Galapagos Islands

On 7 September the *Beagle* sailed for the Galapagos, with a steady gentle wind and a gloomy sky. The fifteenth was spent on a quick survey of the outer coast of Chatham Island (San Cristóbal) at the south-eastern corner of the archipelago. The next morning a whale-boat was left at Hood Island (Española) for the Master Edward Chaffers and the Mate Arthur Mellersh to examine the island and the anchorages around it, and the yawl was sent away under the command of Lieutenant Sulivan with Messrs Stewart and Johnson and ten seamen to survey the central part of the archipelago. In the afternoon FitzRoy and Charles landed for an hour at the south-west end of Chatham, where Charles's first impression was:

> These islands at a distance have a sloping uniform outline, excepting where broken by sundry paps & hillocks. The whole is black Lava, *completely* covered by small leafless brushwood & low trees. The fragments of Lava where most porous are reddish & like cylinders; the stunted trees show little signs of life. The black rocks heated by the rays of the Vertical sun like a stove, give to the air a close & sultry feeling. The plants also smell unpleasantly. The country was compared to what we might imagine the cultivated parts of the Infernal regions to be.[151]

FitzRoy wrote:

> This part of the island is low, and very rugged. We landed upon black, dismal-looking heaps of broken lava, forming a shore fit for Pandemonium. Innumerable crabs and hideous iguanas started in

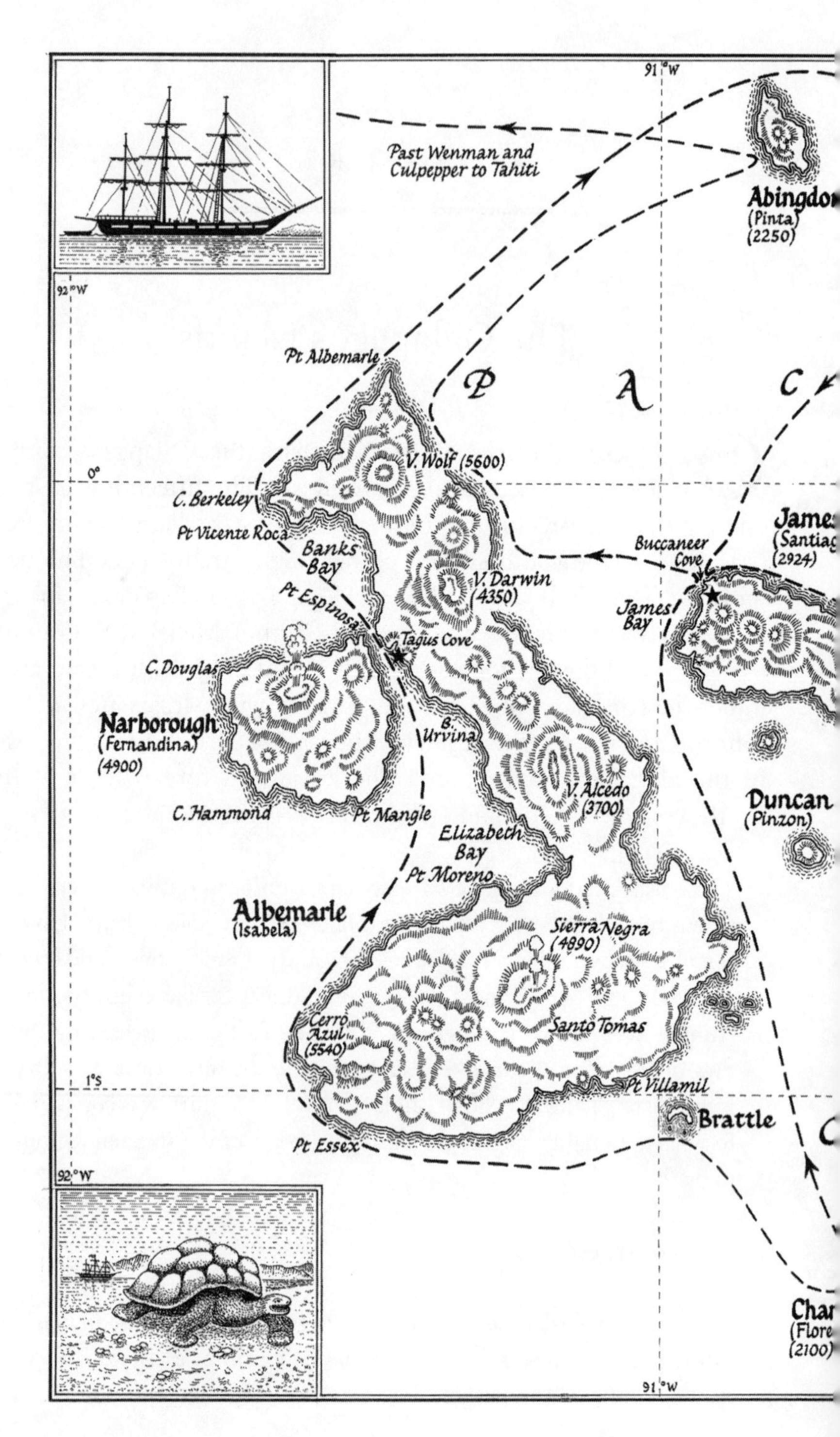
91°W
Past Wenman and
Culpepper to Tahiti
Abingdo
(Pinta)
(2250)
92°W
Pt Albemarle
P A C
0°
V. Wolf (5600)
C. Berkeley
Pt Vicente Roca
Banks
Bay
Pt Espinosa
V. Darwin
(4350)
Buccaneer
Cove
James
(Santiag
(2924)
James
Bay
Tagus Cove
C. Douglas
Narborough
(Fernandina)
(4900)
B.
Urvina
V. Alcedo
(3700)
C. Hammond
Pt Mangle
Duncan
(Pinzon)
Elizabeth
Bay
Pt Moreno
Albemarle
(Isabela)
Sierra Negra
(4890)
Cerro
Azul
(5540)
Santo Tomas
1°S
Pt Villamil
Brattle
Pt Essex
92°W
Char
(Flore
(2100)
91°W

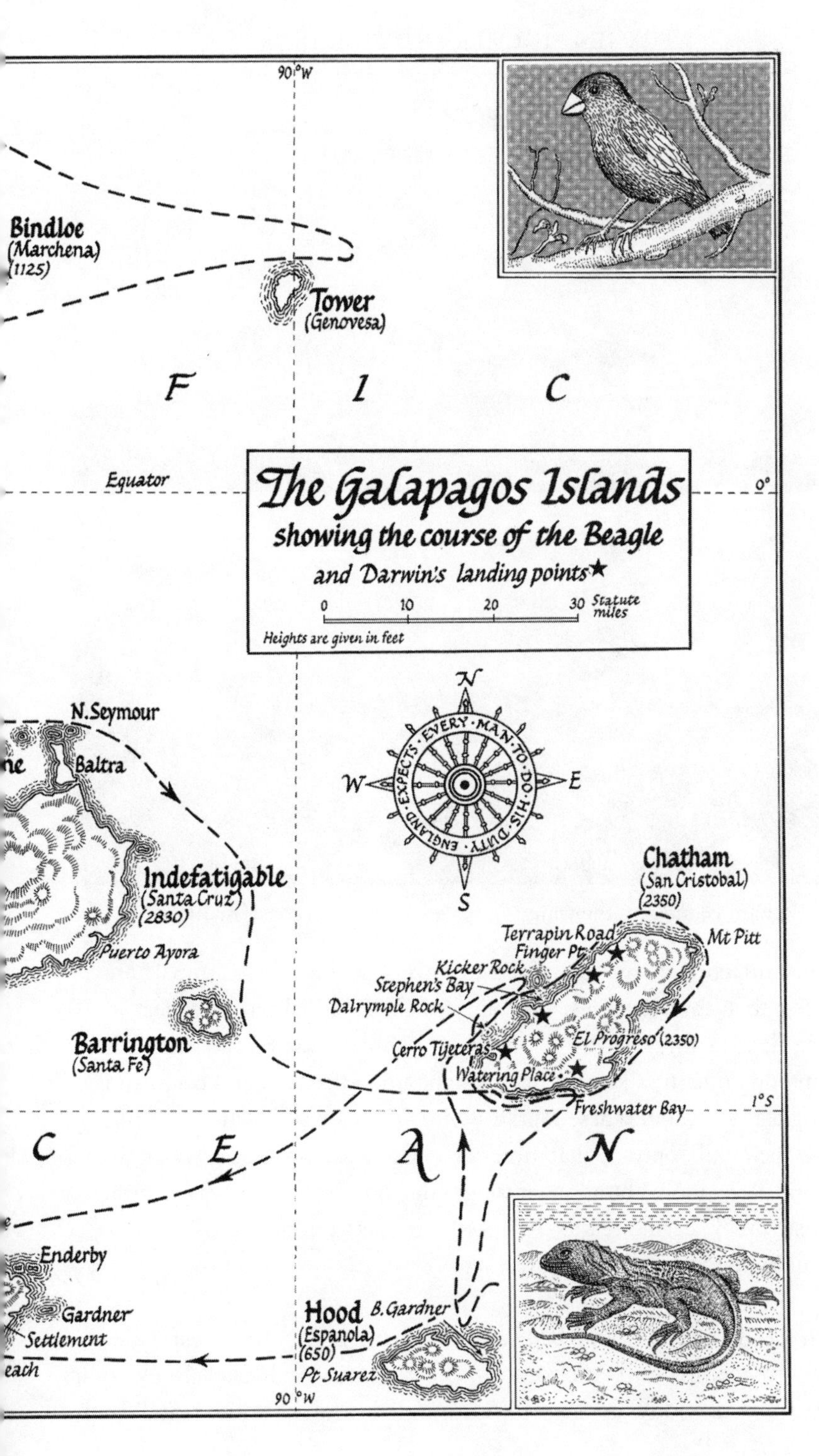

90°W
Bindloe
(Marchena)
(1125)
Tower
(Genovesa)
F
I
C
Equator
The Galapagos Islands
showing the course of the Beagle
and Darwin's landing points★
0
10
20
30
Statute miles
Heights are given in feet
0°
N
W
E
S
EVERY MAN TO DO HIS DUTY
ENGLAND EXPECTS
N.Seymour
Baltra
Indefatigable
(Santa Cruz)
(2830)
Puerto Ayora
Barrington
(Santa Fe)
Chatham
(San Cristobal)
(2350)
Terrapin Road
Mt Pitt
Finger Pt
Kicker Rock
Stephen's Bay
Dalrymple Rock
El Progreso (2350)
Cerro Tijeteras
Watering Place
Freshwater Bay
1°S
C
E
A
N
Enderby
Gardner
Settlement
Hood
B. Gardner
(Espanola)
(650)
Pt Suarez
90°W

September 1835

Chatham Island and the watering place on its south coast, by P.G. King

every direction as we scrambled from rock to rock. Few animals are uglier than these iguanas; they are lizard-shaped, about three feet in length; of a dirty black colour; with a great mouth, and a pouch hanging under it; a kind of horny mane upon the neck and back; and long claws and tail. These reptiles swim with ease and swiftness – but use their tails only at that time. At a few yards from the water we found vegetation abundant, though the only soil seen was a little loose dusty earth, scattered upon and between the broken lava. Walking is extremely difficult.[152]

On 17 September the *Beagle* moved on along the west side of Chatham Island, and anchored slightly south of Kicker Rock at a place known today as Puerto Grande.[153] The bay swarmed with animals, with fish, sharks and turtles popping up their heads everywhere, and Charles collected several fishes, which in due course back at Swaffham Bulbeck were identified by Leonard Jenyns as belonging to new species. Charles went on shore, where the mocking birds (Plate 19) greeted him by picking at his shoes as they still do to visitors, for he wrote in his pocket book, 'The Thenca very tame & curious in these Islands. I certainly recognise S. America in Ornithology. Would a botanist? 3/4 of plants in flower.'

In his commonplace journal he wrote:

> The black Lava rocks on the beach are frequented by large (2–3 feet) most disgusting clumsy Lizards. They are as black as the porous rocks over which they crawl & seek their prey from the Sea. Somebody calls them "imps of darkness". They assuredly well become the land they inhabit. When on shore I proceeded to botanize & obtained 10 different flowers; but such insignificant, ugly little flowers, as would better become an Arctic than a Tropical country. The birds are Strangers to Man & think him as innocent as their countrymen the huge Tortoises. Little birds within 3 & four feet, quietly hopped about the Bushes & were not frightened by stones being thrown at them. Mr King killed one with his hat & I pushed off a branch with the end of my gun a large Hawk.[154]

He later described the marine iguanas *Amblyrhynchus* in greater detail, and revised his initial guess that they fed on marine animals rather than seaweed:

> The Lizard which bears this name is said in the Blonde's Voyage* to have been described from a specimen brought from the shores of the Pacifick. This animal is excessively abundant on all the Islands in the whole group. It never proceeds many yards inland from the rocky sea

* On her way to Hawaii in 1825, HMS *Blonde*, then commanded by the naval 7th Lord George Byron, first cousin of the poetical 6th Lord George Byron, had found at Narborough Island (Fernandina) an 'innumerable host of sea-guanas'.

beach: There, on the large fragments of black Lava, groups may be seen basking with outstreched [*sic*] legs. They are hideous looking animals; stupid & sluggish in their motions. Their color is black, their general size rather more than 2 ft long. On Albermale [*sic*] Is[d] they appear to grow much larger than in any other place, one weighed 20 ℔. I saw *very few* small ones: so that I suppose their breeding season is now coming on: I could not hear any particulars respecting their manner of breeding. These animals have occassionally [*sic*] been seen some hundred yards at sea, swimming. The structure of their bodies points out aquatic habits. Yet it is remarkable, that when shuffling over the tidal rocks it is scarcely possible to drive them into the water. From this reason, it is easy to catch them by the tail, after driving them on a point. They have no idea of biting, & only sometimes when frightened squirt a drop of fluid from each nostril.* Having seized a large one by the tail, I threw it several times a good distance into a deep pool left by the retiring tide. Invariably it returned in the same direction from which it was thrown to the spot where I stood. Its motion was rapid, swimming at the bottom of the water & occassionally helping itself by its feet on the stones. As soon as it was near the margin, it either tried to conceal itself in the sea-weed or entered some hole or crack. As soon as it thought the danger was over it crawled out on the dry stones, & again would sooner be caught than voluntarily enter the water. What can be the reason of this? are its habitual enimies [*sic*] sharks or other *marine* animals? The manner of swimming is singular, consisting solely in a wriggling motion of tail & body, the legs being motionless, collapsed & stretched out behind.

I opened the stomach (or rather duodenum) of several, it was largely distended by quantities of minced pieces of sea-weed, of that kind which grows in *thin* foliaceous expansions of a bright green & dull red color. There was not a trace of any animal matter: Mr Bynoe, however, found a piece of a Crab in them: this might have entered accidentally, in a like manner as I have seen a Caterpillar in the

* The marine iguana rids itself of the excess salt in its diet by means of a nasal salt gland whose secretions account for this fluid.

Portraits of a marine iguana and a land iguana in characteristic poses, by Thalia Grant

stomach of the Tortoise. I conceive the largeness of the intestine is in perfect agreement with its herbivirous appetite. – Capt Colnett* states they go out to sea in shoals to fish: I cannot believe this is the object, nor is it very clear what their object can be. Does such sea-weed grow more abundantly a little way from the coast? They appear to be able to survive a long time without breathing. One was sunk with a weight for nearly an hour, & was then very active in its motions. Their limbs are well adapted for crawling amongst the rough & fissured rocks of Lava, & we have mentioned that with their tail & body they can swim well.[155]

* Captain James Colnett had said in his account of visiting the Galapagos in HMS *Rattler* in 1792 that 'the sea guanas go in herds a fishing, and sun themselves on the rocks like seals, and may be called alligators in miniature'. It was probably Colnett who first erected the post barrel on the north coast of Charles Island.

September 1835

On 21 September Charles and Syms Covington were landed by boat a few miles up the coast to the north-east at an area thick with volcanic cones, and with large circular pits up to eighty feet deep that were not cones, but had been formed by the collapse of the roofs of huge caverns filled with gas during an eruption. The age of the various lava streams could be judged from the amount of vegetation that had grown on them, and on one of the more ancient ones Charles met two very large tortoises.

> One was eating a Cactus, & then quietly walked away. The other gave a deep & loud hiss and then drew back his head. They were so heavy, I could scarcely lift them off the ground. Surrounded by the black lava, the leafless shrubs & large Cacti, they appeared most old-fashioned antediluvian animals; or rather inhabitants of some other planet.[156]

Presumably for practical reasons, Charles did not bring back in his collections a whole marine iguana, nor for that matter a whole tortoise. But it was relatively simple to collect their parasites, so that in his list of 'Specimens not in Spirits' the first entry for Chatham Island was 'Acarus [a mite or tick] from great black Sea Guana or Lizard', and the sixth entry was 'Acarus from Pudenda of common great land Tortoise'. Precisely when these were collected was not recorded.

Charles Island, by P.G. King

September 1835

After Charles had spent two further days collecting and examining the geology of the lava flows, the *Beagle* moved on to the north side of Charles Island (Floreana) and anchored at Post Office Bay. In the highlands of the island there was between June and November a light rain or drizzle carried by the prevailing southerly winds that enabled sweet potatoes and plantains to flourish in cleared areas. A penal settlement, housing two to three hundred prisoners from the Republic of the Equator (Ecuador), had been established there in 1832 with an Englishman, Nicholas Lawson, currently acting as Governor. By a fortunate chance Lawson had come down to Post Office Bay on 24 September to visit a whaling vessel, and he was able to conduct FitzRoy and Charles up to the settlement. He explained that the people under his charge led what Charles described as a sort of Robinson Crusoe life, with ample supplies of vegetables, and woods abounding with wild pigs and goats. But the principal source of meat was still the tortoises, although their numbers had fallen drastically in recent years, and no longer could the ship's company of a frigate capture two hundred tortoises in one day. The oldest of the animals were very large – one caught in 1830 required six men to lift it into the boat, and two men could not turn it over on to its back. Some too large to be carried away had dates carved on their shells, one which Lawson had seen being 1786. He reckoned that the supply would last for another twenty years, but he was already sending parties to James Island, where there was a salt mine, to collect tortoises and salt the meat.* Ultimately the most important piece of information about the tortoises imparted by Lawson was the constancy of their different shapes and sizes between the individual islands, as Charles duly recorded in his notes:

> This animal is, I believe, found in all the Islands of the Archipelago; certainly in the greater number. The Tortoises frequent in preference the high & damp parts, but they occur likewise in the low & arid

* The Floreana tortoise, whose Latin name is *Geochelone elephantophus galapagoensis*, had become extinct by the end of the nineteenth century.

districts. It is said that slight variations in the form of the shell are constant according to the Island which they inhabit – also the average largest size appears equally to vary according to the locality. M[r] Lawson states he can on seeing a Tortoise pronounce with certainty from which island it has been brought. The Tortoises grow to a very large size: there are some which require 8 or 10 men to lift them: The old Males are the largest. The females rarely grow to so great a size. The male can readily be told from the females by the greater length of its tail. The Tortoises which live on those Islands where there is no water, or in dry parts of others, live chiefly on the succulent Cactus: I have seen those which live in the higher parts, eating largely of a pale green filamentous Lichen, which hangs like presses from the boughs of the trees, also various leaves & especially the berries of a tree (called Guyavitas) which are acid & Austere.

The Tortoise is very fond of Water, & drinks large quantities & wallows in the mud. Even those which frequent districts far removed from the water travel occassionally to it; they stay two or three days near the Springs & then return. My informants differed widely in the frequency of these visits. It seems however certain that they travel far faster than at first would be imagined. They ground their opinion on seeing how far some marked animal has travelled in a given time. They consider they would pass over 8 miles of ground in two or three days. One large one, I found by pacing, walked at the rate of 60 yards in 10 minutes, or 360 in the hour. At this pace, the animal would go four miles in the day & have a short time to rest. When thus proceeding to the Springs, they travel by broard & well-beaten tracks, which branch off to all points of the Isl[d]. I should have prefaced that in these Isd[s] there are only a few watering places & these only in the highest & central parts. When first I landed at Chatham Is[d]; the object of these tracks was to me inexplicable. The effect in seeing such numbers of these huge animals, meeting each other in the high-ways, the one set thirsty & the other having drunk their fill, was very curious. When the Tortoises arrive at the water, quite heedless of spectators they greedily begin to drink: for this purpose they bury their heads to above their eyes in the mud & water & swallow about 10 mouthfulls in the

Tortoises with dome-shaped and saddle-back shells, by Thalia Grant

minute. The inhabitants when very thirsty sometimes have killed these animals in order to drink the water in the Bladder, which is very capacious. I tasted some, which was only slightly bitter. The water in the Pericardium is described as being more limpid & pure.

The female Tortoise generally places her eggs in groups of four or five in number & covers them up with earth. Where the ground is rocky she drops them indiscriminately. Mr Bynoe found 7 eggs laid along in a kind of crack. The egg is quite sphærical, white & hard, the circumference of one was 7 & 3/8 of inch. The young Tortoise, during its earliest life, frequently falls a prey to the Caracara [Galapagos hawk], which is so common in these islands. The old ones occassionally meet their death by falling over precipices: but the inhabitants have never found one dead from Natural causes. The Males copulate with the female in the manner of a frog. They remain joined for some hours. During this time the Male utters a hoarse roar or bellowing, which can be heard at more than 100 yards distance. When this is heard in the woods, they know certainly that the animals are copulating. The male at no other time, & the female never, uses its voice. There are now, in the beginning of October, eggs in the ground & in the belly. The people believe they are perfectly deaf; certainly when passing a tortoise, no notice is taken till it actually sees you: then drawing in its head & legs & uttering a deep hiss, he falls with a heavy sound on the ground, as if struck dead.

The people employ the meat largely, eating it both fresh & salt, & it is very good. The meat abounds with yellow fat, which is fryed down & gives a beautifully clear & good oil. When an animal is caught, a slit is made in the skin near the tail to see if the fat on the dorsal plate is thick; if it is not the animal is liberated & recovers from the wound. If it is thick it is killed by cutting open the breast plate on each side with an axe & removing from the living animal the serviceable parts of the Meat & liver &c &c. In order to secure the Tortoises, it is not sufficient to turn them like a Turtle, for they will frequently regain their proper position.[157]

Unfortunately, the full implications of Mr Lawson's observation that each island had its own brand of tortoise were not immediately taken in by Charles, for had they been he would surely have collected one of the shells of the Floreana tortoises, now sadly extinct, that as noted by FitzRoy were being used in 'an apology for a garden, to cover young plants, instead of flower pots'. Perhaps this was because in the islands that he happened to have visited Charles had not seen the extreme examples illustrated in Thalia Grant's drawing (previous page) of the saddle-back tortoises with long necks adapted to reach high into the vegetation for food in the lowest and driest islands, and the dome-shaped tortoises with shorter necks that lived in the highest islands containing extensive damp areas, where there were plenty of low-growing plants on which to feed.

Nevertheless, while the ship was in Chile, Charles had been thinking hard about the distribution of evidently closely related birds and mammals found on the east and west sides of the Andes, and considering possible reasons for their patterns of migration other than their having been created to occupy particular areas. His thoughts had also begun to move in new directions in order to account for the close relationships between the huge extinct animals that he had found, and the much smaller living ones. His notes on first landing in the Galapagos, when he wrote: 'I industriously collected all the animals, plants, insects & reptiles from this Island. It will be very interesting to find from future comparison to what district or "centre of creation" the organized beings of this archipelago must be attached,'[158] suggest that he was still not quite ready at that time to desert the doctrine of Special Creation. But when nine months later he wrote the famous passage in his Ornithology Notes (see pp.370–2) in which he first admitted to suspicions that species might after all be mutable, one of the pieces of evidence that he cited was the variability of the tortoises between the individual islands in the Galapagos, though it was to 'the Spaniards' rather than Mr Lawson that he credited this information.

The *Beagle* next sailed west past Brattle (Tortuga) Island, the largest example in the Galapagos of a tuff-crater,* whose perfect crescent shape had resulted from its long-continued erosion by weather from the south. Having surveyed the southern end of Albemarle (Isabela) Island, they sailed north up the west coast of the island, where Charles was impressed by its array of huge volcanoes. High up on the side of the volcano now known as Sierra Negra he saw a crater from which there was a jet of steam, and recalled that when the *Blonde* had been in the Galapagos in 1825 volcanic activity had been observed. He would

* Tuff is a material produced by a consolidation of volcanic ashes and other erupted material as it cools. Charles had noted that a brown-coloured tuff was typical of many of the craters in the Galapagos.

Albemarle Island, by P.G. King

very much have liked to have gone ashore to climb up one of the volcanoes, but this was an ambition that was never satisfied. On the evening of 30 September the *Beagle* anchored at Banks (Tagus) Cove, an old volcanic crater tucked in between Albemarle and Narborough (Fernandina) Islands. They had hoped to find some fresh water there, but were disappointed. South of the cove Charles found a most beautiful crater, elliptic in shape, nearly a mile in its longer axis, and five hundred feet deep. At its bottom there was water that looked blue and clear, but was as salty as brine. The structure of this crater was unusual in that it consisted entirely of a volcanic sandstone, and he filled his Geological Notes with a detailed description of its special features.

On 2 October the *Beagle* sailed on round the north end of Albemarle, and after a fierce struggle to windward against a strong current reached Buccaneer Cove at James (Santiago) Island on the eighth. Charles, Syms Covington, Mr Bynoe and H. Fuller were then landed with provisions to await the return of the ship after watering at Chatham. They pitched their tents in a small valley close to the beach, after some competition to find a vacant site with the land iguanas of which they had seen numbers on Albemarle. Charles wrote in his Zoology Notes:

> This animal clearly belongs to the same genus as the last, it being a terrestrial, whilst the other is an aquatic species. They are found only

in the central division of the Islands, viz. Barrington, Indefatigable, Albermale & James Is[d]. To the North in Charles, Hood or Chatham, & to the South in Tower, Bindloes & Abingdon, I neither heard of, or saw one. They frequent in the above Islands both the upper, central & damp parts as well as the lower dry sterile districts: in the latter kind of soil their numbers are more especially abundant. I cannot give a better idea of this than by stating we had difficulty in finding a piece of ground free from their burrows large enough to pitch our tents. They are ugly animals, & from their low facial angle have a singularly stupid appearance. Capt. FitzRoy['s] specimens will give a good idea of their size. Their colors are, whole belly, front legs, head "Saffron Y & Dutch orange"* – upper side of head nearly white. Whole back behind the front legs, upper side of hind legs & whole tail "Hyacinth R". This in parts is duller, in others brighter passing into "Tile R". I have seen a few individuals, especially the younger ones, quite sooty on the whole upper side of their bodies.

They are torpid slow animals, crawling when not frightened with their belly & tail on the ground. frequently they doze on the parched ground, with their eyes closed & hind legs stretched outwards. In none of their motions, is there that celerity & alertness which is so conspicuous in true Lacertas [lizards] & Iguanas. Their habits are diurnal: they seldom wander to any distance from their burrows: when frightened they rush to them with a most awkward gait: excepting going down hill their motion, from the lateral position of their legs, is not quick. They are not timorous. When attentively watching an intruder they curl their tails, & raising themselves as if in defiance on their front legs, vertically shake their heads with a quick motion. I have seen small Muscivorous Lizards perform the same gestures. This gives them rather a fierce aspect, but in truth they are far the contrary. When however being caught & plagued with a stick they will bite it severely. Two being placed on the ground close together will fight & bite each other till blood is drawn.

As I have said they all inhabit burrows, these they make some-

* The colours in inverted commas were again taken from Syme's little book.

James Island

times between the fragments of Lava, but more generally in the ground, composed of Volcanic Sandstone. The burrows do not appear deep & enter at a small angle: hence when walking over the "warrens" the soil perpetually gives way. When excavating these holes, the opposite sides of the body work alternately; one front leg scratches the earth for a short time & throws it towards the hind. This latter is well placed so as [to] heave the soil beyond the mouth of hole. The opposite side then takes up the task. Those individuals & they are the greater number, which inhabit the extremely arid land, never drink water during nearly the whole year. These eat much of the succulent Cactus, which is in evident high esteem. When a piece is thrown towards them, each will try to seize & carry it away as dogs do with a bone. They eat however deliberately, without chewing the pieces. The Cactus is in request amongst all animals, I have seen little birds picking at the opposite end of a piece which a Lizard was eating: & afterwards it would hop on with complete indifference on its back. In their stomachs vegetable fibres, leaves of different trees, especially the Mimosa were always found. In the high damp country their chief food is the berry called Guayavitas; it is the same which the Tortoises eat, & has an acid astringent taste. Here also they are said to drink water. To obtain the leaves they climb short heights up the trees: I have frequently seen them clinging to the branches of the Mimosa. Thus their habits are as entirely herbivorous as in the black sea-kind. The meat when cooked is white & esteemed, by those who can bring their stomachs to such a regimen, good food. I observe the pores on under sides of hind thighs are very large. By pressure a cylindrical organ is protruded to the length of some tenths of an inch.

Is any other genus amongst the Saurians [lizards] herbivorous?

I cannot help suspecting that this genus, the species of which are so well adapted to their respective localities, is peculiar to this group of Isd^{s}.[159]

At Buccaneer Cove they found a party of Mr Lawson's men who were salting tortoise meat, and who helped to provide them with some precious water. Later, when the surf broke over the water source and spoiled the fresh water, they were rescued by an American whaler which generously gave them three casks of water and a bucket of onions. Lawson's men took them up for a couple of days to the 'hovels' at an elevation of two thousand feet where the tortoise hunters lived. There they sampled tortoise meat fried in the transparent oil obtained from the fat. Always ready for a new gastronomic experience, Charles commented that 'The Breast-plate with the meat attached to it is roasted as the Gauchos do the "Carne con cuero" (meat on the hide). It is then very good. Young tortoises make capital soup – otherwise the meat is, to my taste, but indifferent food.'

They busily collected specimens of every kind during this their last week on the Galapagos, and Charles summarised their achievements among the small birds with a confidence that proved to be somewhat misplaced:

I believe the collections of birds formed by Mr Bynoe, Capt. FitzRoy & myself will give a nearly perfect series of the birds. At this time of year (end of Septemb & beginning of Octob), from the state in which the birds appeared to be I should imagine the last years produce had nearly attained perfect plumage. In no female of the smaller birds the eggs in the Ovarium were much developed. The Ornithology is manifestly S. American. Far the preponderant number of individuals belongs to the Finchs & Gross-beaks. There appears to be much difficulty in ascertaining the Species. My series would tend to show that only the old Cocks possessed a jet black plumage: but M[r] Bynoe & Fuller have each a small black female bird. Certainly the numbers of brown & blackish ones is immensely great to those perfectly black. Species as in margin are well characterized. I only saw them in James Is[d] & in one place. They were there however numerous, feeding with the various other species. M[r] Bynoe has a much blacker specimen. I should state that all the Species (& doves) feed together in great numbers indiscriminately, their favourite resort being in the dry long grass in the lower & dry parts of Island, where in the soil many seeds are lying dormant. The Icterus like Finch is distinct in its habits: its general resort is hopping & climbing about the Cactus trees, picking with its sharp beak the flowers & fruit. Not infrequently however, it alights on the ground & feeds with the flocks of other species. Out of the *many* specimens which I have seen of this bird, the only one which was black I by good fortune procured. M[r] Bynoe has one other. I have no doubt respecting its identity: for it was shot with the others on the Cactus: This is an illustration of the comparative rarity of the black kinds. The Gross-beaks are very injurious. They will strike seeds & plants when buried 6 inches beneath the ground.

The insectivorous birds are comparatively rare: they are equally found in the low dry country & high damp parts. I was astonished to find amongst the luxuriant damp vegetation an exceeding Scarcity of insects (so much so that the fact is very remarkable) This being the case, it is no wonder that the above order of birds should be scarce.[160]

Almost all of the birds in question belonged to the Geospizinae, the name given to a new family by John Gould when he classified Charles's specimens soon after the *Beagle*'s return to England. But despite the considerable scientific interest that they ultimately proved to possess, they were admittedly not very glamorous when seen in the field. One immediate mistake made by Charles was to assume as may be seen above that the blackness of the individuals was an important feature, for in fact it merely varies with age. Moreover, he

October 1835

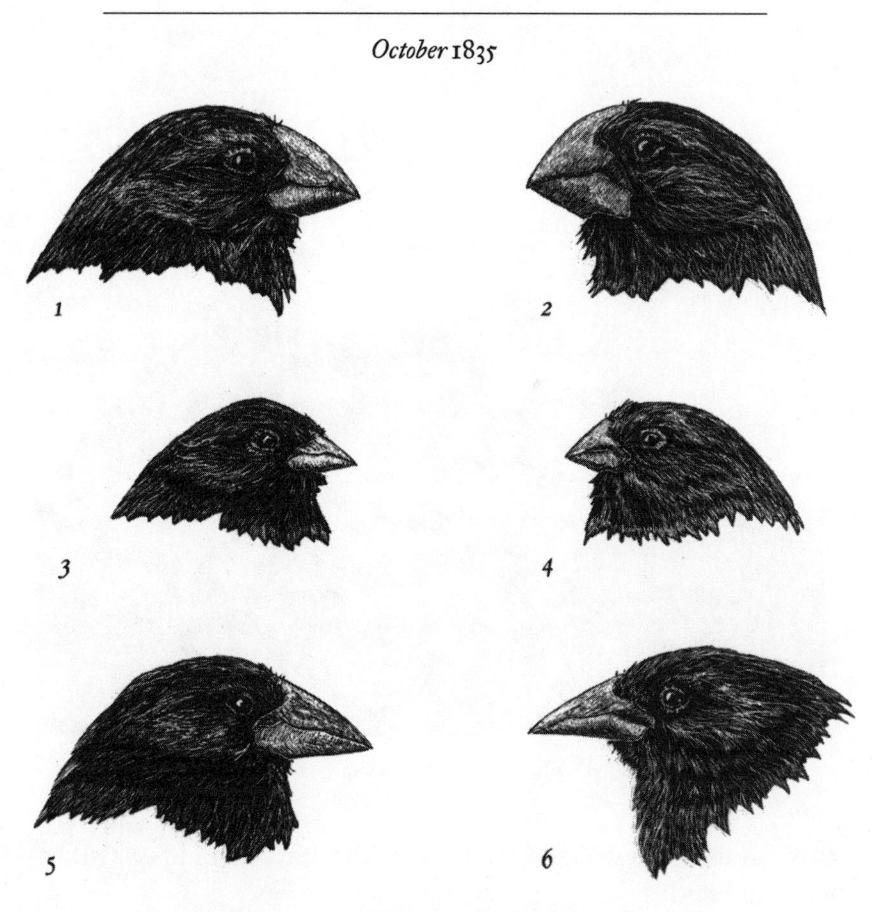

Beaks of six species of Darwin's finches, by Thalia Grant

was slow to appreciate that the vital distinguishing characteristic of the different members of the family would be the size and shapes of their beaks, so that like the shapes of the tortoises mentioned to him by Mr Lawson, the appearance of the birds varied significantly between the different islands. He consequently got himself into serious trouble when he failed for once to label all his specimens properly with the name of the island where they had been shot, and had to appeal to FitzRoy and others for some of their hopefully better labelled birds.

When writing the 1839 edition of the *Journal of Researches* Charles

October 1835

Daphne Major, by Thalia Grant

took advantage of what he had by then learnt from Gould's classifications, and said:

> in the thirteen species of ground-finches, a nearly perfect gradation may be traced, from a beak extraordinarily thick, to one so fine, that it may be compared to that of a warbler. I very much suspect, that certain members of the series are confined to different islands; therefore, if the collection had been made on any *one* island, it would not have presented so perfect a gradation. It is clear, that if several islands have each their peculiar species of the same genera, when these are placed together, they will have a wide range of character. But there is not space in this work, to enter on this curious subject.

FitzRoy's views on the Geospizinae were somewhat different, for having on his marriage in 1836 become a firm believer in the Bible, he wrote: 'All the small birds that live on these lava-covered islands have short beaks, very thick at the base, like that of a bull-finch. This appears to be one of those admirable provisions of Infinite Wisdom by which each created thing is adapted to the place for which it was intended.'

In the 1845 edition of the *Journal of Researches* Charles unwrapped his still strictly private ideas about speciation a little further, and admitted that 'Seeing this gradation and diversity of structure in one small, intimately related group of birds, one might really fancy that from an original paucity of birds in this archipelago, one species had been taken and modified for different ends.'

It turned out in the end that the Geospizinae did not occupy as central a place in Charles's thinking as the legend about them that built up nearly a century later supposed.* It was David Lack[164] who first popularised the christening of the Geospizinae as 'Darwin's finches', and gave an admirable account of his research into the evolution of their beaks. The final word has recently been said by Peter Grant,[165] in describing the observations made with the help of his wife and daughters on the small island known as Daphne Major at the centre of the Galapagos. Living on Daphne are some hundreds of the medium ground finches *Geospiza fortis*, and of the cactus finches *G. scandens*, and the whole population could be ringed and weighed, their crops sampled, and their beaks measured, over a long succession of rainfall cycles governed closely by El Niño with consequent variations in the availability of seeds of different sizes and hardness to feed the birds. The effect of natural selection on the numbers of each

* The origin and fate of the popular legend that on first seeing the Geospizinae in the Galapagos, Charles's theory of evolution came to him in a blinding flash has been authoritatively discussed in two important papers by Frank Sulloway.[161,162] A detailed analysis of the labelling of the specimens collected by Charles himself, supplemented by those of FitzRoy, Bynoe, Fuller and Covington, reveals what can only be described as a taxonomist's nightmare. Charles invariably insisted on having plenty of good evidence in support of any theory worthy of consideration, and perhaps his realisation that he had collected relatively few of the Geospizinae from not many of the islands may explain why they were not mentioned specifically either in his *Transmutation Notebooks*[163] nor in *On the Origin of Species*.

species, and on the evolution of beaks of appropriate sizes, has thus been tested successfully for twenty-seven generations of finches studied under strictly natural conditions. What must be admitted to have been Charles's somewhat indecisive beginning has therefore been succeeded after 166 years by a triumphant demonstration of the mode of evolution of his finches observed in real time.

Charles had nevertheless appreciated that in the case of the mocking birds with which he had become familiar in Argentina and Chile there were consistent variations between the several islands, and wrote in his Zoology Notes:

> This bird which is so closely allied to the Thenca of Chili (Callandra of B. Ayres) is singular from existing as varieties or distinct species in the different Is^{ds}. I have four specimens from as many Is^{ds}. These will be found to be 2 or 3 varieties. Each variety is constant in its own Island.
>
> The Thenca of Albermale Is^{d} is the same as that of Chatham Is^{d}. This is a parallel fact to the one mentioned about the Tortoises. These birds are abundant in all parts: are very tame & inquisitive: habits exactly similar to the Thenca. *runs* fast, active, lively: sings tolerably well, is very fond of picking meat near houses, builds a simple open nest. I believe the note or cry is different from that of Chili.[166]

In his Ornithology Notes written nine months later (see p.370), this significant passage was appreciably extended.

Charles noted that an appreciable number of sea birds had also been collected, but these were of lesser interest, apart from the frigate bird with its spectacular flight, on which he speculated later[167] that its ability to avoid getting its feet wet was linked with the scooping out of the webbing between its toes.

> There is also the Frigate Bird. There is one part of the habits of this bird which has not been sufficiently described; it is the manner in which this bird picks up fish or bits of meat from the surface of the water without wetting even its feet. I never saw one alight on the water. Like an arrow the bird descends from a great height with

extended head, by the aid of its tail & long wings turns with extraordinary dexterity at the moments of seizing its object with its long beak. It is a noble bird seen on the wing, either when soaring in flocks at a stupendous height, or as showing the perfect skill in evolutions when many are darting at the same floating morsel. If the piece of meat sinks above 6 inches deep it is lost.[168]

Captain Beaufort's lengthy Memorandum of instructions for the *Beagle*'s voyage included the following passage:

There are also some remarkable phenomena, which will be announced in the Nautical Almanacks, and which will occur during the Beagle's voyage. Some of these will be highly interesting to astronomers, and if it would not much derange her operations, she should be taken to some convenient anchorage for the purpose of landing the instruments.

If a comet should be discovered while the Beagle is in port, its position should be determined every night by observing its transit over the meridian, always accompanied by the transits of the nearest known stars, and by circum-meridional altitudes, or by measuring its angular distance from three well-situated stars by a sextant. This latter process can be effected even at sea, and the mean of several observations may give very near approximations to its real position.[169]

Halley's comet[170] was due to make one of its reappearances at intervals of about seventy-six years with a perihelion passage (when it was closest to the sun) in November 1835, and might therefore have been visible while the *Beagle* was in the Galapagos that October. But FitzRoy made no mention in the ship's log or anywhere else of having observed it, perhaps because from the *Beagle*'s geographical position the comet would have been very low above the horizon.

The *Beagle* had meanwhile been taking water on board at Chatham Island, and adding thirty more tortoises to the eighteen captured in September. But apart from two small ones kept for a while by Charles and Syms Covington as pets, they were all eaten on the way home. On 17 October a boat was sent to pick up Charles and his party; on the nineteenth the yawl was picked up at Abingdon (Pinta) Island; and on the twentieth after surveying the two small islands a hundred miles north of the group, Wolf (Wenman) and Culpepper, the *Beagle* set sail for her passage of 3200 miles to Tahiti.

CHAPTER 27

Across the Pacific to Tahiti

With studding-sails set on each side, and the trade wind blowing steadily, the *Beagle* travelled westwards at 150 or 160 miles a day in a manner that Charles found most satisfactory. On 3 November a black tern flew past the ship, indicating that land was not far away, for they were approaching the archipelago known as the Low or 'Dangerous' Islands, an extensive group of coral islets and atolls only a few feet above water, though coconut trees grew on some of them. After ten days they were at last clear of the archipelago, and on 15 November they anchored in Matavai Bay at Tahiti in the Society Islands.

Immediately they were surrounded by canoes, and on landing at Point Venus, where Captain Cook and Joseph Banks had observed the transit of Venus* on 3 June 1769,[171] they were greeted by a large crowd of laughing merry faces. It was fortunate that the day was not Sunday, as their calendar told them it was, for since the days of freedom and licence in the South Sea islands seen by the *Endeavour* and *Bounty* in the 1770s, missionaries had imposed new standards of behaviour and a strict injunction against launching a canoe on the

* The prime purpose of the voyage of HMS *Endeavour* to the South Seas in 1768 to 1771 – in the course of which Cook discovered Australia – was to observe at Tahiti the transit of the planet Venus across the face of the sun due to take place in 1769, but then not again until 1874. This was an important observation for determination of the distance between the earth and the sun. Cook's observatory at Tahiti had become the standard centre in the South Pacific for the determination of longitude, and Beaufort's Plan of the *Beagle's* voyage therefore instructed FitzRoy to verify the chronometers at Point Venus.

Sabbath. But the *Beagle* had not yet crossed the International Date Line, and in Tahiti Monday was considered already to have arrived. In the *Beagle*'s log book the following day was accordingly entered, 'owing to our, so far successful, chase of the sun', as Tuesday the seventeenth rather than Monday the sixteenth.

In a letter to his sister Caroline, Charles had written: 'With respect to Otaheite, that *fallen* paradise, I do not believe there will be much to see.' He was pleased with the inhabitants, for there was a mildness in the expression of their faces that at once banished the idea of a savage. But he unexpectedly disapproved strongly of their slowness to adopt a more European style of dress, and of their remaining so 'ridiculously' naked. He preferred the men to the women, for most of them were tattooed, and 'the ornaments so gracefully follow the curvature of the body that they really have a very elegant & pleasing effect'. But he disliked the similar tattooing of the women, very commonly down to their fingers, and the almost universal fashion of shaving their hair 'from the upper part of the head in a circular manner so as only to leave an outer ring of hair. The Missionairies [*sic*] have tried to persuade the people to change this habit, but it is the fashion & that is answer enough at Tahiti as well as Paris.' Although Charles liked the custom of the women to wear a flower at the back of the head or through a small hole in each ear, he concluded that they were in greater want of some becoming costume even than the men.

The following morning he decided to make an excursion into the mountains, and came ashore with some provisions in a bag and two blankets for Syms Covington and himself, lashed to each end of a pole. He had asked Mr Wilson, the missionary in charge of the district around Matavai, to provide him with two guides accustomed to carrying such poles with heavy loads, whom he had firmly instructed to provide themselves with food and clothing. But they maintained that there was plenty of food in the mountains, and that their skins would cover them adequately. Charles and Covington set off up the valley of the river Tia-auru, which arose at the base of the seven-thousand-foot mountains in the centre of the island, and flowed into the sea near Point Venus. The valley soon narrowed to little more than the width of the river bed, with nearly vertical walls several thousand feet high forming a spectacular mountain gorge. On a ledge of rock beneath the lava cliff, the guides provided a lunch of small fish and fresh-water prawns dextrously caught in the stream. Higher up, the river divided into three smaller streams of which two descended down impassable waterfalls. The third had however been found to offer a somewhat precarious access to less precipitous parts above, by climbing from one ledge to another with the aid of ropes placed in position by one of the guides after using the trunk of a dead tree as a ladder. Thus they ascended, sometimes by ledges, and sometimes up knife-edge ridges with profound ravines on either side. Charles had seen much greater mountains in the Andes, but none nearly as abrupt as these.

In the evening the guides quickly built an excellent house from the stems of the bamboos and the large leaves of the bananas, using strips of bamboo bark for twine. After lighting a fire in just a few seconds by rubbing a pointed stick vigorously in a groove in a peculiarly white and very light wood, stones about the size of a cricket ball were well heated, and buried with small parcels wrapped in leaves containing pieces of beef, fish, ripe and unripe bananas, and the tops of the wild arum lily. In a quarter of an hour a most delicious meal was cooked, and served with cool water from the stream in coconut shells. In front of the camp was an extensive brake of wild sugarcane, and nearby there were wild yams (sweet potatoes) and another wild lily called Ti that had huge brown roots that were as sweet as treacle, and served for the party's dessert. Neither of the guides would touch his food before saying a brief grace, and before going to sleep the elder Tahitian fell on his knees and said a long prayer in his native tongue. Charles commented that 'Those Travellers who hint that a Tahitian prays only when the eyes of the missionary are fixed on him, should have slept with us that night on the mountain side.'

The banks of the stream were shaded by the knotted dark green stems of a liliaceous plant called ava, which had been much appreciated by the natives in former days for its powerful intoxicating (and

View of Papiete Harbour, Tahiti, by Martens

possibly hallucinogenic) effects, but had been banished by the missionaries to such remote ravines. Charles chewed a piece, which had an acrid and unpleasant taste. Having with him a flask of spirits, he noticed that when his guides were persuaded to break their resolutions and accept a sip, they put their fingers before their mouths and uttered the word 'Missionary'. The explanation was that after the prohibition of the use of ava, the introduction of spirits took place instead, and drunkenness became prevalent. However, the missionaries had succeeded in persuading all the chiefs and the Queen herself to join a Temperance Society, and with their influential backing, alcohol had joined the ban, at least for the present.

On Sunday, 22 November some of the *Beagle*'s officers and crew attended a service at Papawa, a little cove close to Matavai, where the missionary Mr Nott, who had been longest in Tahiti, and who had just completed a translation of the entire Bible into Tahitian, had built a chapel. FitzRoy took a party to Papiete, the capital of the island seven miles from Matavai, where they attended divine service in the English chapel, a large airy framework of wood, about fifty feet in length and thirty wide, with a high angular roof, filled to excess with people of all ages and both sexes. The service was con-

Mr Nott's old chapel at Papawa, by P.G. King

ducted first in the Tahitian language and then in English by Mr Pritchard, the senior missionary on the island, whose fluency in Tahitian was most impressive, and at whose house they afterwards met three daughters of other missionaries who seemed to speak Tahitian even better than their own language. Both Charles and FitzRoy were concerned at the vehemence with which the activities of the missionaries had sometimes been publicly attacked in recent years, and were anxious to form their own opinion of the situation. In a joint article eventually published in South Africa,[172] supported by a number of extracts from Charles's commonplace journal (*The Beagle Diary*), they came down firmly on the side of the missionaries in denying that the Tahitians had become a gloomy race and lived in fear of them. Respect for the missionaries they certainly felt, as Charles had learnt during his expedition into the mountains, but of fear he saw no trace, and FitzRoy, in his much longer account of the visit to Tahiti,[173] fully agreed. On the vexed question of the lack of virtue of the women in Tahiti, Charles pointed out that the scenes described by Cook and Banks sixty-six years earlier, when Queen Oberea was found happily and unashamedly in bed with her twenty-five-year-old lover, were the consequence of a morality taught to daughters by their mothers in line with the religious concepts of the period. He added, 'it is useless to argue against such men; I believe that disappointed at not finding the field of licentiousness quite as open as formerly, they will not give credit to a morality which they do not wish to practice, or to a religion which they undervalue, if not despise'.

The next day Charles hired a canoe and men to take him out to see the coral reef that fringed the island of Tahiti from half a mile to a mile and a half from the shore. He recorded that:

> We paddled for some time about the reef admiring the pretty branching Corals. It is my opinion that besides the avowed ignorance concerning the tiny architects of each individual species, little is yet known, in spite of the much that has been written, of the structure & origin of the Coral Islands & reefs.[174]

This was in fact the first occasion on which he had seen a coral reef for himself, though it is clear from entries in his pocket-books that his thoughts had first turned to them during the months when he was working on the alternating elevation and depression of the west coast of Chile that must have taken place in past ages. He was familiar with the chapter on the origin of the form of coral islands in the second volume of Lyell's *Principles*,[175] and with Lyell's adoption of the view that the circular atolls in the Pacific were simply the crests of submarine volcanoes lifted above the sea with the rims and floors of their craters grown over with corals. Charles had however preferred the opposite conclusion, that it was more probable that a gradual subsidence of the ocean floor while the corals grew rapidly at the surface of the sea could better account for the formation not only of small circular atolls, but also of barrier reefs fringing much larger islands. What he saw that day at Tahiti, and which could also be seen at the neighbouring island of Eimeo (Moorea), confirmed that he was thinking on the right lines. He returned to the subject four months later when the *Beagle* arrived at the Keeling Islands.

An official task in Tahiti that had been entrusted to FitzRoy by

Queen Pomare's house, by Martens

Commodore Mason in Valparaiso was to pursue negotiations with respect to an incident that had taken place five years earlier when the *Truro*, a small vessel under English colours, had been plundered and her Master and Mate murdered by the inhabitants of the 'Paamotu' or Low Islands (the Tuamotu Archipelago in a modern atlas), which at that time were nominally governed by the Queen of Tahiti. The British government had demanded compensation, and the Queen had agreed to pay the equivalent of 2853 dollars on or before 1 September 1835. FitzRoy was instructed to find out when and how this debt would be paid.

On 23 November, Queen Pomare came over from Eimeo to Papiete, sitting on the gunwale of a whale-boat and without any retinue. On hearing that she was ready to give him an audience, FitzRoy, with Mr Pritchard to act as interpreter, went to her house, where they found her on her own, looking very uncomfortable and evidently expecting a severe lecture about the *Truro*. Her mother, husband and foster-father entered the room and sat down on the floor, and FitzRoy delivered Commodore Mason's letter to her. It was agreed to hold a meeting of the chiefs next day at which she would preside, and FitzRoy and Pritchard took their leave after shaking hands with the royal party. FitzRoy was glad that they had not been expected to rub noses, but felt it was a pity that the missionaries had not done more to encourage the Tahitians to support their Queen 'by some of those ceremonious distinctions which have, in all ages and nations, accompanied the chief authority'.

The following day, FitzRoy took Charles and all the officers who could be spared from duties on the *Beagle* to Papiete, and joined Mr Wilson, Mr Henry (son of another missionary, who had been born in Tahiti) and Hitote (a well-known local chief) at Mr Pritchard's house.

When a messenger told them that the Queen and other chiefs had arrived, they walked over to the English chapel. They expected to find Pomare on the principal pews near the pulpit at the eastern end, but in fact she and the chiefs were sitting humbly on the rough benches at the western end. After everyone's hands had been shaken, Pomare accepted Mr Pritchard as her interpreter, and the proceedings were opened with the reading by FitzRoy of Commodore Mason's letter. When asked why the money had not yet been paid, the chiefs replied that they had not understood how and to whom the payment was to be made, and that it was intended at once to remedy the omission. FitzRoy warned Pomare firmly that if this was not done, her reputation in the eyes of the British government would suffer, for he had been told to take a tough line with her because the Paamotu Isles were rich in pearl oysters, and there was a suspicion that Pomare's relatives had been taking advantage of her youth at the time of the *Truro* incident in order to monopolise the trade in pearls. It was decided to pay the debt at once, using thirty-six tons of pearl oyster-shells belonging to Pomare that were currently lying at Papiete, and to collect the remainder from her friends. FitzRoy pointed out that justice would not necessarily be served if innocent Tahitians were to suffer from the misdeeds of the Paamotu islanders. This was an argument on which the chiefs had divided views, but in the end a satisfactory compromise was reached by an adjustment between the present and former valuations of the shells. Pomare was still worried that the next man-of-war might complain that the murderers had never been given up, but FitzRoy assured her that she would be regarded as having tried them by the laws of Tahiti, and had then exercised her legal right to pardon them.

Having dealt with the *Truro*, FitzRoy delivered a mild rebuke to Pomare for having recently been paying more attention to the views of the young people around her than to the older and more trustworthy counsellors who were now present. This led to a question from one of the seven chiefs about the complicated case of an English whaler, the *Venilia*, which had recently unloaded thirteen of her men on the island, who had proceeded to make serious nuisances of themselves. The Tahitians had been advised by Captain Hill, another English sea-captain often in their waters, to demand 390 dollars from the British government in damages, but could get no help in the matter from the nearest British Consul in the Sandwich (now Hawaiian) Islands. They had therefore written to their beloved King of Britain to ask that Captain Hill, or failing him the Reverend George Pritchard, should be appointed Consul in Tahiti. The eventual outcome was that Mr Pritchard was so appointed. Charles recorded that he was very favourably impressed at this meeting by the extreme good sense, reasoning powers, moderation, candour and prompt resolution that had been displayed by the Tahitians.

His official business thus completed, FitzRoy invited the Queen to visit the *Beagle* next day. Four boats were sent to fetch her, and the ship was dressed with flags and the yards manned to welcome her on board. Charles noted that the behaviour of the chiefs who came with her was very proper, because unlike the Fuegians they begged for nothing, and seemed very grateful for the small presents that were given to them. But deserting his usual kindliness, he said:

> The Queen is an awkquard large woman, without any beauty, gracefulness or dignity of manners. She appears to have only one royal attribute, viz a perfect immoveability of expression (& that generally rather a sulky one) under all circumstances.[176]

After tea, forgetting the prejudices of the missionaries against anything except hymns, FitzRoy proposed that they should hear a few of

Cottages at Tahiti, by Martens

the seamen's songs, for which Mr Pritchard could think of no translation except '*himene*'. 'Rule Britannia' and one or two grave performances passed off successfully, but what went down much best was a very noisy comic song, of which the Queen remarked that it certainly could not be '*hymeni*'. The royal party did not leave the ship until past midnight, apparently well contented with their visit.

Charles noted that apart for the coral reefs, he had seen little of special geological interest on Tahiti. The island was wholly volcanic in origin, and from the character of most of the lavas in the valley of Tia-auru he concluded that they must originally have flowed beneath the sea. He found no evidence that there had been any recent elevation of the land, though geologists have subsequently found marine remains at a considerable height in the mountains of Tahiti, showing that an appreciable elevation must have taken place in the very distant past. He collected a number of fishes in the waters around the island, but little else. On the evening of 3 December, the *Beagle* set sail for New Zealand, and they bade farewell to 'the island to which every traveller has offered up his tribute of admiration'.

CHAPTER 28

New Zealand

After several days of light winds the *Beagle* passed close to the island of Aitutaki, which was a combination of both classes of coral reef, consisting of an irregular hilly mass surrounded by a well defined circle of low reefs. A fortnight later, New Zealand was sighted in the distance, and Charles comforted himself with the thought that now they had passed the Meridian of the Antipodes, at last every league that they covered brought them one league closer to England. On 21 December they anchored in the Bay of Islands at the north-eastern tip of North Island. In several parts of the bay there were groups of tidy-looking houses close to the water's edge. Three whaling ships were at anchor, and there were a few canoes crossing from one shore to another, but only one of them came alongside the *Beagle*. The scene was distinctly less than encouraging in comparison with their joyful and boisterous welcome at Tahiti.

In the afternoon they went on shore to one of the larger groups of houses, due in course to become a village called Paihia. It was occupied by missionaries, with their servants and labourers. Altogether there were two or three hundred English residents around the bay. The next morning, Charles took a walk and quickly found that the hills were largely impenetrable thanks to being densely covered by tall ferns and low bushes growing like cypress. What was more, almost all of them had at one time been fortified by being cut into steps and successive terraces, protected by deep trenches. It was later explained to him by the Revd William Williams that these were the Pas described by Captain Cook in which earlier immigrants had defended themselves from attacks by the native Maoris. After the introduction of fire-arms the Pas had always been built on level ground, and consisted of a double stockade of thick and tall posts, placed in a zigzag line so that every part could be flanked. Within the stockade a mound of earth was built up, behind which the defenders could rest, or from the top of which they could fire their guns. The ubiquity of the Pas at once brought home to Charles the main difference between the natives of Tahiti, who were invariably so friendly towards strangers, and the warlike Maoris with their habitual unfriendliness. Captain Cook had said that their conduct on first seeing a ship was to throw volleys of stones at it, and to say boldly, 'Come on shore, and we will kill and eat you all.' By 1835 there was much less warfare, though in the southern parts of New Zealand the natives had still to be tamed. But the basic natures of the Maori tribesmen had not changed.

In the evening Charles went with FitzRoy and one of the missionaries to Kororareka, the largest village in the neighbourhood. Besides a considerable native population there were many English residents, but these were unfortunately of the most worthless character, including many runaway convicts from New South Wales. There were many spirits shops, and drunkenness and all kinds of vice were rife. Although many tribes in other parts of New Zealand had embraced Christianity, here the greater part were unconverted heathens. In places like Kororareka the missionaries were held in little esteem, and complained far more of the conduct of their own countrymen than of the natives. Charles said that the missionaries felt that the only protection on which they could rely was actually from the native chiefs against Englishmen.

Charles and FitzRoy walked about the village and talked to many of the Maoris, men, women and children. The people with whom they could most readily be compared were of course the Tahitians, but although the Maoris were possibly superior in energy, Charles felt that in every other respect they were definitely inferior. Just from their faces he concluded that the Tahitians were relatively much more civilised, and he reckoned that in the whole of New Zealand one could not find a single man whose face and mien could be compared with those of the old Tahitian chief Utamme. A factor which con-

Pictures of New Zealanders by Robert FitzRoy and Augustus Earle

tributed to this unfavourable comparison was the extraordinary manner in which the Maoris' faces were all tattooed with complicated though strictly symmetrical patterns of lines cut in the skin with a blunt-edged metal tool, and stained black. According to FitzRoy, these patterns were primitive armorial bearings to distinguish between tribes, but they also served to prevent the features from changing with age. When the missionaries tried to prevent some of the women from being tattooed, they pleaded, 'Let us have a few lines on our lips, that they may not shrivel when we are old.' Whatever else, the incisions involved in the tattooing had the effect of destroying the play of the superficial muscles of the face, and giving the Maoris an unattractive air of rigid inflexibility.

The Maoris were tall and solidly built, but contrary to Augustus Earle's opinion in *A narrative of a nine months' residence in New Zealand*[177] did not compare at all well in elegance with the working classes in Tahiti. FitzRoy and Charles were also somewhat incensed by Earle's book because of his criticism of the coldness of the missionaries towards him, for they knew that Earle had always been received with far more civility than his open licentiousness would have deserved. The persons and houses of the Maoris were filthily dirty, and they never thought of washing. The land was divided by well-determined boundaries between the various tribes, each tribe consisting of free men, and of their slaves taken in war; but there was no vestige of any formal law to govern the people. Charles said that if the state in which the Fuegians lived was fixed as zero in the scale of governments, the Maoris would not be much higher, while Tahiti as they had found it would occupy quite a respectable position.

At a place called Waimate, about fifteen miles from the Bay of Islands and halfway between the east and west coasts, the missionaries had bought some land for agricultural purposes, and the Revd William Williams having invited Charles to visit him there, Mr Busby the British Resident in charge of the district had offered to transport Charles in his boat to a creek near Waimate and to provide a guide for the rest of his walk. The road lay along a well-beaten path bordered on each side by the tall fern that covered the whole country.

Although no one could go starving in New Zealand, because the roots of this fern were nutritious, and there were plenty of shellfish on the seashore, the introduction of the rather more palatable potato had been welcome. They passed through a native village where after Charles had been taught the ceremony of pressing noses, they rested for half an hour while the guides smoked their pipes.

At length they reached Waimate, where the sudden appearance of three neat English farmhouses placed there as if by a magician's wand was exceedingly pleasing. The missionaries Messrs Williams, Davis and Clarke lived in the farmhouses, and there were huts for the native labourers near them. On an adjoining slope there were fine crops of barley and wheat in full ear, and of potatoes and clover. There were gardens in which every kind of English fruit, and others for warmer climates, were growing vigorously. There were stables, a threshing barn with its winnowing machine, a blacksmith's forge, and a happy mixture of poultry and pigs in the middle of the farmyard. A few hundred yards away, a stream had been dammed, and a fine watermill had been erected. Since five years earlier, the site had been growing nothing but ferns, and as all the labour had been provided by previously untrained Maoris, the scene provided good support for the policy of civilising the natives more by introducing them to the practical arts than by religious instruction.

The next morning, prayers were read in the native tongue to the whole family. It was market day, and after breakfast Charles saw the natives bringing in their stocks of potatoes, Indian corn or pigs to exchange for blankets and tobacco, and even after some persuasion for soap. Later he walked out with Williams and Davis to a neighbouring forest where the famous Kauri pine was growing, and found one whose circumference at the base was thirty-one feet, and which grew with no branches and almost no change of girth to sixty feet or more. He was introduced to the hemp that was New Zealand's second most important export after timber, and saw the many beautiful tree ferns. But in the woods there were very few birds, and he was not fortunate enough to catch a glimpse of a kiwi, one of the flightless forest-dwelling ratite birds now known to be characteristic of New Zealand. He was indeed struck forcibly by the fact that such a large island, extending across nearly a thousand miles of latitude and in parts 150 miles broad, with a fine climate and land of heights up to fourteen thousand feet, should possess only one indigenous animal, the word he always used for mammals, which was a small rat. Had he arrived in the Galapagos before the whalers had begun to dump their livestock there, he would have observed that the same thing applied. The problem of accounting for the inhabitants of oceanic islands was one that later on greatly exercised him.[178]

FitzRoy had also come to Waimate, and shared Charles's approval of the farming enterprise. The missionaries were very interested to hear about FitzRoy's efforts in Tierra del Fuego, which they regarded as a promising first step and not in the least a failure. It was soon decided that Richard Matthews would remain in New Zealand with his brother, a respectable young missionary mechanic established at the north end of the island, and lately married to Mr Davis's daughter.

On the afternoon of 30 December, the *Beagle* stood out from the Bay of Islands on course for Sydney. Charles thought that all on board were glad to leave New Zealand, for among the natives the charming simplicity that they had found in Tahiti was totally lacking. And only those who had been to Waimate had seen any really attractive country.

CHAPTER 29

Australia

The first port of call in Australia specified in Beaufort's Sailing Instructions for the *Beagle* was Port Jackson, close to Sydney, where the observatory at Parramatta was absolutely determined for longitude, and FitzRoy's chronometers should therefore be checked. While FitzRoy was thus occupied, one of Charles's first errands was to call on Conrad Martens, who, armed with a strong letter of introduction from FitzRoy to Captain P.P. King, had arrived in Sydney on 17 April 1835, and had quickly established a studio in Bridge Street. Soon he had embarked on the splendid series of pictures that quickly made him the leading artist in Australia. Charles ordered from him watercolour developments of pencil drawings that he had made of the *Beagle* in Ponsonby Sound on 5 March 1834, and of the column of men hauling the boat up the Rio Santa Cruz on 3 May 1834.[179]

Charles then hired a guide and two horses to take him across the Blue Mountains to Bathurst, a hundred miles north-west of Sydney, to get a general idea of the appearance of the country.[180] Charles's first impression of the region was of its extreme uniformity, with endless lines of eucalyptus trees that flourished in the dry climate, but whose pale green foliage, not shed annually, and peeling bark did not give an impression of verdure and fertility, but rather of an arid sterility. On the second day he was glad to meet a party of the Aboriginal natives, each carrying bundles of spears and other weapons that they handled with great dexterity. But since they refused firmly to cultivate the ground, or even to take the trouble to look after a flock of sheep, they had to be regarded strictly as hunter-gatherers and he could only classify them as a few degrees higher in civilisation than the Fuegians.

January 1836

After crossing the Nepean River on the Emu Ferry, they took the road recently cut up the side of the sandstone cliffs to the flattish top of the Blue Mountains at a height of some three thousand feet, and after lunching at a small inn called the Weatherboard, Charles walked down to see the spectacular cliffs at Wentworth Falls, which had been painted by Conrad Martens soon after his arrival at Sydney (Plate 20). The next day he went to see the similar scene at Govett's Leap. These were examples of the several remarkable valleys bounded by vertical cliffs that cut into the mountains and opened into wide amphitheatrical depressions. He decided at first that these valleys must have been eroded by surface water, but then came to doubt whether this was compatible with the existing rate of erosion, and in describing the phenomenon in his book on volcanic islands[181] suggested that the area might once have been under the sea with marine erosion at work. Later still when it was realised that the Blue Mountains had been raised as much as 180 million years ago, and that the valleys could have been eroding for much longer than had previously been thought, geologists came round to accept that Charles's original solution was valid.

He then descended to the granite country a thousand feet lower, where he had been invited to visit a large sheep-grazing establishment. Here he found that the labour was provided by assigned convict servants, and his strong feelings about slaves at once came to the surface. He wrote in his journal:

> The Sunset of a fine day will generally cast an air of happy contentment on any scene; but here at this retired farmhouse the brightest tints on the surrounding woods could not make me forget that forty hardened profligate men were ceasing their daily labours, like the Slaves from Africa, yet without their just claim for compassion.[182]

But although some of the men might have been hardened profligates undeserving of compassion, Charles was not necessarily right to assume that they were being unjustly treated on the station.

He was also taken out kangaroo hunting, and although not a

single kangaroo was to be seen, the greyhounds pursued a kangaroo rat into a hollow tree, from which they dragged an animal no larger than a rabbit, but clearly a marsupial. Not long before, the district had abounded with wild animals, but the English greyhounds had effectively wiped many of them out. Nevertheless there were some large flocks of sulphur-crested white cockatoos feeding in a cornfield, handsome black-and-white magpies similar to the European ones, and some exotically coloured parrots. Charles also counted himself particularly fortunate to come across a group of the famous duck-billed platypuses playing about in a stream, though showing very little more of their bodies than water rats. The farm superintendent Mr Browne shot one, and Charles wrote to Captain P.P. King, whom he proposed to visit on his way back to Sydney, saying, 'I consider it a great feat, to be in at the death of so wonderful an animal.' He must also have stuffed the platypus, because he commented that the specimen did not give a good idea of the appearance of the head and beak in life, the beak having become hard and contracted. But the animal was not in the end preserved in his collection.

In his journal that day he wrote a significant entry:

> A little time before this, I had been lying on a sunny bank & was reflecting on the strange character of the Animals of this country as compared to the rest of the World. An unbeliever in everything beyond his own reason, might exclaim "Surely two distinct Creators must have been at work; their object has however been the same & certainly the end in each case is complete". Whilst thus thinking, I observed the conical pitfall of a Lion-Ant: A fly fell in & immediately disappeared; then came a large but unwary Ant; his struggles to escape being very violent, the little jets of sand described by Kirby (Vol. 1, p.425) were promptly directed against him. (*Note in margin*: NB The pitfall was not above half the size of the one described by Kirby.) His fate however was better than that of the poor fly's: Without a doubt this predacious Larva belongs to the same genus, but to a different species from the Europæan one. Now what would the Disbeliever say to this? Would any two workmen ever hit on so beautiful, so simple & yet so artificial

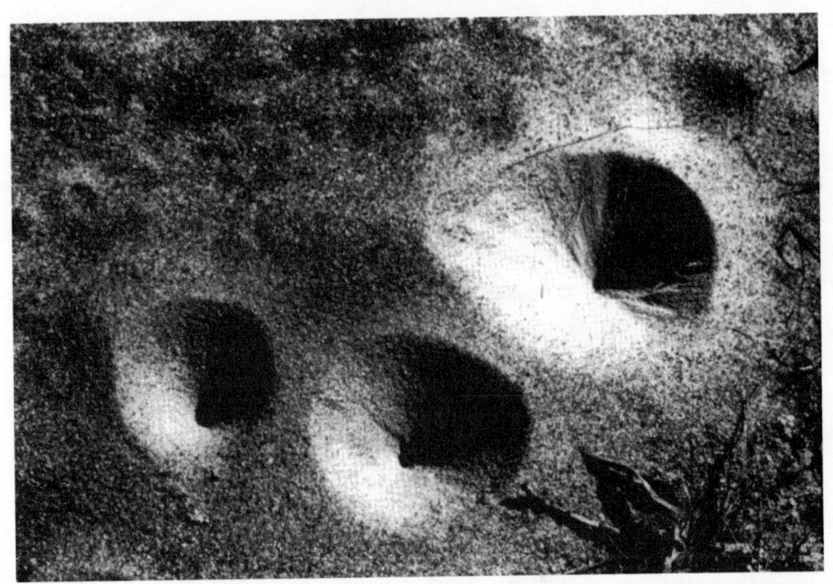

The conical pitfalls of an ant lion

> a contrivance? It cannot be thought so. The one hand has surely worked throughout the universe. A Geologist perhaps would suggest that the periods of Creation have been distinct & remote the one from the other; that the Creator rested in his labour.[183]

It has to be appreciated that at this moment in time Charles's thoughts had definitely begun to move beyond a realisation that Special Creation had difficulties in providing adequate explanations for his observations on the geographical distributions of a number of species of birds and mammals, and on the resemblances between certain existing and extinct species. But it would nevertheless be six months before what he had seen of the mocking birds in the Galapagos would lead him to set down on paper his first doubts on the stability of species. Early in 1836, in this private document intended only for his family, he was still speaking of the Creator, and concluding that a single hand had worked throughout the universe. When in 1839 he was editing the passage for inclusion in the *Journal of Researches*, the situation had changed, for his series of notebooks

An Australian ant lion, the larva of a lacewing of order Neuroptera

on the mechanism of speciation was by then well under way and his intention was to keep his dangerous ideas strictly private, and not to give them away to anyone. So the main changes that he made were to substitute 'sceptic' for 'disbeliever', and to omit the final sentence. In the 1845 edition of the *Journal of Researches*, the episode was confined to a footnote, and the last five sentences were omitted.

Charles rode on into Bathurst, where he had an invitation to visit the Commander of the military barracks. But his impression of the place was not encouraging, thanks mainly to a severe drought that had reduced the Macquarie River to a series of puddles, and had deprived the officers of their sole recreation of shooting quails. He returned by a different road called Lockyers Line which was rather more hilly and picturesque, taking him back across the Blue Mountains, where he slept at the Weatherboard and took another walk to the grand amphitheatre at Wentworth Falls. After crossing the river once more at Emu Ferry, he met Captain King and was taken to the farm at Dunheved that King had been given in 1806 by his father P.G. King (Snr), then Governor of the colony, and to which he had retired in 1831 on leaving the Royal Navy. They spent a pleasant afternoon walking about the farm and talking about the natural history of Tierra del Fuego.

The following day they rode over to Parramatta, where King's brother-in-law, Hannibal Hawkins MacArthur, had just built a sumptuous Greek Revival mansion called Vianaco on his property at Vineyard. There were about eighteen people at lunch in the dining room, and it sounded strange in Charles's ears to hear nice-looking young ladies exclaim, 'Oh we are Australian, and know nothing about England.' All the same, two of the six daughters of the house were later married to Charles's *Beagle* shipmates. One was the amiable First Lieutenant John Wickham, who commanded the *Beagle* from 1837 to 1842 for a third surveying voyage of the north-western coast of Australia, and then retired from the Navy to marry Annie and settle for some years in Australia. The other was Philip Gidley King, who stayed in Australia with his father, married Elizabeth in 1843, and had a long and successful career managing a variety of agricultural and business enterprises. In the afternoon, Charles left this most English-like house and rode by himself into Sydney.

Writing to his sister Susan from Sydney on 28 January 1836, Charles painted a surprisingly favourable picture of the future of Australia, though he could not help disliking the large element of ex-convicts in the population who were intensely acquisitive and knew no pleasure beyond sensuality. He ended his letter with a confession:

> Tell my Father I really am afraid I shall be obliged to draw a small bill at Hobart. I know my Father will say that a hint from me on such a subject is worthy of as much attention, as if it was foretold by a sacred revelation. But I do not feel in truth oracular on the subject. I have been extravagant & bought two water-colour sketches, one of the S. Cruz river & another in T. del Fuego; 3 guineas each, from Martens, who is now established as an Artist at this place. I would not have bought them if I could have guessed how expensive my ride in Bathurst turned out.

At the same time, FitzRoy's choice was a view of Moorea, for which he paid two guineas. Later on he commissioned watercolour developments of quite a number of Martens's *Beagle* drawings that were then engraved as illustrations for *Narrative 1* and 2.

The *Beagle*'s next port of call on Beaufort's list was Hobart in Tasmania, where they anchored at the mouth of the Derwent River on 5 February. Charles went ashore to find the town better supplied with water than Sydney, but otherwise much less impressive. It depressed him to learn that after bitter fighting with the white immigrants, the Aborigines had all been systematically rounded up, and at that time some sixty survivors were kept as prisoners on Flinders Island in the Bass Strait. In 1847 the settlement was moved to an old convict station south of Hobart, but European diseases and exploitation by itinerant sealers and timber-workers continued to take a toll, and the last full-blooded Tasmanian Aborigine died in 1876.

During the next few days Charles took some pleasant walks to examine the geology of Tasmania. With some difficulty because of the thickness of the huge gum trees and flourishing tree ferns, he climbed to the top of Mount Wellington, where at the summit he found large angular masses of naked greenstone. The lower parts of the mountains were encased in basaltic rocks which showed signs of having once flowed as lava, and which contained fossil shells and bryozoans similar to those found in the Silurian, Devonian and Carboniferous strata* in Europe. At Ralphs Bay the eastern and western shores were covered to a height of thirty feet above the high-water mark by broken seashells mixed with pebbles, indicating that there had been only a rather small elevation of the land in more recent times.

While Charles was concentrating primarily on geology, Syms Covington collected some interesting lizards including an oak skink and a blotched blue-tongued lizard belonging to the same genus. Charles also collected a snake whose abdomen burst open during the capture, revealing a developed egg from which a small snake emerged. He had been under the impression that the snake belonged to the family Colubridae with which he was familiar at home, which

* Thus between 300 and 400 million years ago.

was non-venomous and egg-laying, and was puzzled that this snake was evidently viviparous, giving birth to live young. The explanation was that his specimen was not as he supposed a harmless Colubrid, but was a highly venomous member of the family Elapidae, which are indeed live-bearing, being either a black tiger snake or a copperhead. Fortunately his mistake had no disastrous consequences.

One of Charles's favourite topics wherever he went was always the distribution of the coprophagous (dung-eating) beetles that played an important biological role in countries where there were large herds of herbivores whose dung needed to be well distributed over the land to maintain its fertility. But in places like Maldonado in Uruguay he had found the ample repast afforded by the immense herds of horses and cattle to be largely untouched, presumably because there were no native dung beetles to feed on the dung. In Tasmania the only large herbivores were kangaroos, whose dung was pelleted and not obviously suitable for consumption by dung beetles. He therefore carefully examined the dung of the herds that had been introduced only thirty years earlier, and at once found that no fewer than six different species of dung beetle were abundant. He commented that the subject was a curious one that deserved further investigation.

Another interesting find that Charles made beneath a dead and rotten tree in the forest was a family of terrestrial turbellarian flatworms of a previously unknown kind, evidently related to others which he had come across in South America. He kept some specimens alive in his cabin on the *Beagle* in a saucer with rotten wood for eight weeks before they died from the excessive temperature at the latitude that the ship had then reached. They were in due course classified in London as the type specimens of a new genus Tasmanoplana in subfamily Geoplaninae. Before they died he made some valuable observations on the regeneration of bisected specimens, on the death of terrestrial flatworms, and on their photophobia.

On the social front, Charles had been given an introduction to the Surveyor General of Tasmania Mr Frankland, who not only gave him some useful geological specimens from areas that he was unable to visit for himself, but entertained him at home for what Charles described as the most agreeable evening he had spent since leaving England. He decided that Society in Hobart had a considerable advantage over Sydney in not having any wealthy convicts.

The next in the *Beagle*'s chain of meridian distances that had to be established was at a settlement just established at the south-western corner of Australia that ultimately became Albany. After three weeks of steady sailing, for once without any gales, but in a long westerly swell that gave Charles no little misery, the ship anchored in the mouth of the inner harbour of King George's Sound. Here they stayed for eight excessively boring days. At first sight the bright green colour of the brushwood and other plants viewed from a distance seemed to suggest a fertile environment, but this impression was dispelled by Charles's first walk, which led him to wish never to walk again in so uninviting a country. The only cultivated area was around the house of the new Governor, previously a half-pay naval officer in Lyme Regis, who had been bribed to accept the post by the award of a knighthood. On his arrival with his nine children and eleven servants, he found that he had doubled the population of the settlement.

An advantage of the new colony was the cooperative and friendly nature of the Aborigines, who although slightly built, with bearded faces that Charles thought extremely ugly, in disagreement with FitzRoy's view that they had 'rather good countenances', were capable of and willing to perform, unlike the natives of New South Wales, the hardest manual labour. Charles found it impossible not to like such quiet, good-natured men. On the second day of the *Beagle*'s visit, some people from a large tribe called the White Cockatoo Men came to the town, and were asked to hold a corroboree with the King George's men. Tempted according to FitzRoy by an offer from Charles of tubs of boiled rice and sugar, they accepted, and when they had all painted themselves with white spots and lines, the corroboree began. Sometimes running in Indian file, and sometimes sideways, stamping and grunting, beating their clubs and weapons, gesticulating and wriggling their bodies, the two groups merged and separated. Some of the dances apparently represented past wars and victories. In the emu dance each man extended his bent arm to imitate the movement of the neck of the bird, and in another one man imitated an emu grazing in the woods while a second pretended to spear him. All the participants were in the highest spirits, and Charles felt he had never before seen natives so happy and so perfectly at ease.

The geology of King George's Sound was not of particular interest, but with Syms Covington's help Charles made useful collections of fish, where four out of his ten specimens were new species, and a mouse caught among bushes by a trap baited with cheese was later classified as a bush rat new to Australia, but was unexpectedly not a marsupial. Then there were some elegant frogs, an air-breathing limpet still alive after accidentally having been kept in dry paper for twelve days, and a barnacle with curious six-legged larvae. No less than sixty-six species of insects were collected, of which forty-eight were new, and six were in due course named after their discoverer. One trick, sadly, was missed by Charles, for in an area now world-famous for the exceptionally large number of indigenous plants that it houses, he saw none. He might have done better in spring; and he could not of course have known, when forty years later he was working at Down House on insectivorous plants, how impressive the digestive powers of shrubs in the Nepentheaceae like the Albany pitcher plant of West Australia would turn out to be.

The *Beagle*'s departure from King George's Sound was delayed a week by bad weather, when Charles wrote in his journal, 'Farewell Australia, you are a rising infant & doubtless some day will reign a great princess in the South; but you are too great & ambitious for affection, yet not great enough for respect; I leave your shores without sorrow or regret.'

CHAPTER 30

Cocos Keeling Islands

Captain Beaufort had suggested with regard to the *Beagle's* homeward passage that 'perhaps, in crossing the ocean, if circumstances are favourable, she might look at the Keeling Islands, and settle their position'. So on 1 April 1836 the Beagle arrived at the channel entering the outer reef barrier into the central lagoon of the Southern Cocos or Keeling Islands in the east of the Indian Ocean. This remote group of coral islands had first been discovered in 1608, but little notice had been taken of them until in 1826 an adventurer had settled there with some Malay slaves. Not long afterwards an English sea-captain called Ross, together with his Mate Mr Liesk, who greatly impressed Charles by his intelligence, had established the beginnings of an orderly settlement. The future prosperity of the islands depended entirely on the numerous coconut trees growing on them, for their sole product was coconut oil, and in addition the coconuts fed the pigs, poultry and ducks of the Malays. Of less commercial, though of appreciable biological, interest was the population of the remarkable coconut crabs, *Birgus latro* Linn., an interesting species of large land hermit crabs adapted not only to climb up trees, but also possessing claws able to open coconut shells and feed on their contents.

Charles was pleased beyond measure to find himself able at last to take a proper look at the structure of a coral island. The water being unusually smooth, he waded in as far as the living mounds of coral on which the swell of the open sea was breaking. In the gullies and hollows there were beautiful green and other-coloured fishes, and the forms and tints of the corals themselves were admirable, although he felt that some naturalists had perhaps indulged in

April 1836

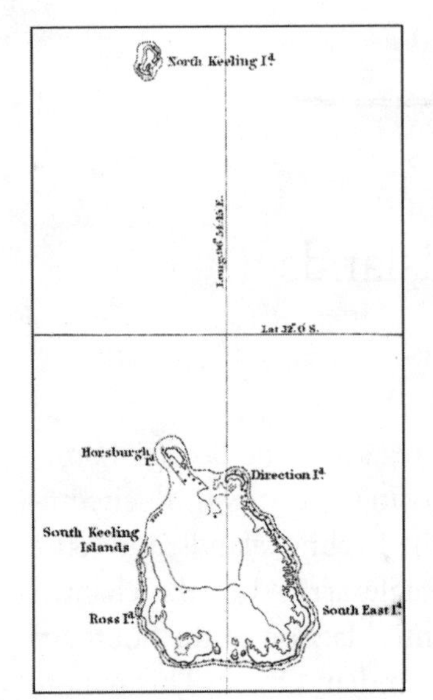

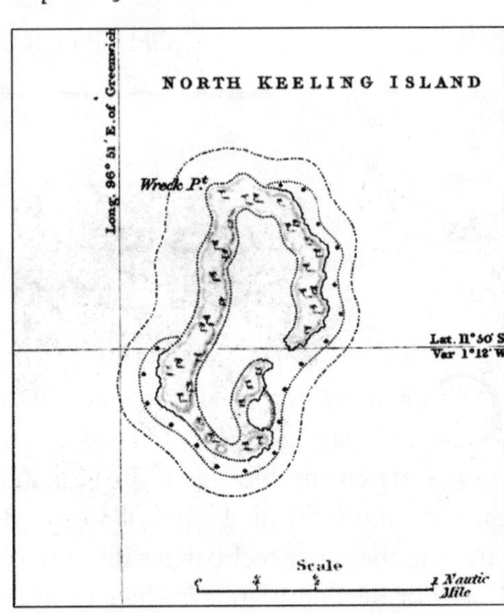

North and South Keeling Islands

excessively exuberant language in describing the beauties of submarine grottoes. A critical point that he said had first been pointed out to him by some of the chiefs in Tahiti, and which was fully confirmed by the observant Mr Liesk, was that even a brief exposure of the living corals to the air and sunlight was fatal. It was therefore difficult to secure a specimen of the truly living coral except when the tide was unusually low, and by using a jumping pole to reach the outermost margin of the reef over which the waves were breaking. However, on two occasions he performed a successful jump, and so found what he called the living 'Porites', forming great irregular rounded masses from four to eight feet broad, and little less in thickness. These mounds were separated from one another by narrow crooked channels about six feet deep, intersecting the reef at right angles. The polyps in the uppermost cells were indeed all dead, but three or four inches lower down on the sides they were alive, forming a projecting border. A few yards further out he could see that the polyps covering the tops of the mounds were still alive, so that he was evidently standing at the exact upward and shoreward limit of the growing corals that formed the outer margin of the reef. From his careful description of the specimens that he collected, the Porites was a true or stony reef-building coral* of order Scleractinia, family Faviidae, and species *Leptoria phrygia*.

The only other coral always found by Mr Liesk at the outer edge of the reefs that seemed capable of standing up to the fury of the breakers was of less importance and belonged to a different order. It was a 'fire coral' or millepore hydrocoral, growing in thick vertical plates with inconspicuous polyps to form a coarse honeycombed mass whose structure Charles described in detail. Charles called it *Millepora complanata*, and its modern name is *M. platyphylla*. It possessed little or no slimy matter on its surface, and yet had a strong and disagreeable odour. It was also capable of producing a stinging sensation on Charles's skin that reminded him of his encounter at the beginning of the voyage with a Portuguese man-of-war (see pp.54–5), like which it might have had protective nematocysts.

In the outer parts of the reef Charles found in addition *Millepora tenella* that too secreted a heavy calcareous skeleton, though it did not sting him. The most abundant corals in the shallow still waters of the central lagoon were further scleractinian species with elegant and more openly branching structures, that included *Pocillopora verrucosa* and *Acropora corymbosa*. There were also scleractinians that Charles described as growing in thin, brittle, stony, foliaceous expansions with stars on both surfaces, and that may have been *Turbinaria bifrons*.

From soundings down to twelve fathoms around the reef it appeared that the commonest coral was living *Millepora alcicornis*. But on the upper margin of the reef, where the top surface of the Porites was dead, three species flourished of what Charles called Nullipora,

* The colonies in this order of colonial hydrozoans containing some 2500 extant species mostly form massive calcareous (stony) exoskeletons. A vitally important characteristic possessed by all reef-building corals is that they house within themselves primitive plants (algae) existing symbiotically, and capable of sharing with their hosts the products of their photosynthesis. Such corals therefore flourish only in shallow and clear water where there is plenty of light.

April 1836

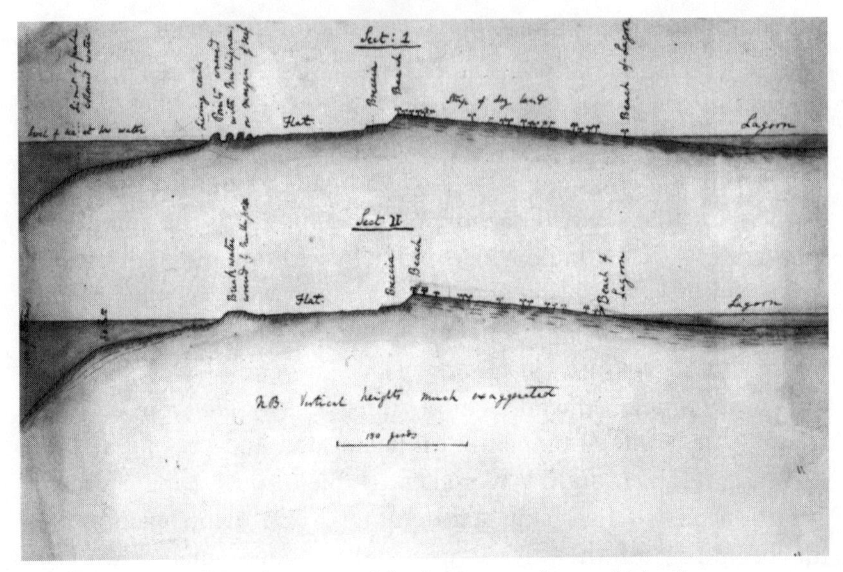

Charles's drawings of the formation of a coral island

and were symbiotic coralline algae such as *Amphiroa* or *Bossea*, green algae such as Chlorophyta, or red algae such as Rhodophyta. These could survive provided that they were immersed in the breaking sea water and exposed to the air for only part of the time, and formed a layer two or three feet in thickness that had a valuable protective effect on the more sensitive reef-building corals.

This visit to the South Keeling Islands was of very great value to Charles, for it enabled him to set his rapidly developing thoughts on the mode of origin of coral islands on an increasingly solid basis by investigating for himself some of the underlying geological facts in this particular example. And at the same time he looked in impressive detail into the biology of the reef-building corals, and examined the internal structure of their remarkable building blocks.[184] On the morning of 12 April he took a last look at the lagoon whose shores were composed of the finest white sand, and from which smaller creeks penetrated the surrounding woods, revealing a field of glittering sand representing water, around the border of which the coconut trees extended their tall waving trunks. He wrote in his journal:

April 1836

12th. In the morning we stood out of the Lagoon. I am glad we have visited these Islands; such formations surely rank high amongst the wonderful objects of this world. It is not a wonder which at first strikes the eye of the body, but rather after reflection, the eye of reason. We feel surprised when travellers relate accounts of the vast piles & extent of some ancient ruins; but how insignificant are the greatest of them, when compared to the matter here accumulated by various small animals. Throughout the whole group of Islands, every single atom, even from the most minute particle to large fragments of rocks, bear the stamp of once having been subjected to the power of organic arrangement. Capt. FitzRoy at the distance of but little more than a mile from the shore sounded with a line 7200 feet long, & found no bottom.

Hence we must consider this Isld as the summit of a lofty mountain; to how great a depth or thickness the work of the Coral animal extends is quite uncertain. If the opinion that the rock-making Polypi continue to build upwards, as the foundation of the Isl^{d} from volcanic agency after intervals gradually subsides, is granted to be true; then probably the Coral limestone must be of great thickness. We see certain Is^{ds} in the Pacifick, such as Tahiti & Eimeo, mentioned in this journal, which are encircled by a Coral reef separated from the shore by channels & basins of still water. Various causes tend to check the growth of the most efficient kinds of Corals in these situations. Hence if we imagine such an Island, after long successive intervals to subside a few feet, in a manner similar, but with a movement opposite to the continent of S. America; the coral would then be continued upwards, rising from the foundation of the encircling reef. In time the central land would sink beneath the level of the sea & disappear, but the coral would have completed its circular wall. Should we not then have a Lagoon Island? Under this view, we must look at a Lagoon Is^{d} as a monument raised by myriads of tiny architects, to mark the spot where a former land lies buried in the depths of the ocean.

CHAPTER 31

Mauritius, Cape of Good Hope, St Helena and Ascension Island

A fortnight later, the *Beagle* rounded the northern end of the island of Mauritius – formerly the Île de France, but since the end of the Napoleonic Wars under British administration – and anchored at Port Louis on its western coast. Charles admired the sloping central plain of the village known as Pamplemousses, 'grapefruit trees', coloured bright green by the many fields of sugarcane, whose output was said to have increased seventy-five-fold since England took possession. Port Louis remained attractively French in its character, the shops were all French, and the English residents spoke in French to their servants. There was a very pretty little theatre in which operas were excellently performed, being – praise be to their tastes, said Charles – much preferred by the inhabitants to common plays. There were surprisingly large and well stocked booksellers' shops, attesting to the adherence of Mauritius to the culture of the Old rather than the New World. Men of many races were to be seen in the streets, for convicts from India had been banished to the island for life. However, they seemed quiet and well conducted, and from their cleanliness and faithful observance of their strange religious practices, it was impossible not to prefer them to the convicts of New South Wales.

To catch a ship sailing for England next day, Charles wrote to his sister Caroline from Port Louis on 29 April:

> Now one glimpse of my dear home would be better than the united kingdoms of all the glorious Tropics. Whilst we are at sea, & the

weather is fine, my time passes smoothly, because I am very busy. My occupation consists in rearranging old geological notes: the rearranging generally consists in totally rewriting them. I am just now beginning to discover the difficulty of expressing one's ideas on paper. As long as it consists solely of description it is pretty easy; but where reasoning comes into play, to make a proper connection, a clearness & a moderate fluency, is to me, as I have said, a difficulty of which I had no idea.

I am in high spirits about my geology, & even aspire to the hope that my observations will be considered of some utility by real geologists. I see very clearly, it will be necessary to live in London for a year, by which time with hard work, the greater part, I trust, of my materials will be exhausted. Will you ask Erasmus to put my name down to the Whyndham or any other club; if afterwards it should be advisable not to enter it, there is no harm done. The Captain has a cousin in the Whyndham, whom he thinks will be able to get me in. Tell Erasmus to turn in his mind for some lodgings with good big rooms in some vulgar part of London. Now that I am planning about England, I really believe she is not at so hopeless a distance. Will you tell my Father I have drawn a bill of 30£. The Captain is daily becoming a happier man, he now looks forward with cheerfulness to the work which is before him. He, like myself, is busy all day writing, but instead of geology, it is the account of the Voyage. I sometimes fear his "Book" will be rather diffuse, but in most other respects it certainly will be good: his style is very simple & excellent. He has proposed to me to join him in publishing the account, that is for him to have the disposal & arranging of my journal & to mingle it with his own. Of course I have said I am perfectly willing, if he wants materials, or thinks the chit-chat details of my journal are in any ways worth publishing. He has read over the part I have on board, & likes it.

I shall be anxious to hear your opinions, for it is a most dangerous task, in these days, to publish accounts of parts of the world which have so frequently been visited. It is a rare piece of good fortune for me, that of the many errant (in ships) Naturalists, there have been few or rather no geologists. I shall enter the field unopposed. I assure you I

> look forward with no little anxiety to the time when Henslow, putting on a grave face, shall decide on the merits of my notes. If he shakes his head in a disapproving manner: I shall then know that I had better at once give up science, for science will have given up me. For I have worked with every grain of energy I possess.[185]

On 1 May, Charles climbed La Pouce, a 2600-foot mountain to the south of Port Louis, and saw how the centre of the island resembled the basin of a huge volcanic crater, with La Pouce and other mountains forming parts of a more or less continuous outer wall. It appeared that the sea had recently reached the base of these mountains, since when the land had risen appreciably, for around much of the circumference of the island there were large blocks of upraised coral, and beds of a conglomerate of shells and corals. There had also been substantial flows of lava round and between the basaltic ring of mountains.

Charles and Mr Stokes were invited by the Surveyor General of Mauritius, Captain Lloyd, well known for his survey across the Isthmus of Panama, to stay at his delightful country house six miles from Port Louis. He showed them some of the geology, the hedges of mimosa, and the avenues of mangoes. He also owned the only elephant in Mauritius, on which they were invited to enjoy a ride in true Indian fashion. Charles was impressed by the noiseless step of the elephant, but felt as others have done that for a long journey the motion would be rather fatiguing.

The *Beagle* then continued on her passage to the west, catching on the way a glimpse of Madagascar, another oceanic island whose unique population of lemurs would eventually make it equal in interest to the Galapagos. But the lemurs received no special attention for a while, because there was no meridian distance to be established there, and the *Beagle* passed by to anchor in Simon's Bay on 31 May. In Cape Town, twenty-two miles to the north, the only hotel was crowded out by passengers on ships from India, and Charles found himself in a set of excessively wealthy Nabobs who were heavy prosers but who could not, poor fellows, have raised between themselves a single healthy liver.

Having taken a room in a boarding house he did not find much of interest either in the town or in the neighbouring country at the village of the Paarl and the surrounding mountains. But during his second week in Cape Town he fared much better, and became acquainted with several very pleasant people. One was Dr Andrew Smith, an army surgeon who became the leading authority on the zoology of South Africa, and who had just returned from an expedition beyond the Tropic of Capricorn where he had found a new species of rhinoceros. Another was Thomas Maclear, who was the Astronomer Royal at the Cape of Good Hope for some years.

Most significantly of all, Charles and FitzRoy met the distinguished astronomer, mathematician, chemist and philosopher John Herschel, whose book on the Study of Natural Philosophy[186] Charles had read during his last year at Cambridge, stirring up in him 'a burning zeal to add even the most humble contribution to the noble structure of Natural Science'. Herschel had come out to the Cape of Good Hope with his twenty-foot telescope in 1834 in order to carry out a systematic survey of the southern skies, and after making some of the best observations on the return of Halley's comet in 1835, had been the last person to see the comet in May 1836 before it disappeared from sight until 1909. Charles later wrote disarmingly of him that 'He never talked much, but every word which he uttered was worth listening to. He was very shy and he often had a distressed expression. Lady Caroline Bell, at whose house I dined at the C. of Good Hope, admired Herschel much, but said that he always came into a room as if he knew that his hands were dirty, and that he knew that his wife knew that they were dirty.'

FitzRoy had been unhappy to find in Cape Town that there were strong prejudices against the activities of the missionaries in South Africa and elsewhere, and through the good offices of John Herschel, submitted to a South African journal his only joint article with Charles,[187] in which the accusations of the critics were refuted by a series of examples taken from Charles's commonplace journal on what they had observed of the current activities of the missionaries in Tahiti and New Zealand.

July 1836

After returning to the *Beagle* at Simon's Bay, where Charles took a last long walk with Bartholomew Sulivan to examine the geology of the immediately surrounding mountains, they put to sea on 18 June accompanied by their usual ill fortune – first a fierce gale of wind, and then scarcely any wind at all. Eleven days later, the *Beagle* crossed the Tropic of Capricorn for the sixth and Charles earnestly trusted the last time, and on 8 July they anchored off St Helena.

This most forbidding of islands rose like a huge castle from the ocean, with a great wall built by successive streams of black lava surrounding it. Around the town, many small forts and guns had been built up, although the natural defences did not appear to need much assistance. Charles found lodgings in a cottage close to Napoleon's tomb in the centre of the island at an elevation of about two thousand feet, primarily because it was convenient as a base for his geological observations. He did not feel that either the tomb itself, or the house a mile up the road where Napoleon had lived, which was in a filthy and neglected state with its walls disfigured with the names and graffiti of visitors, created feelings at all in unison with the imagined resting place of so great a spirit. Syms Covington was surprised that the tomb, of which he made a drawing, was so simple, and noted with approval that the small well beside it had provided Napoleon daily with beautiful clear water, and that an old soldier was stationed beside it to prevent its misuse by sightseers. Whether Covington also approved of the current provision in Napoleon's house of a billiard room for the use of visitors, where wine was sold for their consumption, was not recorded.

Near the coast the rough lava was entirely destitute of vegetation, but further inland the appearance of the island was notable in two respects. First was the extraordinary number of roads and forts that had been built, for although the island had served primarily as a prison, the public expenditure that had at one time been devoted to it seemed quite out of proportion to the extent or basic value of the cultivable land. With the recent withdrawal from its administration of the East India Company, and the departure of many relatively well-paid public servants, the remaining population of emancipated slaves were

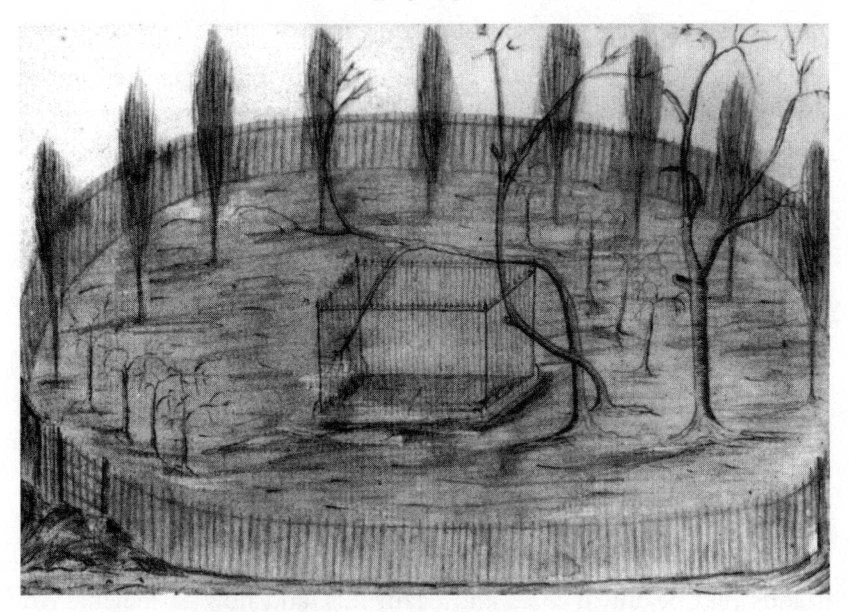

Syms Covington's drawing of Napoleon's tomb

mostly unemployed, and their poverty could only worsen. The fine times when 'Bony' was there, as Charles's old guide, once a goatherd, called them, could never return. Another bequest from those days was the decidedly English appearance of much of the vegetation, with hills crowned with plantations of Scotch firs, sloping banks thickly covered with bright-yellow gorse bushes, hedges loaded with blackberries, and streams bordered by weeping willows. There were said to be four times as many imported species as indigenous ones, many of which evidently preferred the climate of St Helena to that of England.

Situated so remote from any continent, St Helena possessed a unique flora and some insects, but no birds and animals of its own apart from those recently introduced. Among the insects, Charles was once again surprised to find two of his pets, species of dung-eating beetles of family Scarabaeidae that were common under dung. Since the island had originally possessed no large quadrupeds such as goats, horses or cattle, though there might have been a mouse, on what food could the beetles have lived if they were indigenous? It was evident that they could only have arrived accidentally with the cattle or horses, and over the years both here and in Tasmania there must surely have been many occasions when they could have sneaked in with the straw and bedding of the imported animals. Perhaps this had not happened at Maldonado because there the cattle were only being shipped out, and were not being brought in from Europe.

Partridges and pheasants were tolerably abundant, and the island was far too English not to be subject to strict game laws. Charles was shocked to hear of one such law even more unjust than those imposed in England, for it prohibited the burning by poor people, for the export of soda, of a plant growing on the coastal rocks, because the partridges would otherwise have nowhere to build.

There was much resemblance in structure and in geological history between St Jago in the Cape Verde Islands, where Charles's initiation into the birth and subsequent development of volcanic islands had begun in 1832, and Mauritius, and now St Helena. All three islands were examples of 'craters of elevation' where a ring of basaltic mountains had rested on older submarine beds of different compositions. Deluges of more recent lavas had then flowed from the centre of the island towards and between the original mountains, and at St Helena the central platform had been filled up by them. All three islands had been raised in mass. At Mauritius the sea had reached the bases of the mountains in a recent geological period, as it had now done at St Helena. Charles's conclusions were summarised in a lecture given to the Geological Society in 1838,[188] and in the second of his geological books.[189]

The last to be visited by Charles on what might later have been termed his Cook's Tour of Volcanic Islands was Ascension Island, only five days' sailing from St Helena. This was a desolate clump of black volcanic rocks, roughly triangular with sides of about six miles, rising to a central summit 2840 feet high around which there were a few houses and cultivated fields. On the western side a few stone houses and a barracks manned by marines had been built. The only inhabitants in addition to the marines were some Negroes liberated from slave ships who were paid and victualled by the British government.

July 1836

Charles walked up the road to the summit, alongside which there were milestones and carefully maintained cisterns at which every thirsty passer-by could drink good water. A major problem of the island was its restricted rainfall, which meant that there were no trees growing on it, and that the output of the central springs had to be managed at all times with naval discipline. Charles was told that the witty people of St Helena, where there was no shortage of water, used to say with some justice, 'We know we live on a rock, but the poor people at Ascension live on a cinder.'

In the days that followed, he explored much of the island, but even when the sun was shining brightly it could not be called beautiful, for to be just it stared with naked hideousness. The lava streams were everywhere punctuated by rugged hummocks, between which there were spaces covered in some places by pumice, ashes and volcanic sandstone, and in others by curious volcanic bombs varying in size between that of an apple and of a man's body. They were spherical or pear-shaped, with a hardened outer crust of lava encasing a cellular interior, and were supposed to have spun rapidly while cooling during an eruption. Elsewhere there were many different types of volcanic rock containing strikingly laminated mixtures of dark and glassy obsidian with paler minerals. Seen from the sea, the ground at one end of the island was mottled with white patches, which Charles now found to consist of the only living creatures to be seen apart from himself, which were thousands of sleeping sea birds.

> On the beach a great sea, although the breeze was light, was tumbling over the broken lava rocks. The ocean is a raging monster, insult him a thousand miles distant, & his great carcase is stirred with anger through half an hemisphere.[190]

During the long periods at sea of the previous months, Charles had been working hard, first on his Geology Notes, and then on his Zoology Specimen Lists. Here he marked the numbered entries with capital letters in pencil – A for animal (i.e. a mammal), B for bird, C for a crustacean, and so on – showing Syms Covington to which of the separate lists they were to be assigned, and adding various other instructions. In some cases the lists were drawn up in Covington's hand, but for the mammals and birds Charles prepared them all himself, often adding instructive general comments. His Ornithological Notes were probably begun after his departure from the Cocos Keeling Islands, and finished on his arrival at Ascension Island.[191]

The five crucial pages about the Galapagos were written with a new pen and almost without corrections, suggesting that Charles had had time to reflect carefully on what needed to be said:

> These islands are scattered over a space of ocean included between 125 miles of Latitude & 140 of Longitude. They are situated directly beneath the Equator and about 500 miles from the coast of S. America. The constitution of the land is entirely Volcanic; and the climate being extremely arid, the Islands are but thinly clothed with nearly leafless, stunted brushwood or trees. On the windward side however, & at an elevation between one & two thousand feet, the clouds fertilize the soil; & it then produces a green & tolerably luxuriant vegetation. In such favourable spots, & under so genial a climate, I expected to have found swarms of various insects; to my surprise, these were scarce to a degree which I never remember to have observed in any other such country. Probably these green Oases, bordered by arid land, & placed in the midst of the sea, are effectually excluded from receiving any migratory colonists. However this may arise, the scarcity of prey causes a like scarcity of insectivorous birds & the green woods are scarcely tenanted by a single animal. The greater number of birds haunt, & are adapted for the dry & wretched looking thickets of the coast land; here however a store of food is laid up. Annually, heavy torrents of rain at one particular season fall; grasses & other plants rapidly shoot up, flower, & as rapidly disappear. The seeds however lie dormant till the next year, buried in the cindery soil. Hence these Finches* are in number of species & individuals far pre-

* The family of birds ultimately to be classified by the ornithologist John Gould as the Geospizinae.

ponderant over any other family of birds. Amongst the species of this family there reigns (to me) an inexplicable confusion. Of each kind, some are jet black, & from this, by intermediate shades to brown; the proportional number, in all the black kinds is *exceedingly* small; yet my series of specimens would go to show, that that color is proper to the old cock birds alone. On the other hand Messrs Bynoe & Fuller assert they have each a small jet black bird of the female sex. Moreover a gradation in form of the bill appears to me to exist. There is no possibility of distinguishing the species by their habits, as they are all similar, & they feed together (also with doves) in large irregular flocks. I should observe, that with respect to the probable age of the smaller birds, that in no case were any of the feathers imperfect, or bill soft, so as to indicate immaturity, & on the other hand in no case were the eggs in the ovarium of the hen birds much developed. I should suppose the season of incubation would be two or *three months* later.

. . . .

Thenca 3306*: male: Charles Isd –
do 3307: do: Chatham Isd –
These birds are closely allied in appearance to the Thenca of Chile (2169) or Callandra of La Plata (1216). In their habits I cannot point out a single difference. They are lively, inquisitive, active, *run fast*, frequent houses to pick the meat of the Tortoise which is hung up, sing tolerably well; are said to build a simple open nest; are *very* tame, a character in common with the other birds. I *imagined* however its note or cry was rather different from the Thenca of Chile? Are very abundant over the whole Island; are chiefly tempted up into the high & damp parts by the houses & cleared land.

I have specimens from four of the larger Islands: the two above enumerated, and (3349: female. Albermarle Isd.) & (3350: male: James Island. The specimens from Chatham & Albermarle Isd. appear to be the same; but the other two are different. In each Isld. each kind is

* Specimen number in *Zoology Notes*.

July 1836

exclusively found: habits of all are indistinguishable. When I recollect the fact that from the form of the body, shape of scales & general size, the Spaniards can at once pronounce from which Island any Tortoise may have been brought. When I see these Islands in sight of each other, & possessed of but a scanty stock of animals, tenanted by these birds but slightly differing in structure & filling the same place in Nature, I must suspect they are only varieties. The only fact of a similar kind of which I am aware is the constant asserted difference between the wolf-like Fox of East & West Falkland Islds.

If there is the slightest foundation for these remarks the zoology of Archipelagoes will be well worth examining; for such facts would undermine the stability of Species.[192]

In the first of these paragraphs, Charles neatly explains how the shortage of insects on an isolated oceanic island, which he here observes for the first time, might be causally linked with an ecology favouring seed-eating birds. But with his usual honesty he admits to being able to make no sense at all of the mixture of species that are feeding together, noting only in passing that there is a gradation in the form of the bill. It is in the second of these paragraphs that he slightly extends his previous remarks about the Galapagos mocking birds, and commits himself for the very first time in writing to the admission that he is on the track of evidence that existing species are not, as has hitherto been universally believed, immutable.

CHAPTER 32

A Quick Dash to Bahia and Home to Falmouth

The last lap in the *Beagle*'s return to England was a quick dash across the Atlantic to Bahia, for FitzRoy wanted to make a final check on the meridian distance that he had measured there four and a half years earlier.

Charles was glad to find that his enjoyment of tropical scenery had not been diminished in the least by familiarity. He wrote in his journal:

> In the last walk I took, I stopped again and again to gaze on such beauties, & tried to fix for ever in my mind, an impression which at the time I knew must sooner or later fade away. The forms of the Orange tree, the Cocoa nut, the Palms, the Mango, the Banana, will remain clear & separate, but the thousand beauties which unite them all into one perfect scene, must perish: yet they will leave, like a tale heard in childhood, a picture full of indistinct, but most beautiful figures.[193]

He also found time to collect some insects in the woods, and on the beach he took specimens of coralline algae of genera *Melobesia* and *Halimeda*. He noted that when left in sea water in the sunlight the algae exhaled a good deal of 'gaz', and wondered what this gas might be.*

* He was inadvertently repeating a famous experiment first performed by the chemist Joseph Priestley (1733–1804) in 1777, when he showed that under such conditions bubbles of oxygen would be generated.

October 1836

On 6 August the *Beagle* weighed anchor and stood out to sea for home, but the wind was up to its usual tricks, and a week later they were obliged to take shelter for several days at Pernambuco (now known as Recife), still on the coast of Brazil. They crossed the Equator at last on 21 August, this time without any ceremonies, and on 4 September anchored at Porto Praya in the Cape Verde Islands. St Jago had lost its former appeal, although the great baobab tree had improved its appearance by clothing itself with thick green foliage. Charles nevertheless found things of interest in the natural history of the arid lava plains, and collected four small birds that inhabited them. Two weeks later they reached the island of Terceira in the Azores, and Charles was taken to visit a volcanic crater where jets of steam issued from some cracks, and was happy to see some old English friends among the birds. From there, thanks to God, and to the infinite relief of all on board, the *Beagle* was able at last to steer a direct course for England, and anchored at Falmouth on Sunday, 2 October 1836.

That same night, and a dreadfully stormy one it was, Charles took the mail coach home to Shrewsbury, where he arrived after two days on the road. After a joyful reunion with his father and sisters, his first thought was to write to his Uncle Jos at Maer, 'as being my first Lord of the Admiralty', and next he wrote to Robert FitzRoy:

> My dear FitzRoy
>
> I arrived here yesterday morning at Breakfast time, & thank God, found all my dear good sisters & father quite well. My father appears more cheerful and very little older than when I left. My sisters assure me I do not look the least different, & I am able to return the compliment. Indeed all England appears changed, excepting the good old Town of Shrewsbury & its inhabitants – which for all I can see to the contrary may go on as they now are to Doomsday. I wish with all my heart, I was writing to you amongst your friends instead of that horrid Plymouth. But the day will soon come and you will be as happy as I now am. I do assure you I am a very great man at home – the five years voyage has certainly raised me a hundred per cent. I fear such greatness must experience a fall.

October 1836

I am thoroughly ashamed of myself, in what a dead and half alive state I spent the few last days on board, my only excuse is that certainly I was not quite well. The first day in the mail tired me, but as I drew nearer to Shrewsbury everything now looked more beautiful & cheerful. In passing Gloucestershire & Worcestershire I wished much for you to admire the fields woods & orchards. The stupid people on the coach did not seem to think the fields one bit greener than usual, but I am sure we should have thoroughly agreed that the wide world does not contain so happy a prospect as the rich cultivated land of England.

I hope you will not forget to send me a note telling me how you go on. I do indeed hope all your vexations and trouble with respect to our voyage which we now *know* has an end, have come to a close. If you do not receive much satisfaction for all the mental and bodily energy you have expended in His Majesty's Service, you will be most hardly treated. I put my radical sisters into an uproar at some of the *prudent* (if they were not *honest* whigs, I *would* say shabby), proceedings of our Government. By the way I must tell you for the honor & glory of the family, that my father has a large engraving of King George the IV, put up in his sitting Room. But I am no renegade, and by the time we meet, my politics will be as firmly fixed and as wisely founded as ever they were.

I thought when I began this letter I would convince you what a steady & sober frame of mind I was in. But I find I am writing most precious nonsense. Two or three of our labourers yesterday immediately set to work, and got most excessively drunk in honour of the arrival of Master Charles. Who then shall gainsay if Master Charles himself chooses to make himself a fool.

Good bye – God bless you – I hope you are as happy, but much wiser than your most sincere but unworthy Philos. Chas Darwin.

Two weeks later, FitzRoy replied from London:

Dearest Philos

What you will say to me for not having written before I know not – but really I have *not* been idle or forgetful.

I trusted to Fuller [his steward in the *Beagle*] for all immediately necessary information and I will now try to give you the rest.

Captain Beaufort was out of town when my letters & papers reached London (from Falmouth) and the Chart duster put them *away* in a corner (excepting *one* private one) to await Capt. B's *return*!!! Those papers related to the Chronometric results &c &c – upon which the necessity for our going to Woolwich was to be founded.

Orders had been sent to Plymouth for the Beagle to pay off there – but L[d]. Amelius Beauclerk had civility & sense enough to stay proceedings and approve of my going to London to see the Lords & Masters myself – I boarded Sir John Barrow [Secretary of the Admiralty], and then made a stalking horse of *him* while attacking the *others*.

All was satisfactorily settled in a very short time – and they acceded civilly to my proposal of calling at *Portsmouth*. I was delighted to see that the Valp[o]. cargo of charts had not only *arrived* but that they were mostly *Engraved* – or in the Engraver's hands – and on a *large* scale. They have given much satisfaction at the Hyd[l]. Office.

I have promised to give them a short paper for the Geog[l]. Society – a slight *sketch* of our voyage – I will do what I can – according to time and ask you to add and correct.

Rice Trevor [FitzRoy's brother-in-law] & Alex[r]. Wood [FitzRoy's cousin] crossed me on the road – they in one mail – I in another – but I was soon down again & with them at Devonport – Fuller told me you looked very well and had on a *good hat*!

Who the deuce was *my cousin* in a *broad brimmed* hat?

I was delighted by your letter. The account of your family – & the joy tipsy style of the whole letter were *very* pleasing. Indeed Charles Darwin I have *also* been *very* happy – even at that horrid place Plymouth – for that horrid place contains a *treasure* to *me* which even *you* were ignorant of!! Now guess – and think & guess again. Believe it, or not, – the news is *true* – I am going to be *married*!!!!!! to Mary OBrian.*

---

* Maria Henrietta O'Brien, daughter of a retired Major-General living in the country, became FitzRoy's first wife, and mother of his four children.

November 1836

Now you may know that I had decided on this step, long *very long* ago. All is settled & we shall be married in December. Rice Trevor, Alex[r]. Wood & Talbot like her *much*. Pray call on my sister in Stratton Street – she longs to see you, – and ask to see the *children*.

Money matters are better than *you* think. Your's most sincerely Rob[t]. FitzRoy

Beagle. Portsmouth 20th. Oct. (night) – arrived this morning – shall sail tomorrow morning for Woolwich. Sights are *taken*.

On 28 October, the *Beagle* anchored at Greenwich, where her chronometer rates and final observations were recorded, and where appropriately Alexander Usborne and his companions came on board, having just completed their survey of the whole coast of Peru from Atacama to Guayaqil for addition to the *Beagle*'s charts, though without making any additions to Charles's collection of shells. The *Beagle* was paid off at Woolwich on 17 November.

So ended the voyage of the *Beagle* in the cordial state of friendship between the Captain and his gentlemanly scientist in which it had begun. Of course there had been FitzRoy's sudden bursts of temper that were equally quick to subside, and of course there had been the many occasions recalled by Charles in his autobiography when officers coming on duty in the forenoon used to ask 'whether much hot coffee had been served out this morning?' – meaning, how was the Captain's temper? But there were certainly no prolonged or serious disputes between them. The story once in fashion that FitzRoy's main motive in recruiting a scientist had been to find someone who would establish the truth of the Biblical story of the Creation, and that he and Charles had argued vigorously on this issue throughout the voyage, is on FitzRoy's own showing a myth. For in the last chapter of his account of the voyage he wrote:[194]

> While led away by sceptical ideas, and knowing extremely little of the Bible, one of my remarks to a friend, on crossing vast plains composed of rolled stones bedded in diluvial detritus some hundred feet in depth, was "this could never have been effected by a forty days' flood,"

– an expression plainly indicative of the turn of mind, and ignorance of Scripture. I was quite willing to disbelieve what I thought to be the Mosaic account, upon the evidence of a hasty glance, though knowing next to nothing of the record I doubted: – and I mention this particularly, because I have conversed with persons fond of geology, yet knowing no more of the Bible than I knew at that time.

This passage undoubtedly refers to FitzRoy's expedition in 1834 with Charles, hauling the boats up the Rio Santa Cruz across the endless gravel plain studded with oysters in the south of Patagonia. And his point is that at that time he himself knew nothing at all about the teaching of the Bible. It was only after he was married in December 1836 that he was converted by his wife to becoming a religious believer. Charles himself recalled that 'Whilst on board the *Beagle* I was quite orthodox, and I remember being heartily laughed at by several of the officers (though themselves orthodox) for quoting the Bible as an unanswerable authority on some point of morality.' Moreover, he was as has been seen still speaking of a Creator and Centres of Creation early in 1836.

CHAPTER 33

Harvesting the Evidence

Back in England, Charles's most immediate concern was to dispose of his specimens to experts for examination, and he was quickly plunged into a fever of activity that made the busiest time during the voyage seem like tranquillity itself. He fetched the Galapagos plants from the *Beagle* at Greenwich, and took them to Henslow in Cambridge. Henslow warned him that he might find the 'great men' overwhelmed with their own business, and Charles soon found that indeed the Museum of the Zoological Society appeared to be full of unmounted specimens, and 'the Zoologists seem to think a number of undescribed creatures rather a nuisance'.

The geologists were, however, more welcoming, led enthusiastically by Charles Lyell, who had been pressing Henslow since the previous year for news of Charles's return to England. Towards the end of October, Charles and Lyell met for the first time, and at once, said Charles, 'Lyell entered in the *most* good natured manner, and almost without being asked, into all my plans.' Charles was the first and most faithful of Lyell's biological disciples, and they at once became the closest of colleagues, sharing all each other's secrets for the rest of their lives. Lyell was currently President of the Geological Society of London, of which Charles was soon elected a Fellow, though he complained to his old friend and cousin W.D. Fox that letters like F.G.S. and still worse F.R.S. were very expensive. A few days later, dining with the Lyells, Charles met the up-and-coming anatomist Richard Owen, five years older than himself, who had just been appointed Hunterian Professor at the Museum of the Royal College of Surgeons in London, and who immediately offered to look at Charles's fossils.

Soon the cases were being unpacked at the College of Surgeons, and Charles reported to his sister Caroline that 'Some of them are turning out great treasures. One animal, of which I have nearly all the bones, is very closely allied to the Ant Eaters, but of the extraordinary size of a small horse. There is another head, as large as a Rhinoceros, which as far as they can guess, must have been a gnawing animal. Conceive a Rat or a Hare of such a size – What famous cats they ought to have had in those days!'

Before the end of the year, Owen had reached some important conclusions. He rejected Charles's guess that the osseous polygonal plates from Punta Alta belonged to the *Megatherium*, and instead assigned them to a huge extinct armadillo from the family Glyptodontidae named *Hoplophorus*. But Charles's bones nevertheless made valuable additions to what was known about the giant ground sloth, and with them was a new and closely related animal that Owen named *Scelidotherium*. There were in addition fragments from the commonest notoungulate, i.e. a hoof-bearing grazing mammal, *Toxodon*, which was built like a short-legged rhinoceros and resembled a gigantic capybara. When the second shipment from the *Beagle* arrived, there were more bones of *Toxodon*, and a most interesting addition was the animal that Charles had found in the cliffs at Port St Julian, which Owen named *Macrauchenia** and initially classified as an extinct species of camel related to the existing guanaco, though further studies then led him to assign it to a different order of extinct quadrupeds. Lastly there were bones of an extinct

* It suited Charles rather well that *Macrauchenia* should be related to the guanaco existing in South America, but the argument rested on special features of the particular bones available for study, and in the end it mattered most that the animal was long-necked and had three toes on its forefeet. *Macrauchenia* was therefore classified in the end as a litoptern, a thick-skinned quadruped which did not chew the cud, and was distantly related to the tapir, which after all was found elsewhere in South America. The family of Camelidae were nevertheless quite well represented in North America, and in addition to the guanacos at least one, the paleollama whose footsteps were found at Pehuen-Co (see p.113), had reached South America.

rodent, *Ctenomys*, related to the existing guinea-pig, and the tooth of an extinct species of horse.

Owen was a thrusting and ambitious man who in due course became Director of the Natural History Museum,* and as a disbeliever in the transmutation of species, was later a bitter enemy of Charles and his associates. But his first venture into palaeontology, with a demonstration in South America of what he called a 'persistence of type' in which the fossils resembled in type though not in size the animals currently living in the area, was highly successful, and earned him the Wollaston Medal of the Geological Society in 1838. At that period he was critical of the moribund state of the Natural History Museum, and recommended to Charles that he should give his collections of birds and mammals to the Zoological Society of London. His advice was sound, and the taxonomist John Gould agreed to look after the birds, George Waterhouse the insects and mammals, and Thomas Bell the reptiles.

Gould classified and exhibited most of Charles's birds at meetings of the Zoological Society during the first two months of 1837, and quickly reached the unexpected and even startling conclusion that the Galapagos 'finches' belonged to an entirely new group previously unknown to science and found only in the Galapagos Islands. On 13 March Charles moved with Syms Covington to work in London at rooms in 36 Great Marlborough Street, and was then able to discuss the Galapagos birds more closely with Gould. It appeared that like the mocking birds, the different species of finch were restricted to their own particular island, though not all the specimens were well enough labelled for final proof of the point. Charles was already aware of the importance of such evidence in connection with his new but still strictly private ideas that a transmutation of species must have taken place in the Galapagos. So he had hastily to go begging to FitzRoy and his servant Fuller for

* The official title of the museum in South Kensington was until recently British Museum (Natural History), but for the sake of simplicity it will be referred to here as the Natural History Museum.

further specimens from their collections that might help to reinforce the case.[161] On 14 March, his thinking received another nudge from Gould when at a lecture at the Zoological Society he learnt that the 'Avestruz Petise' shot by Conrad Martens at Port Desire was not merely a geographic variation of the ordinary rhea, but on anatomical grounds was definitely a different species that replaced *Rhea americana* in the south of Patagonia.

In the private journal in which for the rest of his life Charles recorded the dates of his various activities, he wrote:

> In July [1837] opened first note Book on "transmutation of species". Had been greatly struck from about month of previous March on character of S. American fossils, & species on Galapagos Archipelago. These facts origin (especially latter) of all my views.[195]

So while Charles occupied himself publicly first with publishing the results of his scientific observations on geology and natural history made on the *Beagle*, and then with making a detailed taxonomic study of the whole order of extinct and living barnacles, in private he was simultaneously working hard on the series of notebooks on geology, the transmutation of species, and metaphysical enquiries[196] that led eventually to the publication of *On the Origin of Species* in 1859.

Charles's first published work, apart from the article with FitzRoy and the extracts from his letters to Henslow printed privately for members of the Cambridge Philosophical Society in 1835, was the third of the volumes entitled *Narrative of the Surveying Voyages of H.M. Ships Adventure and Beagle between the years 1826 and 1836* that were published in 1839. Volume 1 was written by Captain P.P. King, and Volume 2 by FitzRoy. Volume 3 had been written by Charles between March and September 1837, drawing about half of the material from his commonplace journal, now *The Beagle Diary*, and the other half from his Geology and Zoology Notes. But Captains King and FitzRoy were slower in completing their volumes. Volume 3 immediately outsold the other two, and it was therefore reprinted as *Journal of Researches into the Geology and Natural History of the various Countries visited by H.M.S. Beagle from 1832 to 1836*. Charles afterwards regarded it as the favourite among his printed works, saying that 'the success of this my first literary child always tickles my vanity more than that of any of my other books'. A revised version was published in 1845, and the *Journal of Researches* has since been reprinted many times in many languages.

Having found expert taxonomists to classify his thousands of specimens, Charles next succeeded in persuading the Chancellor of the Exchequer to grant £1000 for the publication in five parts under his joint editorship of *The Zoology of the voyage of H.M.S. Beagle, under the command of Captain FitzRoy, during the years 1832 to 1836*. Part I on *Fossil Mammalia* by Richard Owen was published in 1838–40, Part II on *Mammalia* by George Waterhouse in 1838–39, Part III on *Birds* by John Gould in 1839–41, Part IV on *Fish* by Leonard Jenyns in 1840–2, and Part V on *Reptiles* by Thomas Bell in 1842–43. Hundreds of new species were first described in these volumes, together with comments on their behaviour and habits, and pictures of many of them. The harvest also included identifications that appeared elsewhere of many new plants[197] and insects.[198]

An area to which Charles had actually devoted more time than any other was the study of marine invertebrates of many kinds caught in his plankton net, this having been something that he could do when the *Beagle* was at sea and he could not add to his other collections. He also confessed that working on the internal anatomy of these tiny creatures under his microscope, and making surprisingly good drawings of what he saw, gave him particular pleasure. In the course of these studies he sorted out Linnaeus's controversial zoophytes into bryozoans, coralline algae and true corals, and made pioneering observations on a variety of crustaceans, tunicates, turbellarian flatworms and other creatures. Most of this work remained unpublished in his lifetime, but a transcription of his notes and specimen lists has recently appeared.[199]

Charles was now free to pursue his geological interests. On 4 January 1837 he read a paper to the Geological Society on his evidence for the recent occurrence of elevation on the coast of Chile. This was followed up on 31 May by an account of his theory of the mode of formation of coral reefs. On 16 February 1838 he reluctantly accepted an invitation from William Whewell to become Secretary of the Geological Society. On 8 March he read an important paper to the Society entitled 'On the connexion of certain volcanic phenomena in South America; and on the formation of mountain chains and volcanos, as an effect of the same power by which continents are elevated'. It was received with respect, for his arguments on the succession of elevations and declines of the level of the land that must have taken place were convincing. But it did not make a permanent impression, in the absence at that time of any evidence as to what precisely the underlying forces affecting the rise and fall of the earth's crust might be. It was not until the 1960s, when the theory of plate tectonics caused a revolution in the thinking of geologists, that it could be appreciated how close to the truth Charles's geological theorisation had in fact been.

In October 1838, Charles noted that he had begun to write his book on *The structure and distribution of coral reefs*, and that it would require much reading. It was published in May 1842, followed in January 1844 by *Geological observations on the volcanic islands visited during the voyage of H.M.S. Beagle*, and in October 1846 by *Geological observations on South America*. In his *Autobiography*, Charles said he calculated that the writing of his three geological books had consumed four and a half years' steady work, and he was afraid that his intermittent illness must have lost quite a bit more.

However, he was very far from idle in his intensive speculation and constructive thinking about the transmutation of species that was recorded in the series of notebooks begun on board the *Beagle* in May 1836 with the 'Red' notebook.[200] Its first 112 pages were confined to geological topics, but subsequent entries were written in London early in 1837 after Richard Owen's identification of the *Macrauchenia* bones from Port St Julian and John Gould's identification of a second species of rhea in the south of Patagonia, by which time Charles was firmly launched on considering the ways in which new species might arise.

The sorting out of the subsequent notebooks written between 1837 and 1844 later presented scholars with a difficult problem, because although throughout his life Charles preserved all his papers with meticulous care, he had a somewhat tiresome habit, after the notebooks had originally been filled, of going back to them and cutting out an appreciable number of 'useful' pages to be kept elsewhere in quite a different order. Nevertheless, they have finally been transcribed in the order in which they were probably written.[196] As will be seen, their existence was carefully concealed from the outside world for many years to come, partly because Charles was well aware of the opposition that he would encounter in many quarters to such a radical line of research, and partly because of his invariable and time-consuming insistence on assembling wholly trustworthy evidence in support of all his arguments. To begin with, Lyell was the only person with whom his species work was closely discussed, though Henslow, Jenyns, Waterhouse and W.D. Fox were aware of his interest in the origin and variation of species. But in 1843, Charles embarked on a vigorous correspondence with Joseph Dalton Hooker, who had just returned as the botanist on HMS *Erebus* with James Clark Ross's Antarctic expedition (1839–43). Soon they were deep in discussions of the botany of Tierra del Fuego and Charles's volcanic islands, and Hooker had accepted an invitation to look at the Galapagos plants. When in 1844 they met at Down House in Kent, Hooker became the first person to whom the prime secret of the principle of Natural Selection was revealed, and thereafter he remained the closest of all Charles's helpers.

Notebook A was devoted wholly to geology, and therefore to the issues covered in the three books that Charles wrote between 1838 and 1846. Notebook B was headed *Zoonomia*, the title of Erasmus Darwin's huge treatise on medicine and biology published in 1794, whose importance had once been dismissed rather summarily by Charles (see pp.7–8). It conveyed the basic message set out by Charles as:

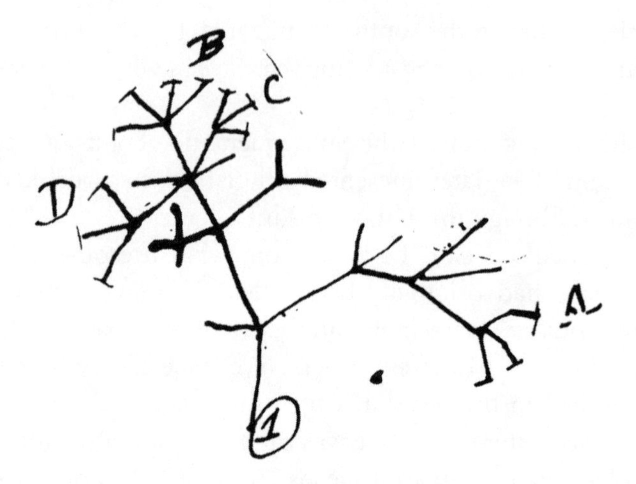

Charles's branching diagram from Notebook B

> The Grand Question, which every naturalist ought to have before him, when dissecting a whale, or classifying a mite, a fungus, or an infusorian,* is "What are the laws of life".[201]

After explaining how species might adapt to changing circumstances through modification of the two types of reproduction described in *Zoonomia*, Charles switched to the relationship between transmutation and systematics, and quickly arrived at the first of his branching diagrams showing how several different groups of animals might be descended from a single ancestor.

The idea that he thus introduced was what became known as the Theory of Common Descent, which proposed that evolution had taken place by the modification of animals from their starting pattern in a stepwise fashion. It was not new, for it had previously occurred in slightly different forms to Erasmus Darwin, to the eminent French naturalist Georges-Louis Leclerc, comte de Buffon (1707–88), and to Lamarck. There were flaws in the version put

* A single-celled protozoan animal equipped with cilia for locomotion and the generation of feeding currents.

forward in *Zoonomia* that Charles not unreasonably disliked, but it cannot be denied that his grandfather, in his best and posthumously published poem *The Temple of Nature*,[202] had arrived at the most elegant presentation of the theory, even if it was purely speculative and was accompanied by very little supporting evidence. In Canto I of the poem, lines 227 to 314 run:

> Ere Time began, from flaming Chaos hurl'd
> Rose the bright spheres, which form the circling world;
> Earths from each sun with quick explosions burst,
> And second planets issued from the first.
> Then, whilst the sea at their coeval birth,
> Surge over surge, involv'd the shoreless earth;
> Nurs'd by warm sun-beams in primeval caves
> Organic Life began beneath the waves.
> . . .
> Hence without parent by spontaneous birth
> Rise the first specks of animated earth;
> From Nature's womb the plant or insect swims,
> And buds or breathes, with microscopic limbs.
>
> ORGANIC LIFE beneath the shoreless waves
> Was born and nurs'd in Ocean's pearly caves
> First forms minute, unseen by spheric glass,
> Move on the mud, or pierce the watery mass;
> These, as successive generations bloom,
> New powers acquire, and larger limbs assume;
> Whence countless groups of vegetation spring,
> And breathing realms of fin, and feet, and wing.
>
> Thus the tall Oak, the giant of the wood,
> Which bears Brittania's thunders on the flood;
> The Whale, unmeasured monster of the main,
> The lordly Lion, monarch of the plain,
> The Eagle soaring in the realms of air,

> Whose eye undazzled drinks the solar glare,
> Imperial man, who rules the bestial crowd,
> Of language, reason, and reflection proud,
> With brow erect who scorns this earthy sod,
> And styles himself the image of his God;
> Arose from rudiments of form and sense,
> An embryon point, or microscopic ens!*

But although Charles accepted and improved on the theory of common descent with enthusiasm, he never referred to his grandfather or anyone else as introducing it, perhaps because he thought its truth so obvious as to need no acknowledgement.

In notebook C, Charles conducted a masterly survey of the available literature in search of the laws governing such questions as the hereditary transmission of form and the effect and stability of crosses, the distribution of local and wide-ranging species, and the relation between behaviour and structure. In notebook D he was concentrating on the role of reproduction, when as he wrote in his *Autobiography*:

> In October 1838, that is fifteen months after I had begun my systematic enquiry, I happened to read for amusement Malthus on *Population*,[203] and being well prepared to appreciate the struggle for existence which everywhere goes on from long-continued observation of the habits of animals and plants, it at once struck me that under these circumstances favourable variations would tend to be preserved, and unfavourable ones to be destroyed. The result of this would be the formation of new species.[204]

He had now arrived at the Theory of Natural Selection, the second of the main principles on which *On the Origin of Species* was based,[205] providing a viable mechanism for the formation of new

* An 'ens' is an old word describing a being or entity as opposed to an attribute or quality.

species. He therefore settled down to assemble evidence that would bear on both the extent of variation in natural stocks, and on the nature of the selective process that would enable the individuals best adapted to a changing environment to survive. In 1842 he wrote a preliminary sketch of his species theory, concluding with a memorable sentence that faintly echoes Erasmus's poem, and that remained almost unchanged in the final book published in 1859:

> There is a simple grandeur in the view of life with its powers of growth, assimilation and reproduction, being originally breathed into matter under one or a few forms, and that whilst this our planet has gone circling on according to fixed laws, and land and water, in a cycle of change, have gone on replacing each other, that from so simple an origin, through the process of gradual selection of infinitesimal changes, forms most beautiful and most wonderful have been evolved.

In 1844 he wrote a more polished version of the 1842 sketch, which was then set aside, the only people knowing about his views on the transmutation of species being Lyell, Hooker and Jenyns. In 1846, encouraged by Hooker's remark that 'no one has hardly a right to examine the question of species who has not minutely described many', Charles set to work for eight years on the painstaking research into the anatomy and taxonomy of the living and fossil barnacles that was described in his two monographs.[206]

When in 1854 the barnacles were at last out of the way, and thousands of borrowed specimens had been returned to their owners, Charles was free to resume work on his species theory. Searching for evidence on possible mechanisms for the geographical distribution of animals and plants, he carried out experiments on the germination of seeds that had been soaked in sea water, concealed in the carcases of small animals fed to seagoing birds, or carried in the mud picked up on the feet of birds. He hybridised garden vegetables like peas and beans, and having set up a network of experts on variation in domestic animals, himself became a pigeon fancier at Down House. Early in 1856 he at long last began on Lyell's advice to write the 'big book' on the origin of species that never appeared as such, for it was overtaken by events. He also began a correspondence with the naturalist Alfred Russel Wallace, who was collecting specimens in the East Indies, about domesticated birds in the region.

On 18 June 1858, Lyell's warning that Charles might be forestalled came all too true, when he received a letter from Wallace asking for advice about the publication of Wallace's own theory of natural selection. Charles was shattered, for as he explained to Lyell, 'I never saw a more striking coincidence. If Wallace had my M.S. sketch written out in 1842 he could not have made a better short abstract! Even his terms now stand as Heads of my Chapters.' There was some agonised discussion with Lyell and Hooker as to the best action to be taken, for Charles insisted that 'I would far rather burn my whole book than that he or any man shd think that I had behaved in a paltry spirit', while Lyell and Hooker knew well for how much longer than Wallace he had been working on the subject, and what a huge mass of evidence he had accumulated. In the end it was decided that the fairest solution would be for Lyell and Hooker to present at the next meeting of the Linnean Society a joint contribution comprising extracts from the 1844 Sketch, an abstract of a letter from Charles to Professor Asa Gray of Harvard, and Wallace's paper. The Darwin–Wallace papers were duly read to the Society on 1 July 1858.

The cat was now out of the bag as far as the transmutation of species was concerned, so Charles hurriedly produced a shortened version of the big book that he had begun too late to write. *On the Origin of Species* was released to the booksellers on 22 November 1859, and immediately sold out.* Charles later published many other books, but this concludes the story of his greatest work, for which the seeds had been sown during those four and three-quarter venturesome years that he spent on board the *Beagle*.

* 1250 copies were printed initially, and the book was reprinted many times.

CHAPTER 34

Farewell to Robert FitzRoy

Writing to his sister Caroline about Captain FitzRoy early in the *Beagle* voyage, Charles said:

I never before came across a man whom I could *fancy* being a Napoleon or a Nelson. I should not call him clever, yet I feel convinced nothing is too great or too high for him. His ascendancy over everybody is quite curious: the extent to which every officer & man feels the slightest rebuke or praise would have been, before seeing him, incomprehensible. It is very amusing to see all hands hauling at a rope, they not supposing him on deck, & then observe the effect when he utters a syllable: it is like a string of dray horses, when the waggoner gives one of his aweful smacks . . . His many good qualities are great & numerous: altogether he is the strongest marked character I ever fell in with.[207]

Near the end of the voyage, he wrote to his sister Susan:

Thank God the Captain is as home sick as I am, & I trust he will rather grow worse than better . . . I have been for the last 12 months on very Cordial terms with him. He is an extra ordinary but noble character, unfortunately however affected with strong peculiarities of temper. Of this, no man is more aware than himself, as he shows by his attempts to conquer them. I often doubt what will be his end, under many circumstances I am sure it would be a brilliant one, under others I fear a very unhappy one.[208]

Charles's closeness to Robert FitzRoy, and his indebtedness to him, were greater than has often been recognised, and a brief account of FitzRoy's career after he and Charles parted at Woolwich deserves to be included. In the summer of 1841, Lord Melbourne's Whig government fell, and a general election followed. With no further naval appointment in sight, FitzRoy was glad to accept an offer from his maternal uncle, the Tory Lord Londonderry, of candidature for one of the County of Durham's seats. Unfortunately a second Tory, Mr William Sheppard, was also nominated, though he later withdrew after a complicated and public quarrel with FitzRoy, accusations from both sides that honour had been infringed, and challenges to a duel. But when FitzRoy had safely been elected, tempers had still not cooled. Outside the United Services Club in the Mall one day, Sheppard, brandishing a whip over his head, announced in a loud voice, 'Captain FitzRoy! I will not strike you. But consider yourself horsewhipped!' FitzRoy retaliated with his umbrella, but the contestants were then separated by their friends. Neither of them emerged from the affair with a great deal of credit.

While serving as an MP, FitzRoy was appointed as an Elder of Trinity House,* and to be Acting Conservator of the River Mersey, on whose problems he produced a valuable report. He was also involved in launching the Mercantile Marine Act requiring both captains and first mates of all boats to have certificates of competence, and suggested how the examinations for these certificates might be conducted. Lastly, he was entrusted with the task of looking after the Archduke Frederick of Austria for three months while his Austrian warship was repaired. The tour, which included calls on the Queen Dowager, the Prime Minister, the Duke of Wellington and other grandees, visits to Woolwich Dockyard and Arsenal, and viewing naval gunnery and military parades, passed off most successfully. At

* By Act of Parliament in 1836, all privately operated lighthouses, lightships, beacons and buoys in England and Wales were bought and transferred to the management of an autonomous non-governmental agency known as Trinity House.

Christmas the Archduke departed in his warship, and FitzRoy went home to his family.

In 1842 the first Lieutenant Governor of New Zealand, William Hobson, died after serving the new colony for less than three years, leaving some intractable problems for his successor to solve. These mainly arose because the population of land-hungry white immigrants was rising fast, and they had to buy their land from its Maori owners under laws that had been incompletely and very poorly drafted. The appointment was offered to FitzRoy, who already had a slight knowledge of New Zealand. He accepted it with rash optimism, and as he wrote to Captain P.P. King, 'trusting in a superintending Providence – and anxious to raise the New Zealanders. I anticipate no great difficulty with them – but abundant trouble with the whites.' In December 1843, Captain and Mrs FitzRoy arrived at Auckland in New Zealand's North Island with their three children.

Earlier that year there had been a serious dispute over the sale of some land at Wairau, near Nelson at the northern end of South Island, which had ended with the massacre by the Maoris of nineteen white settlers, including a brother of their fiery leader Edward Wakefield. At a levée held by FitzRoy at Nelson, he delivered a rebuke in his best hot coffee style to Wakefield's son Jerningham, which was hardly calculated to calm the feelings of the whites. Then at a hearing on the facts underlying the incident he determined that the law as far as the sale was concerned had in fact been on the Maoris' side; but he felt that since the Maoris so far outnumbered the whites, it was simply not practicable to take revenge on their chiefs for the murders. It was not a promising start, for he had neither impressed nor pleased anyone on either side.

With no goodwill behind him, FitzRoy now not only had to solve the land problem, but also to rescue his exchequer from impending bankruptcy. Neither the British Treasury nor any Australian bank would lend him any money, so ignoring his instructions as Governor he issued paper money on which 5 per cent interest was payable. To raise this interest, he cancelled the embargo on the selling of land by anyone except his government, and put a high tax on further land sales, which was often higher than the value of the land itself, infuriating the Maoris. Next he was faced with a revolt by some though not all of the Maoris at Kororareka in the Bay of Islands, egged on by discontented whites of the kind of whom he had been critical in 1835. A government flagstaff flying the Union Jack was cut down, re-erected, and cut down again despite protection by the guns of HMS *Hazard* and some badly-led soldiers sent from Sydney to help. The soldiers' powder magazine accidentally blew up, and Kororareka was burnt to the ground. At Nelson, FitzRoy's effigy was burnt.

The Secretary of State for the Colonies in London, Lord Stanley, received from the local government offices in New Zealand what was inevitably a very one-sided picture of what was going on, which together with some justified accusations that FitzRoy had taken several important decisions without prior consultation, led to his dismissal from the office of Governor. He had in fact done his best throughout to do right in his own eyes, though it could be said that because of his inflexibility he was the unfortunate victim of his own high principles.

During 1849 and 1850, having been given the post of Superintendent of Woolwich Dockyard, FitzRoy was occupied with the trials of the Navy's first screw-driven steamship, perhaps appropriately named HMS *Arrogant*. But although he found the ship 'a pleasure to turn', he was irritated by its too frequently suffering from mechanical faults such as foreign matter getting under the valves, and he remained at heart an unconverted sailship man. In 1850 he resigned from active service in the Navy.

In 1851 he was elected a Fellow of the Royal Society, supported by Beaufort, Charles and eleven others, as distinguished for his eminence in the sciences of hydrography and nautical astronomy, and for his accurate measurements of a chain of meridian distances during the *Beagle*'s circumnavigation of the globe. In 1854 the chief maritime powers held at Brussels a conference on meteorology at sea that recommended some urgent action in this field to their governments. The Board of Trade in London was voted some funds, and turned to the Royal Society for advice, with the result that FitzRoy was officially appointed as Meteorological Statist – i.e. statistician – with a staff of three to help him. This was his last and, apart from his captaincy of the *Beagle*, most successful job, in which he justly earned the title of father of meteorology in Great Britain.

FitzRoy at once settled down to circularise the captains of all ships, both naval and maritime, who when provided with suitable instruments would provide him with good data on winds, atmospheric pressure, temperature and humidity, on which weather charts for a given area could be based. Remembering that it was the little men in little boats who suffered most of all from unpredicted storms, this search for data was soon extended to the provision of what became known as FitzRoy barometers to fishing towns and villages all round the coast, together with a *Barometer Manual* of fifty pages, costing one shilling. FitzRoy's basic message was from the start that 'It should always be remembered that the state of the air *foretells coming* weather, rather than indicates weather that is *present*,' and soon he used the phrase that he made his own, the *forecasting* of weather. He was quick to appreciate the value of the newly introduced electric telegraph for obtaining the latest weather information, and set up a chain of eighteen stations on the coasts of England, Scotland and Ireland, and six further afield in western Europe, which reported at frequent intervals to him in London. He then produced synoptic charts of weather including wind speeds and directions, isobars showing atmospheric pressures, and isotherms showing temperatures, together with weather forecasts for the next two days. These were published in *The Times* and other newspapers just as they still are today. He also instituted a system of warning cones which were hoisted at ports and harbours and fishing villages when a gale was to be expected. At long last in 1857 he was promoted to Rear Admiral, and in 1863 to Vice Admiral.

FitzRoy's daughter Laura related to Nora Barlow in 1934 how one day in 1860, when as a young girl she was living with her parents in Kensington, she heard the bell ring, and opened the front door to find the Queen's Messengers come to enquire at the Admiral's house about the weather reports and storm warnings for the following day,

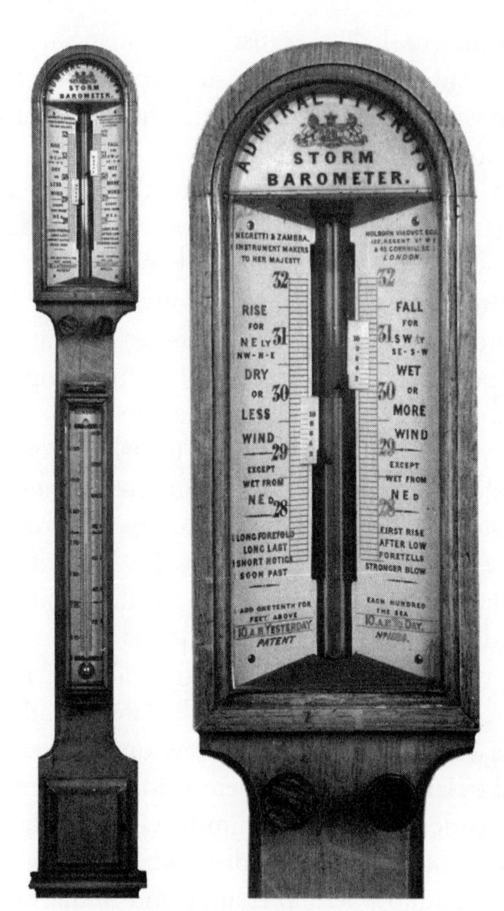

Admiral FitzRoy's storm barometer

when Her Majesty was intending to cross to Osborne on the Isle of Wight. But the story did not relate whether or not the forecast on that occasion was correct.

It was on 30 June 1860 that at a meeting of the British Association for the Advancement of Science in Oxford, the Bishop of Oxford, known as 'Soapy Sam' Wilberforce from his habit of wringing his hands as if washing them whilst preaching or lecturing, delivered a celebrated attack on Charles that was energetically refuted by his principal champions T.H. Huxley and Hooker. Admiral FitzRoy had on the previous day been reading a paper about British Storms, and it has sometimes been claimed that after Huxley had spoken the Admiral rose to his feet in a fury, and flourishing a bible briskly denounced his onetime gentlemanly companion. However, this would appear to be an exaggeration, for the official report on the meeting merely said that Admiral FitzRoy regretted the publication of Mr Darwin's book, and denied Professor Huxley's statement that it was a logical arrangement of facts. Moreover, FitzRoy's reaction immediately after publication of the *Origin* in 1859 had not been anger, but alarm at the extreme views of his 'poor friend and five years messmate Charles Darwin', to whom he had written, 'My dear old friend – I, at least, *cannot* find anything "enobling" in the thought of being a descendant of even the *most* ancient *Ape*.' No reply from Charles has survived. Over the years FitzRoy and Charles corresponded but did not meet very often, the last occasions being when FitzRoy was admitted as a Fellow of the Royal Society in 1851, and when he came to Down House in 1853 and 1857.

Despite the initial success of FitzRoy's weather forecasts, his final years were not happy ones. There was steadily increasing public criticism when inevitably some of the forecasts proved to be wrong, as of course is still happening a century later despite the enormously greater amount of information available to his successors. The great American pioneer M.F. Maury, who had been instrumental in setting up the international conference to which FitzRoy owed his appointment, became unfairly critical of the forecasts made in England and France. Even *The Times* joined in, though at the same time it urged FitzRoy not to give up. He was always prone to depression, but what his doctors prescribed was not unreasonably a period of total rest. He responded by working even harder, until on the morning of Sunday, 30 April 1865, at the age of fifty-nine, he went into his dressing room, took up his razor, and cut his throat. Charles's premonition had tragically turned out to be correct.

Epilogue

In his *Autobiography*, Charles stated that 'The voyage of the *Beagle* has been by far the most important event in my life and has determined my whole career'; but as has been seen in this account of his activities during the voyage, it was only at its very end that his first written doubts were expressed about the stability of existing species. It may therefore be asked how, after more than four years of such painstaking geological study and extensive collecting of animals and plants, he had arrived at making his most important discovery, throwing light on the manner in which new species had arisen?

In his *Autobiography* he also said, perhaps to be in line with current fashion, but far from accurately, that he 'worked on true Baconian principles, and without any theory collected facts on a wholesale scale'. The truth was that the central element in his scientific approach was a passion for theorising, and although he admitted that reasoning was a serious fault while making an observation, he always maintained that it was essential both beforehand and afterwards. This was abundantly clear in the extensive and extremely well organised notes that he made from the outset on all of his geological and zoological observations.

For his first geological project early in the voyage at St Jago in the Cape Verde Islands he was guided not only by what he had learnt five months earlier during his expedition with Adam Sedgwick in North Wales, but also by his careful reading of Volume 1 of Lyell's *Principles*. He proceeded first to theorise, and then after conducting some experiments on rock samples that he had collected, he crossed out the arguments that he already considered to be faulty. A year later he made further corrections and additions to his initial conclusions. In the manner in which he had thus started, so he continued, with a long series of observations on the slow rise of the coast of Patagonia; on the sequences of big rises and falls involved in the creation of the Andes; on the origin of coral reefs; and on the structure of volcanic islands. He also dug up fossils of huge animals that turned out to be related to modern South American species. The final bulk of his geological and palaeontological notes was four times greater than those on natural history.

On the zoological front there was less literature to guide him, though he often referred to the works of the famous French school of '*encyclopédistes*' of which there were copies in the *Beagle*'s library. His first practical step was to invent a new type of net for collecting the plankton living at the surface of the sea, details of whose anatomy he studied under his dissecting microscope while he was at sea. On the seashore at St Jago he then encountered an octopus, the doyen of invertebrates, and made the first of many highly perceptive and original observations not just on the features of this and other specimens that were of importance for their taxonomic classification, but also on their behaviour and the manner in which they moved, and on their relationship with their environment. Whenever possible he performed little experiments to test out his theories, and argued critically with himself about the proper interpretation of the results. His intensely analytical way of thinking, coupled with the orderly efficiency with which he kept his notes, and an exceptional memory for detail, were important keys to his success.

During the voyage Charles was exposed to a challenging array of new scientific facts, and was exceptionally fortunate not only in his timing as far as earthquakes were concerned, but also in being taken by the *Beagle* to what became the prime example of a set of isolated oceanic islands where the development of new species could be observed. His talent for observation, and his genius for finding logical explanations for what he had seen by hard thinking, were properly rewarded by his arrival at *On the Origin of Species*.

Charles had embarked on the *Beagle* as an admirer of William

Paley's philosophy, and until early in 1836 he continued to speak about 'centres of creation' at which the indigenous species had first made their appearance. Some of the areas of biology in which he interested himself, for example the discovery of new species of turbellarian worms or the vexed question of whether corallines were animals or plants, had no immediate bearing on how the species concerned might have originated. But when he began to think about how the distribution of particular creatures had arisen in the first place, or how they had come to interact with one another and their environment, he quickly appreciated that mechanisms other than special creation would have to be taken into account to explain how it came about that 'Nature when she formed these animals & these plants knew that they must reside together'. In the Falklands in 1833 he further pursued this line of thought and started to think about the role of geographical separation, concluding two years later that the animal and bird populations found on the two sides of the huge barrier to migration imposed by the Andes required either that the same species had been created in two different places, or that species were not immutable and might be modified over a long period of time. Although in the Galapagos he missed for the time being the significance of the variation of the beaks of the finches, he did take note of the variation in the mocking birds between the different islands, and he perceived for the first time the 'very remarkable' scarcity of insects on an isolated island, and evidently drew the correct conclusion.

When therefore Charles arrived back in England, he was already poised to take the critical step of admitting that species were not immutable, and to explore the logical consequences. Early in 1837, he had been told by Richard Owen that in South America as in Australia the huge extinct mammals were clearly related to existing much smaller species, and by John Gould that the finches belonged to a remarkable new family of birds with beaks of different sizes, evidently adapted to different diets. He then set out in the privacy of his study to work on the series of Transmutation Notebooks that in the fullness of time led to the publication of *On the Origin of Species*. Owing to his invariable insistence on supporting his arguments with all the evidence that could be gathered, many of Charles's works went through a long period of incubation, and in some cases proof of the correctness of his case did not emerge until long after his death.

His first major scientific theory was thus the 'Theory of the Earth' presented to the Geological Society in 1838,[188] though its impact was limited because it postulated underlying movements of the crust of the earth for whose existence and causation there was at that time no direct evidence. In the 1960s, however, a new unifying theory for geological science was introduced by the advent of plate tectonics, which explained how the earth's outermost shell is broken into about a dozen large rigid plates riding on the weak and partially molten region of the mantle. The collision between the South American Plate covering the land, and the Nazca Plate covering the eastern part of the South Pacific, can now be seen to provide very nice support for Charles's observations.

One of the two main planks supporting Charles's ideas on evolution was the Theory of Common Descent, but although it seemed so obviously valid, it lacked scientific proof until, again in the 1960s, the genetic code was proved to be virtually identical in all living organisms.

After Gregor Mendel had in 1866 published his famous paper about the basic mathematics of the inheritance of characteristics of peas planted in the garden of his monastery at Brünn, Charles could have read it in the library of either the Royal or the Linnean Society. But he seldom went up to London for such a purpose, and worked on steadily in his study at Down House, so remaining unaware of its existence for the rest of his life, as did almost everyone else until it was rediscovered in 1900. Although Mendel read and wrote notes in his copy of *On the Origin of Species*, he made no attempt to call at Down House when he visited England in 1862.[209] In closing this book, it could be said that one of Charles Darwin's most remarkable achievements was to have got so much right without ever knowing anything about Mendelian genetics or the molecular mechanism of heredity.

Notes

Most of the information in this book is drawn from the following sources:

Narrative 1: *Narrative of the surveying voyages of His Majesty's Ships Adventure and Beagle between the years 1826 and 1836, describing their examination of the southern shores of South America, and the Beagle's circumnavigation of the globe. Volume I. Proceedings of the first expedition, 1826–1830, under the command of Captain P. Parker King, R.N., F.R.S.* Henry Colburn, London, 1839

Narrative 2: *Narrative of the surveying voyages . . .of the globe. Volume II. Proceedings of the second expedition, 1831–1836, under the command of Captain Robert Fitz-Roy, R.N.* Henry Colburn, London, 1839. (Substantial extracts from this volume are to be found in David Stanbury (ed.). *A Narrative of the Voyage of H.M.S. Beagle*. The Folio Society, London, 1977.)

Journal of Researches 1: *Narrative of the surveying voyages . . .of the globe. Volume III. Journal and remarks. 1832–1836*. By Charles Darwin Esq., M.A. Sec. Geol. Soc. Henry Colburn, London, 1839

Journal of Researches 2: *Journal of Researches into the Natural History and Geology of the countries visited . . .under the Command of Capt. Fitz Roy, R.N.* By Charles Darwin, M.A., F.R.S. John Murray, London, 1845

Origin: *On the origin of species by means of natural selection, or the preservation of favoured races in the struggle for life*. By Charles Darwin, M.A., F.R.S., F.G.S., F.L.S. &c. John Murray, London, 1859

Notebooks: *Charles Darwin's Notebooks, 1836–1844*. Edited by Paul H. Barrett, Peter J. Gautrey, Sandra Herbert, David Kohn and Sydney Smith. British Museum (Natural History) and Cambridge University Press, 1987

Autobiography: *The Autobiography of Charles Darwin, 1809–1882*. Edited by Nora Barlow. Collins, London, 1958

Beagle Record: *The Beagle Record: Selections from the Original Pictorial Records and Written Accounts of the Voyage of H.M.S. Beagle.* Edited by Richard Keynes. Cambridge University Press, 1979

Beagle Diary: *Charles Darwin's Beagle Diary*. Edited by Richard Keynes. Cambridge University Press, 1988, 2001

Zoology Notes: *Charles Darwin's Zoology Notes and Specimen Lists from H.M.S. Beagle*. Edited by Richard Keynes. Cambridge University Press, 2000

Correspondence 1: *The Correspondence of Charles Darwin. Volume 1. 1821–1836*. Edited by Frederick Burkhardt and Sydney Smith. Cambridge University Press, 1985

Biography: Janet Browne. *Charles Darwin, Vol. 1: Voyaging.* Jonathan Cape, London, 1995

1. *Autobiography* pp.76–7
2. *Journal of Researches 1* and 2
3. Alan Moorehead. *Darwin and the Beagle.* Hamish Hamilton, London, 1969
4. *Origin* p.192
5. Richard Keynes and Hiss Martins-Ferreira (1953) Membrane Potentials in the Electroplates of the Electric Eel. *Journal of Physiology* 119 pp.315–51
6. *Narrative 2*
7. Richard Keynes (ed.). *Charles Darwin's Beagle Diary*. Cambridge University Press, 1988, 2001
8. Diary of observations on the zoology of the places visited during the voyage of H.M.S. Beagle. CUL MSS DAR 30.1, 30.2, 31.1, 31.2
9. Diary of observations on the geology of the places visited during the voyage. CUL MSS DAR 32.1, 32.2, 33
10. Notes on the geology of the places visited during the voyage, Parts I–V. CUL MSS DAR 34, 35, 36, 37, 38
11. CUL MS DAR 118
12. Nora Barlow (ed.). *Darwin and Henslow: The Growth of an Idea. Letters 1831–1860.* John Murray, London, 1967. See also *Correspondence 1*
13. *Zoology Notes*

14. *The zoology of the voyage of H.M.S. Beagle during the years 1832–1836*, Parts I-V by Richard Owen, George Waterhouse, John Gould, Leonard Jenyns and Thomas Bell, edited by Charles Darwin. Reprinted in facsimile by Nova Pacifica, Wellington, New Zealand, 1970
15. James Secord. 'The Discovery of a Vocation: Darwin's Early Geology'. *British Journal of the History of Science* 24, pp.133–57
16. Sandra Herbert. 'Charles Darwin as a Prospective Geological Author'. *British Journal of the History of Science* 24, pp.159–92
17. Frank Rhodes. 'Darwin's Search for a Theory of the Earth: Symmetry, Simplicity and Speculation'. *British Journal of the History of Science* 24, pp.193–229
18. *Biography* pp.167–340
19. Desmond King-Hele. *Erasmus Darwin: A Life of Achievement.* Giles de la Mare Publishers Ltd, London, 1999
20. See P.H. Jesperson, 1948–9. 'Charles Darwin and Dr Grant'. *Lychnos* pp.159–67
21. *Autobiography* p.49
22. Ibid. p.28
23. *Correspondence 5* p.63
24. John Maurice Herbert (1808–82) was later a county court judge on the Monmouth and Cardiff circuit
25. *Autobiography* pp.62–3
26. Alexander von Humboldt, *Personal narrative of travels to the equinoctial regions of the New Continent during the years 1799–1804, by A. de Humboldt, and A. Bonpland; . . .translated into English by H.M. Williams.* 7 vols, London, 1814–29. Vols. 1 and 2, in one, 3rd edn, inscribed 'from J.S. Henslow to CD on his departure, September 1831', were in the *Beagle* library
27. John Frederick William Herschel. *A preliminary discourse on the study of natural philosophy*. Published in D. Lardner's *Cabinet Cyclopædia.* London, 1831
28. *Correspondence 1* pp.125–6
29. *Autobiography* pp.69–70
30. George Bellas Greenough, *Geological Map of England and Wales*, London, 1820. There was a copy in the *Beagle*'s library of Greenough's book *A*

Critical Examination of the First Principles of Geology, London, 1819

31. The notes are transcribed by P.H. Barrett in 'The Sedgwick-Darwin Geological Tour of North Wales' *Proceedings of the American Philosophical Society* 118, pp.146–64 (1974)
32. *Narrative 1* p.xi
33. Ibid. p.385
34. *Narrative 2* loose map
35. *Narrative 1* p.444
36. Nick Hazlewood. *Savage: The Life and Times of Jemmy Button.* Hodder & Stoughton, London, 2000
37. See Fergus Fleming. *Barrow's Boys*. Granta Books, London, 1999
38. *Narrative 2* pp.4–10
39. Ibid. pp.24–41
40. Ibid. p.18
41. *Correspondence 1* pp.127–8
42. Ibid. pp.128–30
43. *Narrative 2* pp.18–19
44. *Correspondence 1* p.131
45. Ibid. p.132
46. Ibid. pp.132–3
47. Ibid. pp.133–5
48. *Correspondence 3* pp.359–60
49. *Narrative 2* p.49
50. *Zoology Notes* p.3
51. *Beagle Diary* p.21
52. *Zoology Notes* p.9
53. See Sandra Herbert (1991). 'Charles Darwin as a Prospective Geological Author'. *British Journal of the History of Science* 24, pp.159–92 and James Secord (1991) 'The Discovery of a Vocation: Darwin's Early Geology'. *Brit. J. Hist. Sci.* 24, pp.133–57; and MS DAR 32.1 pp.15–20
54. *Narrative 2* p.56
55. A copy of P.G. King's account of the voyage written in 1890 is included in CUL MS DAR 107
56. *Autobiography* pp.73–4
57. See *Narrative 2* pp.61–2

58. Ibid. pp.67–72
59. *Beagle Diary* p.139
60. *Correspondence 1* pp.225–8
61. Albert Friendly. *Beaufort of the Admiralty*. Hutchinson, London, 1977
62. *Zoology Notes* p.47
63. *Correspondence 1* pp.250–3
64. *Zoology Notes* pp.37–8
65. Ibid. pp.38–9
66. Ibid. pp.50–2
67. See David Stanbury (ed.). *A Narrative of the Voyage of H.M.S. Beagle*. The Folio Society, London, 1977
68. C. Darwin (1844). 'Observations on the Structure and Propagation of the Genus *Sagitta*'. *Annual Magazine of Natural History* 13, pp.1–6. Reprinted in *Collected Papers 1*, pp.177–82
69. *Zoology Notes* p.73
70. Down House Notebook 1.10
71. CUL MS DAR 262.33
72. *Beagle Diary* pp.107–8
73. *Correspondence 1* pp.279–82
74. See Alfred Sherwood Romer. *Vertebrate Paleontology*. 3rd edition. University of Chicago Press, Chicago and London, 1974
75. Silvia A. Aramayo and Teresa M. de Bianco (1987). *Hallazgo de una icnofauna continental (Pleistoceno tardio) en la localidad de Pehuen-Co (Partido de Coronel Rosales), Provincia de Buenos Aires, Argentina. Parte I: Edentata, Litopterna, Proboscidea.* IV Congreso Latinoamericano de Paleontologia, Bolivia (1987) 1: 516–531.
76. Silvia A. Aramayo and Teresa Manera de Bianco (1996). *Edad y nuevos hallazgos de icnitas de mamiferos y aves en el yacimiento paleoicnologico de Pehuen-Co (Pleistoceno tardio), Provincia de Buenos Aires, Argentina.* Asociación Paleontológica Argentina de Icnología, Publicación Especial 4, 1º Reunión Argentina de Icnología: 47–57. Buenos Aires, 1996.
77. *Zoology Notes* pp.106–9
78. Ibid. pp.120–1
79. See E. Lucas Bridges. *Uttermost Part of the Earth*. Hodder & Stoughton, London, 1948

80. *Journal of the right hon. Sir Joseph Banks during Captain Cook's first voyage in H.M.S. Endeavour in 1768–71 to Tierra Del Fuego, Otahite, New Zealand etc*. Edited by Sir Joseph D. Hooker. London, 1896
81. *Beagle Diary* pp.122–5
82. Henry Norton Sulivan (ed.). *Life and Letters of the Late Admiral Sir Bartholomew James Sulivan, K.C.B. 1810–1890.* John Murray, London, 1896
83. Thomas Bridges. *Yamana-English: A Dictionary of the Speech of Tierra del Fuego*, edited by F. Hestermann and M. Gusinde, privately printed at Mödling, Austria, in 1933
84. *Narrative* 2 pp.228–40
85. Down House Notebook 1.14; and see also Nora Barlow (ed.). *Charles Darwin and the Voyage of the Beagle*. Pilot Press, London, 1945
86. *Beagle Record* pp.131–3
87. Admiralty Records, Record Office ADM/1/1819
88. Roger McDonald. *Mr Darwin's Shooter*. Transworld, London, 1998
89. René Primevère Lesson. *Manuel d'ornithologie* Vol. 2, p.385 Paris, 1828. There was a copy in the *Beagle*'s library
90. *Correspondence 1* p.326
91. Ibid. pp.334–6
92. *Beagle Diary* p.200
93. Ibid. p.202
94. Nat Rutter, Enrique J. Schnack, Julio del Rio, Jorge L. Fasano, Federico L. Isla and Ulrich Radtke (1989). 'Correlation and Dating of Quaternary Littoral Zones Along the Patagonian Coast, Argentina'. *Quaternary Science Reviews* 8, pp.213–34
95. *Beagle Diary* p.209
96. Knut Schmidt-Nielsen. *Desert Animals: Physiological Problems of Heat and Water*. Clarendon Press, Oxford, 1964
97. *Zoology Notes* p.183
98. Michael Organ (ed.). *Conrad Martens: Journal of a Voyage from England to Australia 1833–35.* State Library of NSW Press, Australia, 1994
99. *The Zoology of the Voyage of H.M.S. Beagle . . . 1832 to 1836. Edited by Charles Darwin. Part III. Birds. By John Gould.* pp.123–5. Smith Elder & Co., London, 1836

100. *Correspondence 1* pp.368–72
101. For map and history of Port Desire see Felix Reisenberg. *Cape Horn*. Robert Hale Ltd, London, 1941
102. *Narrative 2* p.320
103. Richard D. Keynes (1997). 'Steps on the Path to the Origin of Species'. *Journal of Theoretical Biology* 187, pp.461–71
104. *Beagle Diary* p.227
105. Nick Hazlewood. *Savage. The Life and Times of Jemmy Button*. Hodder & Stoughton, London, 2000
106. Bridges. *Uttermost Part of the Earth*. op. cit.
107. Patrick Armstrong. *Darwin's Desolate Islands: A Naturalist in the Falklands, 1833 and 1834*. Picton Publishing Ltd, Chippenham, 1992
108. *Zoology Notes* pp.214–15
109. *Narrative 2* p.348
110. *Beagle Diary* p.237
111. *Narrative 2* p.350
112. *Zoology Notes* pp.227–8
113. Ibid. p.xvi
114. *Beagle Diary* p.245
115. Ibid. pp.249–50
116. Ibid. p.250
117. Ibid. p.254
118. *Correspondence 1* p.406–7
119. Ibid. pp.410–12
120. George Pickering. *Creative Malady: Illness in the Lives and Minds of Charles Darwin, Florence Nightingale, Mary Baker Eddy, Sigmund Freud, Marcel Proust and Elizabeth Barrett Browning*. George Allen & Unwin Ltd, London, 1974
121. *Zoology Notes* pp.253–4
122. From the archives of the Hydrographic Department, Taunton
123. *Autobiography* p.75
124. *Correspondence 1* pp.417–20
125. See Down House Notebook 1.8 on pp.229–30 of Nora Barlow (ed.). *Charles Darwin and the Voyage of the Beagle*. Pilot Press Ltd, London, 1945
126. *Journal of Researches 1* p.352

127. Ibid. p.341
128. *Beagle Diary* p.273
129. Henry Norton Sulivan. *Life and letters of the Late Admiral Sir Bartholomew James Sulivan . . . 1810–1890.* London, 1896
130. *Beagle Diary* p.277
131. *Zoology Notes* pp.274–6
132. *Beagle Diary* pp.292–3
133. *Narrative* 2 pp.406–8
134. *Beagle Diary* p.309
135. Charles Lyell. *Principles of Geology, being an attempt to explain the former changes of the earth's surface.* Vol. 2. John Murray, London, 1832
136. *Journal of Researches* 2 pp.327–8
137. Down House Notebook 1.13 and *Beagle Diary* p.315
138. *Beagle Diary* p.315
139. Ralph Colp. *To be an Invalid: The Illness of Charles Darwin.* University of Chicago Press, 1977
140. *Correspondence 1* p.442
141. *Journal of Researches 1* pp.406–7
142. Charles Lyell. *Principles of Geology . . .* Vol. 3, p.131. John Murray, London, 1833
143. *Correspondence 1* p.458
144. *Narrative* 2 pp.428–80
145. H.E.L. Mellersh. *FitzRoy of the Beagle.* Rupert Hart-Davis, London, 1968
146. Duccio Bonavia. *Los Gavilanes.* Corporacion Financiera de Desarrollo S.A. Cofide, Lima, 1982
147. *Beagle Diary* pp.349–50
148. Admiralty Records, Record Office, ADM/1/3848
149. *Correspondence 1* pp.464–5
150. Letter in CUL MS DAR 207
151. *Beagle Diary* pp.351–2
152. *Narrative* 2 pp.486–7
153. Gregory Estes, K. Thalia Grant and Peter R. Grant (2000). 'Darwin in Galapagos: His Footsteps Through the Archipelago'. *Notes Rec. Royal Society, London* 54, pp.343–68

154. *Beagle Diary* p.353
155. *Zoology Notes* pp.293–4
156. *Beagle Diary* p.354
157. *Zoology Notes* pp.291–3
158. *Beagle Diary* p.356
159. *Zoology Notes* pp.294–6
160. Ibid. p.297
161. Frank J. Sulloway (1982). 'Darwin and his Finches: The Evolution of a Legend'. *Journal of the History of Biology* 15, pp.1–53
162. Frank J. Sulloway. 'Darwin's Conversion: The *Beagle* Voyage and its Aftermath'. *Journal of the History of Biology* (1982), 15, pp.325–96
163. Paul H. Barrett, Peter J. Gautrey, Sandra Herbert, David Kohn and Sydney Smith (editors). *Charles Darwin's Notebooks, 1836-1844*. Cambridge University Press, 1987
164. David Lack. *Darwin's Finches: An Essay on the General Biological Theory of Evolution*. Cambridge University Press, 1947.
165. Peter Grant. *Ecology and Evolution of Darwin's Finches*. Princeton University Press, 1999. See also Jonathan Weiner. *The Beak of the Finch: Evolution in Real Time.* Jonathan Cape, London, 1994.
166. *Zoology Notes* p.298
167. *Origin* p.185
168. *Zoology Notes* p.300
169. *Narrative* 2 pp.36–7
170. Peter Lancaster-Brown. *Halley and his Comet.* Blandford Press, Poole, Dorset, 1985
171. Patrick O'Brian. *Joseph Banks: A Life*. Collins Harvill, London, 1987
172. R. FitzRoy and C. Darwin (at sea, 28 June 1836). 'A letter, containing remarks on the moral state of Tahiti, New Zealand, &c'. *South African Christian Recorder* 2 (1836), pp.221–38. Reprinted in Paul H. Barrett (ed.). *The Collected Papers of Charles Darwin* Vol. 1 pp.19–38. University of Chicago Press, Chicago and London, 1977
173. *Narrative* 2 pp.506–63
174. *Beagle Diary* p.378
175. Charles Lyell. *Principles of Geology* . . . Vol. 2, Chapter XVIII. John Murray, London, 1832

176. *Beagle Diary* p.379
177. Augustus Earle. *A narrative of a nine months' residence in New Zealand, in 1827; together with a journal of a residence in Tristan d'Acunha*. London, 1832
178. *Origin* pp.388–406
179. *Beagle Record*
180. F.W. and J.M. Nicholas. *Charles Darwin in Australia.* Cambridge University Press, Sydney, 1989, is a detailed and well illustrated account of this journey
181. Charles Darwin. *Geological Observations on the Volcanic Islands visited during the voyage of H.M.S. Beagle, together with some brief notices of the geology of Australia and the Cape of Good Hope.* Smith Elder & Co., London, 1844
182. *Beagle Diary* p.401
183. Ibid. pp.402–3
184. *Zoology Notes* pp.305–10
185. *Correspondence 1* pp.495–6
186. Sir John Herschel. *A preliminary discourse on the study of natural philosophy*. Published in D. Lardner's *Cabinet Cyclopaedia*, London, 1831
187. R. FitzRoy and C. Darwin (at sea, 28 June 1836). 'A letter containing remarks on the moral state of Tahiti, New Zealand, &c.' loc. cit., pp.221–38
188. Charles Darwin. 'On the connexion of certain volcanic phenomena in South America; and on the formation of mountain chains and volcanos, as the effect of the same power by which continents are elevated' (read 7 March 1838). *Transactions of the Geological Society of London*, 2nd series, pt 3, 5, pp.601–31, 1840
189. Charles Darwin. *Geological Observations on the Volcanic Islands visited during the voyage of H.M.S. Beagle . . .* op. cit.
190. *Beagle Diary* p.432
191. Frank J. Sulloway (1982) 'Darwin's Conversion: The *Beagle* Voyage and its Aftermath'. *Journal of the History of Biology* 15, pp.325–96
192. Nora Barlow (ed.) (1963). 'Darwin's Ornithological Notes'. *Bulletin of the British Museum (Natural History) Historical Series* Vol. 2 No. 7, pp.261–2
193. *Beagle Diary* p.434
194. *Narrative 2* pp.658–9
195. *Correspondence 2* p.431

196. Paul H. Barrett, Peter J. Gautrey, Sandra Herbert, David Kohn and Sydney Smith (eds.). *Charles Darwin's Notebooks, 1836–1844*. British Museum (Natural History) and Cambridge University Press, 1987
197. Duncan M. Porter (ed.) (1987). 'Darwin's Notes on *Beagle* Plants'. *Bulletin of the British Museum (Natural History) Historical Series* 14 (2), pp.145–233
198. Kenneth G.V. Smith (ed.) (1987). 'Darwin's Insects'. *Bulletin of the British Museum (Natural History) Historical Series* 14 (1), pp.1–143
199. *Zoology Notes*
200. Sandra Herbert (ed.). *The Red Notebook of Charles Darwin*. British Museum (Natural History) and Cornell University Press, New York, 1980
201. *Notebooks* p.228
202. Erasmus Darwin. *The Temple of Nature*. J. Johnson, London, 1803
203. Thomas Robert Malthus. *An Essay on the Principle of Population*. 6th edn. London, 1826
204. *Autobiography* p.120
205. Richard D. Keynes (1997). 'Steps on the Path to the Origin of Species'. *Journal of Theoretical Biology* 187, pp.461–71
206. Charles Darwin. *A monograph on the sub-class Cirripedia, with figures of all the species*. Vol. 1 *The Lepadidae; or peduncuted species*. The Ray Society, London, 1851. Vol. 2 *The Balanidae, or sessile Cirripedia; The Verrucidae, etc., etc., etc.* The Ray Society, London, 1854
207. *Correspondence 1* pp.226–7
208. Ibid. p.483
209. Robin Marantz Henig. *A Monk and Two Peas: The Story of Gregor Mendel and the Discovery of Genetics*. Weidenfeld & Nicolson, London, 2000

Index

Page numbers in *italic* refer to illustrations